Developments in System Safety Engineering

Related Titles

The Safety of Systems
Proceedings of the Fifteenth Safety-critical Systems Symposium, Bristol, UK, 2007
Redmill and Anderson (Eds)
978-1-84628-805-0

Improvements in System Safety
Proceedings of the Sixteenth Safety-critical Systems Symposium, Bristol, UK, 2008
Redmill and Anderson (Eds)
978-1-84800-099-5

Safety-Critical Systems: Problems, Process and Practice
Proceedings of the Seventeenth Safety-critical Systems Symposium, Brighton, UK, 2009
Dale and Anderson (Eds)
978-1-84882-348-8

Making Systems Safer
Proceedings of the Eighteenth Safety-critical Systems Symposium, Bristol, UK, 2010
Dale and Anderson (Eds)
978-1-84996-085-4

Advances in Systems Safety
Proceedings of the Nineteenth Safety-critical Systems Symposium, Southampton, UK, 2011
Dale and Anderson (Eds)
978-0-85729-132-5

Achieving Systems Safety
Proceedings of the Twentieth Safety-critical Systems Symposium, Bristol, UK, 2012
Dale and Anderson (Eds)
978-1-4471-2493-1

Assuring the Safety of Systems
Proceedings of the Twenty-first Safety-critical Systems Symposium, Bristol, UK, 2013
Dale and Anderson (Eds)
978-1481018647

Addressing Systems Safety Challenges
Proceedings of the Twenty-second Safety-critical Systems Symposium, Brighton, UK, 2014
Dale and Anderson (Eds)
978-1491263648

Engineering Systems for Safety
Proceedings of the Twenty-third Safety-critical Systems Symposium, Bristol, UK, 2015
Parsons and Anderson (Eds)
978-1505689082

Developing Safe Systems
Proceedings of the Twenty-fourth Safety-critical Systems Symposium, Brighton, UK, 2016
Parsons and Anderson (Eds)
978-1519420077

Mike Parsons • Tim Kelly
Editors

Developments in System Safety Engineering

Proceedings of the Twenty-fifth Safety-critical Systems Symposium, Bristol, UK, 7th-9th February 2017

Safety-Critical Systems Club

SCSC-135

The publication of these proceedings is sponsored by BAE Systems plc and Jaguar Land Rover

Editors

Mike Parsons
NATS CTC
4000 Parkway
Whiteley, Fareham
PO15 7FL
United Kingdom

Tim Kelly
Department of Computer Science
University of York
Deramore Lane, York
YO10 5NG
United Kingdom

ISBN-13: 978-1540796288
ISBN-10: 1540796280

Preface

This book contains the papers presented at the twenty-fifth Safety-critical Systems Symposium (SSS'17). For this year's Silver Jubilee symposium authors produced important material covering topics relevant to safety-critical systems practitioners; we are very grateful to them for their contributions.

The first day themes were New Challenges and New Techniques. John McDermid opened the symposium on a challenging theme: "Playing Catch-Up: The Fate of Safety Engineering?" Talks on safety of systems of systems and assuring unconventional software concluded the first morning. The afternoon featured a keynote talk from Audrey Canning entitled "Functional Safety: Where have we come from? Where are we going?" followed by presentations on autonomy, and the certification of multi-core architectures. The day finished with an amusing special talk from Tom Anderson, "What can I say?".

The two themes of the second day were Software, and Safety and Security. Robin Bloomfield started the day with a keynote talk "Confidence in a connected world: safe, secure, resilient and autonomous". This was followed by talks on handling of hardware errors by software, a description of a formally verified optimizing compiler and a talk on the use of formal proof to meet DO-333. The afternoon featured a keynote talk from Nancy Leveson: "My 36 Years in System Safety Engineering: Looking Backward, Looking Forward". The issues of Reduced Crew Operations on aircraft were considered, followed by a security talk considering the nature of insider threats to organisations. The day finished with a lively "Practitioners' Question Time" where the eminent panellists responded to questions from the audience.

Safety Analysis and Data Safety were the themes of the final day. The morning started with a keynote talk from Dewi Daniels "From the IBM 29 Card Punch to the Boeing 787 Dreamliner (and Beyond)". The first session papers considered interference and sneak analyses, followed by talks on the issues of data in safety systems including one by Harold Thimbleby covering a court case which focussed on data handling in hospitals. Ron Bell started the final session with a retrospective keynote talk: "A brief history of the development of guidelines and standards". The symposium concluded with a topical and entertaining keynote address by Les Hatton "Balancing safety with rampant software feature-itis".

We are grateful to our sponsors for their valuable support and to the exhibitors at the Symposium's tools and services fair for their participation. And we thank Joan Atkinson at Newcastle and Alex King at York for laying the event's foundation through their exemplary planning and organisation.

MP & TK

A message from the sponsors

BAE Systems and Jaguar Land Rover are pleased to support the publication of these proceedings. We recognise the benefit of the Safety-Critical Systems Club in promoting safety engineering in the UK and value the opportunities provided for continued professional development and the recognition and sharing of good practice. The safety of our employees, those using our products and the general public is critical to our business and is recognised as an important social responsibility.

The Safety-Critical Systems Club

organiser of the

Safety-critical Systems Symposium

Safety-critical systems and the accidents that don't happen

When an aircraft crashes, it makes headlines. That hundreds of thousands of flights each week do not crash is accepted as routine. Airliners, air traffic control systems, railway signalling, car braking systems, defence systems, nuclear power stations and medical equipment are some of the critical systems in use, on which life and property depend. New autonomous systems that will affect our daily life are coming on-stream soon, including delivery drones and self-driving road vehicles. That safety-critical systems do work well is because of the expertise and diligence of professional systems safety engineers, regulators and other practitioners who work to minimise both the likelihood that accidents will occur, and the consequences of those that do. Their efforts prevent untold deaths and injuries every year. The Safety-Critical Systems Club (SCSC) has been actively engaged for over twenty-five years to help to ensure that this continues to be the case.

What is the Safety-Critical Systems Club?

The SCSC is the UK's professional network and community for sharing knowledge about safety-critical systems. It brings together engineers and specialists from a range of disciplines working on safety-critical systems in a wide variety of industries, academics researching in the field, providers of the tools and services that help develop the systems, and the regulators who oversee safety. It provides, through publications, seminars, tutorials, a website, working groups and, importantly, at the annual Safety-critical Systems Symposium, opportunities for them to network and benefit from each other's experience in working hard at the accidents that don't happen. It focuses on current and emerging practices in safety engineering, software engineering, and product and process safety standards.

What does the SCSC do?

The SCSC maintains a website (www.scsc.uk), which includes directories of tools and services that assist in the development of safety-critical systems. It publishes a regular newsletter, Safety Systems, three times a year. It organises seminars, workshops and training on general matters or specific subjects of current concern, which are prepared and led by world experts. Since 1993 it has organised the annual Safety-critical Systems Symposium (SSS) where leaders in different aspects of safety, from different industries, including consultants, regulators and academics, meet to exchange information and experience, with the papers published in a proceedings volume. The SCSC supports industry working groups, such as the Data Safety Initiative Working Group (DSIWG) that are addressing concerns raised about data in safety-related systems. New working groups on Autonomy and Service Assurance are starting in 2017 and others are planned. The SCSC carries out all these activities to support its mission:

> *... to raise awareness and facilitate technology transfer in the field of safety-critical systems ...*

History

The SCSC began its work in 1991, supported by the Department of Trade and Industry and the Engineering and Physical Sciences Research Council. The Club has been self-sufficient since 1994, but enjoys the active support of the Health and Safety Executive, the Institution of Engineering and Technology, and the BCS - The Chartered Institute for IT; all are represented on the SCSC Steering Group.

Membership

Membership may be either corporate or individual. Individual membership, which costs £95 a year, entitles the member to Safety Systems three times a year, other mailings, and discounted entry to seminars, workshops and the annual Symposium. Frequently individual membership is paid by the employer.

Corporate membership is for organisations that would like several employees to take advantage of the benefits of SCSC programmes. The amount charged is tailored to the needs of the organisation.

Individual membership can be obtained online at: http://www.scsc.uk/

For more information about membership, please contact: Alex King, Department of Computer Science, University of York, phone: 01904 325402 email: alex.king@scsc.uk or mike.parsons@scsc.uk

Contents

Playing Catch–Up:
The Fate of Safety Engineering?

John A McDermid

University of York

Abstract *Scientific and engineering disciplines advance by establishing, refining, and occasionally rejecting theories. A key part of this process is the constant use of experience (more particularly the results of 'experiments') to challenge orthodoxy. One marker of the vigour of an academic discipline is how much of this challenge comes from within. Safety engineering appears to advance mainly by "playing catch up" i.e. responding to the challenges presented by advances in other disciplines including psychology, human factors, and organisational science. This paper considers whether the trend is inevitable, or whether advances in safety engineering can be more systematic and pro-active.*

1 Introduction

Philosophers of Science propose different models for how scientific disciplines advance. Karl Popper (Popper 1959) was famous for his observation that scientific theories can be falsified, but never proven; the importance of counter-evidence in safety can be seen as a direct reflection of this insight. Imre Lakatos (Lakatos 1974) criticised this model and suggested that scientists 'lived with anomalies' and theories were only replaced when better ones were developed. Thomas Kuhn (Kuhn 1962) suggested that there was a cycle: 'normal science - crisis - revolution - normal science', and others have proposed that this applies in engineering as well. But how does safety engineering advance?

It could be argued that safety engineering follows a similar cycle to that proposed by Kuhn, where the crises are either accidents or major changes in technology. Turner (Turner 1976) suggested that some large-scale accidents are 'sufficiently unexpected and sufficiently disruptive to provoke a cultural reassessment of the artefacts and precautions available to prevent such occurrences'. Piper Alpha (Cullen 1990) is an example of an accident that led to the widespread focus on safety cases (although the safety case concept predates the accident) and an earlier accident led to the introduction of the principle that risks should be reduced As Low As Reasonably Practicable (ALARP) (Edwards

v. National Coal Board 1949). The advent of programmable electronics could be seen as an example of a technological challenge, with the adoption of Safety Integrity Levels (SILs), and cognate concepts, being the response. Put more prosaically, safety engineers are "just playing catch-up" when the orthodoxy (normal engineering) in application domains is overtaken by events.

It is interesting to ask if this is inevitable. If it is not, what can the discipline do to get "on the front foot"? If it is inevitable, then what does this mean about the nature of safety engineering as a discipline? If accident investigations are the source (rather than merely a by-product) of advances in safety thinking, should we be conducting and interpreting them in fundamentally different ways?

The paper explores these questions, seeking to provide some motivation for a more systematic approach to advancement in safety engineering. Section 2 briefly discusses some of the more classical models of advances in science, and considers their relevance to engineering in general and to safety engineering in particular. Section 3 identifies some advances in safety engineering and their "triggers", and speculates on what changes will emerge in the next few years. Section 4 returns to the question in the title. The more specific questions identified above are considered in section 5, and some conclusions are also presented.

2 Theories of Scientific Advances

There are several well-recognised theories of scientific progress, generally concerned with how science (especially physics) advances. Some researchers have tried to apply those theories to engineering, but others have developed independent theories of engineering progress. Some of the seminal work on theories of science and engineering is briefly surveyed before considering whether or not there is, or can be, a similar theory for safety engineering.

2.1 Philosophy of Science

Philosophies of science set out what it means for scientific theories to be valid, but also consider the nature of progress in science. Whilst Popper was not the first to consider such issues, he is generally viewed as producing some of the seminal ideas on the subject.

Popper stated (Popper 1959) that one of the most fundamental requirements is that a theory should be consistent; if it is not then one could draw arbitrary conclusions. This is well understood in the domain of mathematical logic, but is more general.

Popper also introduced the idea that scientific theories are falsifiable – in other words that it is possible to tell if they are wrong, whereas previous models had focused on the role of evidence in confirmation rather than refutation of theories. In practice this means that theories must enable predictions to be made, and that these predictions can be evaluated empirically. For example, we can use Newtonian mechanics to predict the movement of the planets, and we can measure their movement to check the theory. Of course there is a need to be able to distinguish an incorrect theory from a wrong or misleading prediction or experiment, e.g. not including all the planets in predicting planetary motion.

Later Popper (Popper 1977) introduced his "three worlds" model. In this model world 1 is physical entities and phenomena; world 2 is mental or subjective states and events; world 3 is objective knowledge: abstract objects of thought that are explicitly recorded. In this model, predictions made in world 3 can be evaluated in world 1, but world 2 is not amenable to such validation. In the above example, the planets are in world 1, and Newtonian mechanics in world 3.

Popper's model is often interpreted as saying that any falsification (counter-evidence) should lead to a theory being rejected (immediately). But scientists are human. In practice scientists can seek to explain counter-evidence as exceptions, or to refine the model to resolve the anomalies. Lakatos proposed an alternative model, where progress is only made where a better (new) theory is produced and that scientists 'live with anomalies' and don't reject theories on the first failed experiment, as Popper said should be done. Lakatos argues that many theories that we now accept would never have survived if the Popperian model were adopted. Perhaps more importantly he sees there as being competing "projects" (or "research projects") and that one project eventually "wins", i.e. becomes accepted as the "mainstream", when the problems with the other projects (competing theories) makes them too hard to sustain, due to the number of anomalies that have to be accommodated.

There have been a number of treatments of scientific progress that view it more as a social phenomenon. For example Kuhn (Kuhn 1962) proposed that there is a cycle: 'normal science - crisis - revolution - normal science'. He and others suggest that a crisis is needed before the community will break away from the accepted theory, i.e. the current 'normal science'. This view is reflected in some commonplace terminology – for example we talk about the Copernican revolution, and paradigm shifts.

It is not possible to do this complex and subtle subject full justice here, but it is hopefully sufficient to act as a basis for discussing theories of engineering and of safety engineering.

2.2 Philosophy of Engineering

Some researchers have tried to apply philosophies of science to engineering, but many would argue that there is a fundamental difference, e.g. that 'the scientist builds in order to study; the engineer studies in order to build' (Brooks 1996). This suggests that theories of engineering should be different from those of science – but as engineering has a scientific underpinning, they should not be disjoint.

One of the more interesting approaches is that due to Mark Staples (Staples 2014, 2015) using the terminology 'critical rationalism'. He identifies five different roles for philosophies of engineering, some of which build on Popper's three worlds model. In particular he identifies two roles for philosophies of engineering that link between Popper's world 1 and world 3:

'Role 1: characterisations of changes required to be brought about in usage situations, represented (i.e., formalised, documented) as requirements specifications;

Role 5: characterisations of artefacts and their behaviour, represented (i.e., formalised, documented) as designs and descriptions of behaviour.'

The notion of 'system identification' from control theory aligns directly with role 5, whereas role 1 is about prediction – how a system or a design will behave in the world. This is *not* the same as Popper's philosophy of science, as it is about the satisfaction of requirements, not the validity of a theory. Further roles 1 and 5 can be seen as complementary; role 1 specifying what is desired, and role 5 evaluating what has been achieved. Of course role 5 can be used to 'falsify' a system or design – the system does not do (in the real world) what we specified it should do. Thus a system or a design can or should be rejected if it doesn't meet its requirements. But engineers are human. In practice engineers can seek to explain counter-evidence as exceptions, usually in preference to refining the design to resolve the anomalies, due to the cost involved. Indeed this can be viewed as 'normal engineering' and is often quite acceptable, although it may not be if there are safety consequences of the problems. Of course, this does not invalidate theories – in fact the role of theories in engineering is rather different to their role in science.

It is often said that 'all models are wrong, but some are useful'. Put another way, theories are not universal and are perhaps best thought of as being of limited applicability, rather than false. This can be seen pragmatically as, for example, we might use a steady-state model of a system if that is adequate for design, extending it to deal with transients if that is necessary, or perhaps considering through-life degradation of components, if that is necessary to solve the problem at hand. So in engineering we should ask what the model or theory is for – and it is *never* scientific truth – and whether or not it is fit for that purpose, in the context of use. However this means that we need to know the limits of applicability of a theory – for example, in aerodynamics, under what conditions boundary layers separate and the models of laminar flow are not applicable.

Advances in engineering are well documented in some areas, e.g. by Petroski (Petroski 1992), but in terms of anecdotal histories, rather than in more philosophical terms. However Petroski talks about the role of failures (and accidents) in making advances in engineering – and this is very similar to the notion of a crisis in Kuhn's model. However, these advances can also be explained in Lakatos' terms of competing "projects", but where the failures are stimulating the development of new competing theories (producing new "projects").

2.3 Theories and Safety Engineering

Safety engineering is (or at least should be) a special case of engineering and we discuss this in terms of Staples' roles for engineering theories.

We can see role 1 as being about prediction of how a system or design will behave in the world, in safety terms; this is most obviously thought of in terms of quantification of risk. Role 5 is about evaluation of systems in the real world – in a sense in terms of risk, but also more qualitatively. For example we might question a theory – or its application – if behaviour includes failure modes that we had not predicted (whether or not they involve harm).

But what are the theories? For some specific techniques, such as Leveson's STAMP (Leveson 2012) there is an explicit link to cybernetics and systems theory. More widely used techniques such as Failure Modes and Effects Analysis (FMEAs) and its variants or Fault Tree Analysis (FTA) rest on theoretical developments in reliability and statistical analysis, but are applied well beyond the original real world situations described by these theories. These methods are generally interpreted as showing causal dependencies, although the causal theory is not well developed, however fault tree structures resemble the Insufficient but Necessary part of Unnecessary but Sufficient (INUS) conditions model of causality due to Mackie (Mackie 1965).

As with other engineering theories they (the underlying methods) are not universal and have limited applicability; but do we know where they apply? This comes back to the issue of fitness for purpose. We might use FMEAs with bulk failure rates at the level of subsystems where the situation is not demanding; but detailed piece-part FMEAs discriminating the rate of occurrence of failures in different failure modes where precise characterisation of failure rates is needed (or it is hard to show that safety requirements are met otherwise). In the case of FTA we might discover that they are inadequate to model a particular system due to time-dependency between events, or state-dependent failure behaviour.

Taking Lakatos' view, we can describe changes to these theories (e.g. modifications to FTA to include sequencing and non-independence) either as refinements or as responses to an accumulated burden of anomalies. But can we identify 'crises' that lead to Kuhn-style revolutions? In some senses we can – but often these crises come from advances in other domains, e.g. digital electronics, that push the then current theories and methods beyond their domain of applicability. There is at least one other important perspective on advances in safety engineering. As the theories advance, equivalent systems should get safer (or reach the same level of safety more cost-effectively); we return to this point in §4.3.6 when we consider evaluation.

One important aspect of theories has not been mentioned yet – that is the notion of consistency. In what sense, if any, can we say that theories of safety engineering are consistent (or inconsistent)? This is a very significant question, and one that cannot be done justice here. However if safety engineering is to become more mature as a discipline, this is an issue that does need to be addressed.

3 Some Advances in Safety Engineering

The discussion above describes theories – but what happens in practice? Of course it is not possible to discuss every advance in safety engineering, but it is possible to identify some broad changes in capability in the discipline, although the old methods are still used (hopefully) within their domain of applicability. Some of the advances can be thought of as 'crises', in Kuhn's sense, where the advance of technology drove the need for new safety engineering methods. The section sets out a broad chronological progression, and the later sub-sections refer to current and to emerging issues that are likely to (hopefully will) lead to advances in the discipline. The final sub-section considers the role of standards.

3.1 Space – The First Frontier?

Good engineers think about how systems may fail, as well as how they work in normal situations. Whilst it is hard to prove, it can be conjectured that many early systems were "safe enough" because good engineering ensured that failure conditions were considered and/or that design codes ensured safety without the need for analysis. A combination of complexity and criticality was perhaps the first 'crisis' for safety engineering – seen most clearly in the space and nuclear industries.

The response to this – in Kuhn's terms the revolution – can be viewed as the development of systematic methods – the FMEAs and FTAs mentioned above – which embody the causal and risk analyses which are the fundamentals of safety engineering. They have been the "normal engineering" for many decades.

3.2 Process Industries

Generally the challenges to safety engineering relate to technology change or increases in complexity. The development of HAZOP to assess designs of chemical process plant (Kletz 1999) might be viewed as pragmatic improvement – it is easier to think about process plant in flow terms – rather than the response to a crisis. HAZOP models are still firmly based on causal theories, although the focus is more on identifying hazards, than analysing known hazards. Arguably it has a broader scope than FTA and FMEA explicitly considering the human element (i.e. operability) but it doesn't really embody a fundamentally new theory.

The Piper Alpha accident (see figure 1 overleaf) can be viewed as a crisis. Whilst the concept of safety cases – now viewed as a structured argument supported by evidence (Kelly 1998) – predates the Cullen report (Cullen 1990) into the accident, they became more widely used as a consequence of that event. They have become the "normal engineering" in some domains, but not in all. The reason for take-up in some arenas rather than others may be cultural, but it may be because the underlying theory is not so clear. It might be thought that a safety case embodies a "theory of evidence" and some of the academic work on safety cases, e.g. (Weaver 2003), implicitly reflects this view. However, it is hard to find a solid "theory of evidence" in the sense of being able to argue how elements of evidence combine, and that may be one of the reasons for their limited adoption. Another might be that it is a "competing project" in the sense used by Lakatos, and that safety cases are competing with "standards compliance" as a model, see §3.9.

Fig. 1: Piper Alpha

The accident to Nimrod XV230 (Haddon-Cave 2009) led to the role of safety cases being questioned. However the learned Queen's Counsel (QC) (now Judge) concluded that the problem was with the specifics of Nimrod, not the concepts of safety cases, *per se* (and there were many other issues, not just the misleading safety case). In other words this did not constitute a crisis, for safety cases as a method or theory (although clearly it was for the families of those who died in the accident, and others affected by its aftermath). In Lakatos' terms, this was an anomaly for the "safety cases project" but not one that was fatal for the project.

Fig. 2: A Nimrod Aircraft showing the refuelling probe

3.3 Complex Systems

Further growth in complexity has been seen in many sectors. A factor in the growth of complexity has been the ever-increasing use of digital electronics and software, and more recently the "explosion" in the volume of data being processed. Complexity will be considered first, with a focus on the aerospace industry, as that is where the response is perhaps most apparent.

The hypothesis advanced in §3.1 is that designers needed systematic methods such as FMEA and FTA as their designs were too complex for "good engineering" (intuition, perhaps) to suffice to produce adequate designs. However, producing a design and then analysing it with FMEA, FTA, etc. – especially if this was done later in a separate department – might reveal design weaknesses and result in design revision. If the level of change were small, then this could be accommodated, but major issues could lead to multiple redesign cycles. This is a challenge, if not a crisis, and has certainly escalated the costs of some development projects.

Aerospace standards, for example Aerospace Recommended Practice (ARP) 4754 (SAE 1996A) and 4761 (SAE 1996B) introduced the notion of Preliminary System Safety Assessment (PSSA). PSSA is based on a system architecture, is predictive, and will conclude that a design is likely to meet its safety requirements, or not, at an early stage, enabling cheaper design revision, before systems are realised and subject to (full) FMEA and FTA. A later issue (ARP4754A (SAE 2010)) introduced Preliminary Aircraft Safety Assessment (PASA), extending this notion earlier in the life cycle. The underlying theory is no different, but there is an emphasis on "designing to meet" requirements rather than conducting retrospective analysis to see if designs are adequate. Like HAZOP it can be seen as a pragmatic enhancement rather than a new theory. However it is an important enhancement – an improvement in Staples' role 1 – and without it aircraft design processes would be prone to major rework with consequential costs or delays to entry into service.

There was, however, a new "theory" to address problems of complexity, in response to the need to deal with systematic errors, as opposed to random errors that are addressed by tools such as FTA and FMEA. In a sense the crisis was that the existing theories could not analyse systematic errors, and hence were incomplete. Many standards introduced the concept of Safety Integrity Levels (SILs) to address the crisis. The civil aerospace community introduced Development Assurance Levels (DALs); although DALs are different to SILs, at the detailed level, they are aimed at solving the same problem.

It is quite difficult to characterise the underlying "theory". It is perhaps that "greater rigour, and greater independence in verification reduces the probability of residual systematic errors" although that sounds more like a hypothesis than a theory. The variation in the detailed "prescriptions" for SILs and DALs between standards also suggests immaturity in (or absence of a consistent) theory.

Again, there is an argument to say that they are a pragmatic (management) tool and don't really have a strong underpinning. This is not to say that SILs and DALs are not useful, just to query their maturity, and therefore the ability to test their soundness.

3.4 Socio-Technical Systems

Just as it became clear that theories based on random failure were insufficient to address the complexities of software-intensive systems, it also became apparent that treating systems as purely technical was inadequate. The realisation that non-technical factors – the wider social and human systems – were an important part of safety dates back many years, and the issues have been "rediscovered" many times. However, the underpinning theories of safety engineering did not have room for the complex feedback structures inherent in socio-technical systems.

Attempts were made to incorporate human factors in FMEAs, FTAs and safety cases, but the validity of representing humans in the same way as technological components was controversial. This became even more difficult when concern was extended beyond individuals to encompass teams, organisations, and developments such as organisational drift (Snook 2000) and normalisation of deviance (Vaughan 1996).

A range of new techniques was developed to reflect the extent to which safe (or otherwise) behaviours depended on the interaction between the system and the wider social and human systems. This is consistent with Lakatos' view of theoretical progress as a response to an accumulation of difficulties. Perhaps the most notable of the new models proposed to deal with the challenges of socio-technical systems are those developed by Eric Hollnagel and Nancy Leveson.

The Functional Resonance Analysis Method (FRAM) (Hollnagel 2012) introduces a new notion (theory) of "functional resonance" to complement causal models or theories. Loosely functional resonance might be thought of as (not necessarily intended) interactions or dependencies between functions, and moves away from a time-sequence model of causality. Leveson's System Theoretic Accident Methods and Processes (STAMP) (Leveson 2012) builds on classical systems theory and systems thinking, but expanding the scope of influences on systems out to the political level. In some senses this is not a new theory but it is a new application. Also the view of failures here is of "control failures" so this is another shift in the models of failure and causality to deal with socio-technical systems.

Although the references for FRAM and STAMP are relatively recent, the ideas have been evolving for some time. During the same period work was published on "resilience engineering" (Hollnagel et al 2006) setting out the need to make systems "resilient" to make them safe, and the notion of Safety-I and Safety-II was articulated, to define progress in safety engineering. As indicated here, in the author's view there are more than just two "stages" in safety engineering. Further, I would distinguish robustness as designing a system (in the broadest sense) to respond (safely) to known/anticipated failure modes, and resilience as an extension to deal with unanticipated failure modes. This extension is normally achieved by generalising recovery mechanisms and is well established in systems engineering practice. Thus I remain somewhat sceptical about the extent to which resilience engineering really reflects an advance in theory or thinking – although it certainly does in articulation of approaches and precepts.

3.5 Autonomy and Systems of Systems

One of the obvious challenges for safety engineering – and perhaps a crisis – is the growing pressure to deploy autonomous systems, most notably on our roads. One of the (unstated) assumptions behind most safety analyses is that the behaviour of a system is known prior to operation, so it can be assessed – and the assessment is valid throughout operation. Where systems learn or adapt, this assumption is broken. Not all autonomous systems are necessarily adaptive, and the adaptation may just be optimisation within bounds that can be analysed but "full autonomy" is not readily assessable with current techniques. There is some published work in this area, but no real consensus on how to address the problems. To a large extent the safety and artificial intelligence communities have not worked together on these issues (although there was a recent White House-sponsored meeting on the topic) but much remains to be done to develop new theories and methods in this area.

Systems of Systems (SoS) are complex for technical reasons, but also in the author's view because of the lack of a central authority. Put another way, individual systems in the SoS have different objectives and owners and this is a challenge for safety assessment, as there may be multiple perspectives on risk, amongst other difficulties (Rae and Alexander 2011).

SoS and autonomy are linked here because there are situations where it is necessary to address both aspects at once. In the UK, for example, there is a large body of work on connected autonomous (road) vehicles (CAV). CAV will, by definition, be autonomous and through their connectivity will be able to act as an SoS – for example agreeing on speeds to maximize road throughput. Whilst there has been some work on SoS safety including on road systems (Arlow et al 2006) current proposed automotive systems are beyond the state of the art for safety assessment and might, for example, require dynamic safety cases, i.e. updates to the assessment of safety as systems operate. It seems reasonable to say that there is no mature/accepted theory for dealing with such situations, at present.

3.6 Service Oriented Architectures

A further challenge, and one that might prove a crisis for some organisations, is the application of service-oriented architectures (SOA) in safety-related or safety-critical applications. Engineering is about money – or at least subject to cost-constraints; for large-scale, complex, information systems there is great pressure to use commercial-off-the-shelf (COTS) technologies, including very powerful and flexible computer systems (including but not limited to "the cloud"). SOA are usually managed by service level agreements (SLAs) or OLAs (O for Operational). These resemble commercial contracts and a key concept is compensation for failure (to meet the SLA). Such a commercial model does not obviously help with safety – indeed it can be viewed as inimical to safety, and it could cause a crisis.

Arguably this area is even less mature than autonomy and SoS. It may be that the growing focus on "data safety" and the ongoing work of the Data Safety Initiative Working Group (DSIWG) which is updated periodically (DSIWG 2017) will form part of the solution framework for SOA, but it is unlikely to provide a complete theory – as SOA are wider in scope than data, although a "data perspective" may provide a useful abstraction.

3.7 Cyber Security

Cyber security has received more attention than SOA since it is more obviously a problem, as it potentially affects all systems – and most particularly those that are network connected. Here the focus is on the interplay between safety and cyber security, in particular the way in which cyber vulnerabilities might be exploited to impact safety. This issue has been studied for more than a decade, e.g. in the SafeSec project (Lautieri et al 2005) and by the Department of Homeland Security (DHS), and it is interesting to consider what a new theory would look like.

A "harmonised" theory for safety and security – or of cyber-influenced safety – would need to unify some of the concepts and precepts of the two domains, and resolve the (apparent) divergence between unintentional (safety) and intentional (security) causes of harm. There have been several attempts to do this, producing both abstract models, for example by Firesmtih (Firesmith 2003), and practical applications, e.g. to analyse railway systems. Whilst there is merit in this work, it serves primarily to show how difficult it is to merge the concepts, and it is not clear (at least to the author) what a new theory of "cyber-influenced safety" would look like. However one of the more successful avenues might be to look at trying to start with one of the socio-technical theories, rather than the more traditional safety models. (The Safety Critical Systems Club (SCSC) discussed some of these issues in their 'New Challenges' seminar (SCSC 2016).)

3.8 Standards

Standards should embody the understanding (in a domain) of what is needed to achieve and assure safety. In the terms set out here they should reflect theories that are rich enough to provide a sound basis for engineering and assessing systems. Before reflecting on this directly, it is interesting to consider an incident some 25 years ago. When giving a talk at a conference on safety of aircraft I said 'you have to meet the standards, and make it safe'. Having realised what I had said, I was in some trepidation as some of my research sponsors were in the audience (would they become ex-sponsors before the day was out?). But the first one I met said to me 'oh, so you do get it then'. As well as providing relief, this made me start to think about the nature of the standards-making process.

Standards are the result of group consensus and, over time, they evolve – in some cases to reflect challenges or crises. However there are practical limits to this sort of evolutionary process. First, it is necessary to get consensus – and it is easier to agree on the "historic" things than on the new ideas that address challenges. Second, they take a long time to produce or update, with more than a decade between issues, for several important standards. Material gets "frozen" so people are reluctant to go back to reconsider agreed text and out-of-date material can remain in published standards. Third, standards are politics (with a small "p"). If a company believes they have "the edge" over competitors they may wish to get a standard written so they can apply their advantageous methods, but without giving away to others their precious know-how. Fourth, practice in using standards often has relatively little influence on their updates – whilst standards committees are often open to the community, only those with time and (usually) company support can afford to attend, often over several years. And so on.

Despite these limitations there is a strong emphasis on "standards compliance" in many domains – based on my own experience, this seems to be particularly prevalent for railway systems. One of the reasons for this is apparent – it removes a lot of the uncertainties about how to run a development project. This contrasts with safety cases that give flexibility in how to reason about safety – but at the cost of *increased* uncertainty. In fact it is possible to see the contrast with safety cases in Lakatos' terms – with safety cases and standards compliance being "competing projects" (although some have tried to reconcile the two approaches). Both these "competing projects" have their weaknesses, or limitations – but neither project has yet been dealt a "fatal blow" by problems that have arisen.

To be more specific about "challenges" to (crises for) the use of standards, most standards focus on analysis and management of failures. It is well understood in the automotive community, for example, that the safety of normal functionality needs to be considered for autonomous vehicles, and analysis is likely to have to include the interaction between normal functions in connected vehicles. Work is under way to update the automotive industry standard (ISO 2011) but, at the time of writing, it does not address these issues and it may be some time before it does. This is not to criticise ISO 26262 or its developers – just to make concrete the general problem that standards often lag the state-of-the art in system development. Similar phenomena can be seen in other domains, where standards lag industrial (development) practice, and there are unresolved problems in dealing with systems currently being deployed – forever playing catch-up, perhaps?

4 The Inevitability of Playing Catch-Up?

This section considers what it would mean for the safety community to "get on the front foot" then discusses some possible ways in which that might be achieved.

4.1 Getting on the Front Foot?

What would it mean for the safety community to "get on the front foot"? In essence it would be that it was prepared for evolving technology and applications, so what might otherwise have been crises are, instead, simply met by applications of known principles and methods.

To an extent, §3 has set out the difficulties in "getting on the front foot", but there are other pragmatic issues. For example, in some industries, it is difficult to make a business case for work on safety unless it can be shown to reduce project costs or risks. Often the safety of the end-product is taken as a "given" by senior management, so investing to consider the safety of emerging technologies is hard to justify (although the consequences of not doing so can be very expensive).

4.2 Opportunities

There are several ways in which the safety community might "get on the front foot"; this section discusses some of the possibilities and opportunities.

4.3.1 Theories

In principle, if a "universal" theory of safety engineering were produced then it would be robust against new situations and systems. Whilst this sounds like the "holy grail" it might be possible for some aspects of the discipline – humans and organisational frailties change much less rapidly than technology and systems, so it might be that an approach such as FRAM or STAMP could be a stable part of any theory, focusing on the "socio" part of any socio-technical system.

An important question that any such theory must address is the relationship between safety engineering, safe design, and social acceptance of technology. If, as Rae and Alexander suggest (Rae and Alexander 2017), there is a fundamental goal conflict between providing assurance and improving safety, what is the role of safety engineering in negotiating this conflict?

4.3.2 Principles

There are already sets of principles, including the "4+1" principles (Hawkins et al 2013) for software safety assurance. Principles aren't really the same as a "theory" as they don't have the explanatory power, but they can be very useful in guiding practice. They can perhaps be thought of as having a similar lifecycle to theories of engineering – although it is perhaps more likely that "crises" will lead to the identification of limits to the use of the principles, not their rejection.

There are many design principles, e.g. redundancy, control and monitor, etc. which can be applied widely, albeit with constraints on their valid application. Some of these can be applied to autonomous or adaptive systems, e.g. control and monitor, where the adaptation is optimising within a known envelope, but this is unlikely to be possible where the adaptation deals with previously unmanageable situations, e.g. flying an unmanned air vehicle (UAV) following battle damage. In a sense there is nothing new here – principles for products are rather like design patterns with known limits to their application.

Similarly, one can think of operational principles, including those dealing with human behaviour. Principles of rostering for (safety-critical) shift work are well understood, in some domains. At a finer grain, there are principles for design of user interfaces to minimise human error. Again there are limits; for example there has long been evidence, e.g. from the Federal Aviation Administration (FAA), that excessive (reliance on) automation can increase human error (FAA 1996), and this should influence design. Considering proposed levels of autonomy in cars there is a need for principles related to the "handover problem" when control is "given back" to the driver after long periods of autonomous driving, so the driver regains situation awareness before accepting control. Other domains, e.g. semi-automated trains, could benefit from the same principles. Perhaps this is an area where the safety community can get ahead – or at least not be lagging so far behind, if it works on the issues before widespread deployment of autonomous vehicles.

Further, it seems that there is value in principles for design and assessment, or assurance although, again, these may have limits and consequently may need extending. Indeed, there is work on evolving the "4+1 principles" for software assurance to deal with non-traditional software (Ashmore and Lennon 2017), these principles are reflected in the data safety guidance (DSIWG 2017), and perhaps there would be value in similar extensions for SoS, etc. Such an approach might not yield "concrete" methods of analysis, but could provide an agreed approach to assurance.

An important aspect is how the principles are articulated – it is not practicable to go into detail here, but the use of patterns may be appropriate in some cases.

4.3.4 Risk Analysis

Risk analysis is core to safety engineering – yet the evidence is that QRA is not accurate, in practice, and it is not evident that it is more effective than qualitative analysis (Rae et al 2014). That paper proposes a model for improving the maturity of QRA, but there are other opportunities for improvement.

The principle that risks should be reduced ALARP is set out in British law. This requires a risk versus benefit argument to be made if risks can't be made "broadly acceptable". But this ignores the "upside". For example, autonomous driving reduces the risks due to human error – the primary cause of automotive accidents. By embracing an assessment of risk versus benefit for *all* capabilities, this would enable safety engineers to have a much more constructive involvement in design (see below for more on this). This could be encoded in two ways: that there is a net risk reduction, taking all factors into account; that no one sector of those at risk is disadvantaged for the benefit of others, e.g. pedestrians should not be put more at risk to achieve a net reduction in risks on the roads.

A further shift in risk analysis would be a move to (or towards) dynamic risk assessment. Arguably an additional problem for QRA is the implicit assumption that it is possible to evaluate risk, prior to system deployment, for the whole of the system's life. This may lead to worst-case assumptions being made but, more fundamentally, it is a theoretical "flaw" assuming that the statistics are time invariant. To enable or exploit such a shift, system behaviour would have to be "risk-aware", again meaning that safety engineers would have to work more closely with designers to optimise capability. This perhaps sits best with autonomy – humans adjust their behaviour based on (intuitive) assessment of risk, so shouldn't autonomous systems also do so?

Such approaches would probably involve the establishment of new principles, if not new theories – perhaps DRABTO (Dynamic Risk And Benefit Trade-Off) as opposed to ALARP. They would also have technological challenges, but the notion of designs "explaining" their reasoning, as was done for early expert systems, may be a good model to aspire to.

However there is a further problem – there is a difference between "risk" as perceived by engineers, and how the term is used by sociologists, and perhaps more generally by the public. Often here risk relates to uncertainty – which the advocates of QRA seek to remove (or ignore). There are eminent Professors of the "public understanding of risk" – but maybe we need Professors of "understanding how the public perceive risk". For example the attitudes to autonomous cars seem to be bimodal – very high risk ('I wouldn't use one'), or a net benefit to society as human drivers (and especially young males) are a major cause of road accidents. Maybe DRABTO could help to resolve the dichotomy and gain public acceptance.

4.3.5 Emphasis

Another opportunity is to recognise that there could be (should be?) more commonality in process rather than product. Put another way, the community can produce theories that emphasise know-how rather than know-what, which is inevitably more domain dependent. The result of embracing this emphasis might be a "theory of safety-informed design", noting that there are inevitably trade-offs and although safety is important it is not always the ultimate arbiter in system design trade-offs (ALARP tells us this, let alone DRABTO).

The variance in approaches to SILs (and DALs) between domains and standards suggests that producing a "unified" theory for safety-informed design will not be easy, or perhaps will only be done at an abstract level. However it should be noted that the more systems are interconnected or integrated, e.g. using the so-called Internet of Things (IoT), the more problematic it is to have different processes, rules for SILs, etc. across domains. So harmonization might help the safety community to "get on the front foot" for the introduction of the IoT!

4.3.6 Learning, Evaluation and Education

Another opportunity is through learning. Almost by definition, learning is a lagging not a leading process, but do we need to lag so far? The above-mentioned FAA work on the difficulties posed by automation is now twenty years old – but it seems to have had little influence. There are two aspects to this.

First, accidents and incidents tell us about the limits of our knowledge and theories, as well as about mistakes in their application. Thus accident and incident investigation gives an opportunity for Learning from Experience (LfE) that goes far beyond the individual system or incident. Arguably, LfE is good in some sectors, e.g. civil aerospace as seen by declining accident rates (Boeing 2016), but there is a need for the community to focus more strongly on LfE, and potentially to modify the way accident and incident investigation is done to enable this.

Second, the safety community could adopt a much stronger focus on evaluation – actively seeking to determine whether or not our approaches or theories work. The author has previously said that most standards are 'unverified hypotheses' and this, of course, leads us back to the discussion of progress in science and engineering – if we don't look to see whether or not our theories work, then we are unlikely to progress. There are difficulties in evaluation – for example in civil aerospace the target rate for catastrophic hazards is 10^{-9} per flight hour, so such hazards should not be seen in the life of an aircraft type, if the safety work has been done well. However there are more "directly testable" criteria that we can apply to theories and methods (Rae et al 2010), and the community would benefit from taking a more systematic approach to evaluation.

Further, the rather philosophical issues raised here should be seen as pointing out the need for better education (rather than training). In teaching about safety the boundaries of applicability of methods should be made clear – what they are good for, and where they (may) give misleading results. Some courses (including those from my own University) seek to do this, but this is not universal and it is hard to get across this sort of understanding. A lot of teaching is done from the viewpoint of a particular "project" in Lakatos' terms, e.g. standards compliance. Should we not be educating safety engineers about the strengths and limitations of (all the) techniques, enabling them to decide what tools to use when?

These ideas are dawn together, as LfE and evaluation are related – and both provide the critical insights necessary to provide education rather than training.

5 Conclusions

The introduction posed two questions, intended to make the general question about 'just playing catch-up' more specific. The essence of the two more specific questions is 'what can we as a discipline do to get "on the front foot?"' and 'what is the most effective way we can learn from events?'

Realistically, I do not believe we can ever get ahead of design engineering, at least not entirely. Engineering includes numerous examples of practice preceding theory, e.g. the progress of aviation from 'fly-fix-fly' to being able to predict handling qualities (Vincenti 1990). Also the experience of Sir James Dyson making 5,126 unsuccessful prototypes before producing a successful bagless vacuum cleaner (Dyson 2014) shows the difficulty of evolving commercial products. But if the designers don't know how it will work, what can safety engineers do? To put it another way, if we don't have a pre-existing theory for design, it is hard to see how we can produce one for safety analysis. But it seems to me we are 'playing catch-up' from too far behind – to use an analogy, we should be running on the shoulders of the designers not several laps behind. How can we do this? Some of the ideas in §4 are intended to help. Focusing on principles should enable the community to adapt to new problems more quickly; the same is true for the proposed emphasis on process. The suggested move towards DRABTO should enable designers and safety engineers to work together – to a shared objective of producing an effective design that is safe as designed. Even running together, perhaps?

The idea of working more closely with design is a cultural shift, but one that seems to be necessary. This brings us to the second question – which also needs a cultural shift. Caricaturing safety engineering it seems we are content to operate "open loop" – assessing designs, but only being concerned about the effectiveness of methods when accidents occur (and even then often being disinclined to abandon weak theories). This seems to be a very unsatisfactory culture, but consistent with what Lakatos said about "projects". It therefore seems to me that the community needs to consider evaluation and LfE as a core part of what it does – this would enable us to make more rapid progress and to be able to reject, rectify or replace unsound theories much more quickly than we do now.

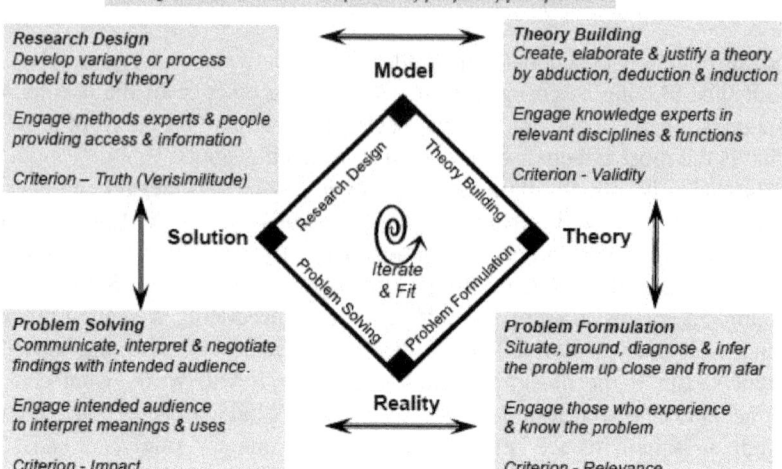

Fig. 3: A Model of 'Engaged Scholarship' from Van de Ven (Van de Ven 2007)

There is perhaps a more profound shift needed – that of accepting that safety engineering is more like social science than science. The model of 'Engaged Scholarship' (Van de Ven 2007) seems to me to be much more representative of how we work in safety than the scientific model. It also perhaps indicates why it is hard to transfer results between domains, because tangible progress is actually made embedded in particular industrial or organisational settings, and findings don't necessarily translate easily to other settings (let alone domains). This may help to explain why Lakatos' views of competing "projects" is so appropriate for safety as evidence from one domain (even counter-evidence which Popper says should refute a theory) is not necessarily very compelling in another, so it is very hard to accumulate compelling evidence to replace a theory (the fact that the ideas discussed in §3 overlap hugely in time also supports this). To me the "self-critical and learning community", that evaluates what it does, using the sorts of engagement schemata outlined above, is a better model than the paradigm revolution called for by Hollnagel and others under the banner of 'Safety-II'.

Acknowledgments Thanks are due to Mike Parsons and Drew Rae both of whom made helpful comments on an earlier draft of this paper; in particular Drew helped me sharpen the main message about progress in the field. I also owe thanks to the many people I have worked with in the safety community over more than 30 years who have helped to shape my understanding of the subject. Of course, the responsibility for any errors of understanding and excessively wild hypothesising remains mine.

References

Arlow A, Duffy C J and McDermid J A (2006) Safety Specification of the Active traffic Management Control System for British Motorways, in proceedings of the 1ˢᵗ IET International Conference on System Safety.

Ashmore, Lennon (2017) Progress Towards the Assurance of Non-Traditional Software in Developments in System Safety Engineering, SCSC 2017

Boeing (2016) Statistical Summary of Commercial Jet Airplane Accidents | 1959-2015 http://www.boeing.com/resources/boeingdotcom/company/about_bca/pdf/statsum.pdf (last accessed 10th October 2016)

Brooks, Jr, FP (1996) The computer scientist as toolsmith II, Communications of the ACM, 39(3):61-68.

The Hon Lord Cullen (1990) The Public Inquiry into the Piper Alpha Disaster, Her Majesty's Stationery Office.

DSIWG (2017) http://scsc.uk/p133 (accessible from February 2017)

Dyson   J   (2014)   http://www.theglobeandmail.com/report-on-business/innovators-at-work/james-dysons-road-to-success-paved-with-failure/article20074990/   (last   accessed 10th October 2016)

Edwards v. National Coal Board (1949) All ER 743

FAA Human Factors Team (1996) The Interfaces between Flightcrews and Modern Flight Deck Systems.

Firesmith DG (2003) Common Concepts Underlying Safety, Security, and Survivability Engineering, Technical Note CMU/SEI-2003-TN-033, Software Engineering Institute.

Haddon-Cave C (2009) The Nimrod Review: An independent review into the broader issues surrounding the loss of the RAF Nimrod MR2 Aircraft XV230 in Afghanistan in 2006, The Stationery Office.

Hawkins RD, Habli I, and Kelly TP (2013) The Principles of Software Safety Assurance, In Proc of International System Safety Conference, System Safety Society.

Hollnagel E., Woods, DD. and Leveson NC (Eds.) (2006) Resilience engineering: Concepts and precepts. Ashgate.

Hollnagel, E. (2012) FRAM – The Functional Resonance Analysis Method. Farnham, Ashgate.

ISO (2011) Road vehicles – Functional safety, ISO 26262.

Kelly TP (1998) Arguing Safety – A Systematic Approach to Managing Safety Case, PhD Thesis, University of York.

Kletz TA (1999) HAZOP and HAZAN: Identifying and Assessing Process Industry Hazards, Institute of Chemical Engineers.

Kuhn TS (1962) The Structure of Scientific Revolutions, University of Chicago Press.

Lakatos I (1974) The role of crucial experiments in science, Studies in History and Philosophy of Science, 4 (4): 344-355.

Lautieri S, Cooper D and Jackson D (2005) SafSec: Commonalities Between Safety and Security Assurance, in Proceedings of the Thirteenth Safety Critical Systems Symposium, Springer Verlag.

Leveson NC (2012) Engineering a Safer World: Applying Systems Thinking to Safety, MIT Press.

Mackie JL (1965) Causes and Conditions, American Philosophical Quarterly, 12: 245–65

Petroski H (1992) To Engineer is Human: The Role of Failure in Successful Design, Vintage Books.

Popper K R (1959) The Logic of Scientific Discovery. Routledge.

Popper K R (1977). The worlds 1, 2 and 3. In: Popper, K R and Eccles, J C. (eds) The Self and Its Brain: An Argument for Interactionism, Routledge.

Rae AJ, Alexander RD (2011) Is the "System of Systems" a Useful Concept for Hazard Analysis, 29th International System Safety Conference

Rae AJ, Alexander RD and McDermid JA (2014) Fixing the Cracks in the Crystal Ball: A Maturity Model for Quantitative Risk Assessment, Reliability Engineering & System Safety, 125: 67-81

Rae AJ, Alexander RD (2017) Probative blindness and false assurance about safety, Safety Science)to appear, March 2017)

Rae AJ, Nicholson M and Alexander RD (2010) The state of practice in system safety research evaluation, 5th IET International Conference on system safety.

SAE (1996A) Certification Considerations for Highly-Integrated or Complex Aircraft Systems, ARP 4574.

SAE (1996B) Guidelines and Methods for Conducting the Safety Assessment Process on Civil Aircraft Systems and Equipment, ARP 4761.

SAE (2010) Guidelines For Development Of Civil Aircraft and Systems, ARP 4754A.

SCSC (2016) http://scsc.uk/e444, accessed December 2016

Snook SA (2000) Friendly Fire, Princeton University Press

Staples M (2014) Critical rationalism and engineering: ontology. Synthese, 191(10): 2255-2279

Staples M (2015) Critical rationalism and engineering: methodology, Synthese, 192 (1): 337-362

Turner BA (1976) The organizational and interorganizational development of disasters, Administrative Science Quarterly, 21(2): 378-397

Van de Ven AH (2007) Engaged Scholarship: A Guide for Organisational and Social Research, Oxford University Press

Vaughan D (1996) The Challenger launch decision, University of Chicago Press

Vincenti WA (1990) What engineers know and how they know it: analytical studies from aeronautical history, Johns Hopkins University Press

Weaver RA (2003) The Safety of Software – Constructing and Assuring Arguments, PhD Thesis, University of York

The "rise of the machine" and the need for a System-of-Systems safety methodology?

Mike Brownsword, Andy German and Ian Mitchell

Atkins

Abstract *The approaches used today for assuring safety of Systems-of-Systems have evolved and matured over many years. Governments are increasingly encouraging automation and businesses are progressively digitising our System-of-Systems, which may ultimately result in the removal of skilled people from positions of control. Examples of these digitised System-of-Systems include driverless trains, planes/drones, submersibles and cars; however this trend extends into all industry domains including the medical and defence areas. The "rise of the machine" cannot be stopped and therefore results in the need for designers and safety engineers to think about: (1) significant System-of-Systems issues and epochs; and (2) how safety can be designed using a common methodology or codes of practice. This paper draws on experience from work undertaken in a number of domains including energy, rail, defence, aerospace and information technology systems. This paper explores System-of-Systems problems including: epoch identification, risk ownership, disparate legislation and regulation, and why bottom-up summing of component safety arguments fails to address the problem. It proposes the need to imagine and develop safe systems of operation and test these. It considers how techniques such as Soft Systems Methodology, Systems Engineering, Systems-Theoretic Accident Model and Processes (STAMP) and Hybrid Development Lifecycles may aid System-of-Systems safety understanding and may help to expose potentially harmful emergent properties.*

1 Introduction

A System-of-Systems (SoS) refers to systems that contains two or more independently managed elements with different parts being subject to different management, investment and control policies (Boardman and Sauser, 2006) although experience is that the System-of Systems will be constrained by existing safety legislation and regulation. The SoS overall socio-technical safety control elements are also subject to differing investment and risk control systems.

Today's SoS safe systems of operation have been organically grown over decades, if not the past couple of centuries, examples being the civil air traffic management system, road transport system or the London underground. These safe systems of operation have often been developed following an accident or investigation, and have undergone course correction through legislation, regulation, standards and codes of practice. Governments are encouraging automation and businesses are digitising our SoS with the aim of removing skilled people from positions of control, such as driverless trains, planes and automobiles. This results in the need for designers and safety engineers to think about significant SoS epochs and how safety can be designed in a common and standard way. In addition, McKinsey (Chui, Manyika, and Miremadi 2015) suggests that as many as 45 percent of the activities individuals are paid to perform today can be automated by adapting currently demonstrated technologies. The adoption of workplace automation has the ability to change the socio-technical controls and hence safety management and safety culture of organisations, which in turn affects SoS safety. We may end up with socio-technical machines managing the safety of SoS machines sooner than we think.

2 Separating revolution from evolution

The need to address SoS safety necessitates careful consideration, as it requires socio-technical and activity modelling across disparate independently managed elements and networks; this can be difficult and time consuming.

Most system safety engineers work on elements of a SoS that requires them to meet extant legislation, regulation, standards and codes of practice with existing safe systems of work, that maybe explicit or implicit, having been developed through custom and practice. The safety analysis is considered within this given socio-technical system and network and is largely cause-hazard-accident centric. Here the SoS and their management evolve over time usually resulting in a downward trend in harm.

When a SoS epoch is started, those wanting to make the change are unlikely to have control over all the system's associated interactions and interfaces, and may not realise that what is being proposed is revolutionary. Here it would be useful if a set of judgement criteria were available to aid the classification of a SoS as either evolutionary or revolutionary. It is suggested that this classification can be achieved based on questions around both the socio-technical systems and the disruptive technologies. This can then be used to justify the investment required to ensure safety is adequately addressed for both the technology and socio-technical elements of the SoS.

From our experience on a number of SoS projects, the following questions are used to aid classification:

1. Are significant changes (including using new technologies) being made to an existing SoS or a new SoS being introduced?
2. Is the proposed SoS covered by current legislation and regulation in terms of its *overall* scope?
3. Do industry product and operational safety standards and design codes of practices define how risks should be managed?
4. Does the existing "safe system of work" (both implicit and explicit) provide risk management that is socially acceptability?

The typical responses include:

1. For some projects, it is easy to identify that it is revolutionary such as Digital Railways. Ben Dunlop, Director of Digital Railway at Atkins *"It shakes to the core the existing working practices and methods that the industry has relied upon for the past 180 years, and it challenges almost every facet of the industry"*. (Dunlop 2016) However, this was not as clear for other projects such as the digitisation of medical records.
2. Goal based legislation and regulation allows for considerable innovation and is therefore not a discriminator, however some regulation is proscriptive, and deliberately constrains innovation. For example, the operation of Unmanned Aerial Systems (UAS) and driverless cars are required to meet the safe operating requirements developed over the past 100 years for systems with crews and drivers. Legislators and Regulators are rightly conservative in their approach unless significant safety gains are demonstrable. The Office of Road and Rail (ORR) has been in consultation about its approach to innovation including Digital Railways, including how it will affect different aspects of the industry (Brown 2015). It has been found that it can be difficult to classify the SoS based on legislation and regulation.

3. Characteristically, industry product and operational safety standards and safe design codes of practices lag behind current technologies partly because these are created by committees, however, if the SoS technology is not addressed and is disruptive then this is clearly a revolutionary change.

4. Our experience in this area is that existing safe systems of work tend to be implicit (custom and practice) and when made explicit aid hazard classification, safe design and the mitigation of risks.

In summary, the provision of a common question set may aid understanding of whether a SoS is evolutionary or revolutionary and help justify the safety engineering approach and associated investment that should be made in understanding the technology and socio-technical safety attributes.

3 Risk Ownership

Responsibility for SoS safety and hazard management remains a problem as the people affected by the SoS may not be in a position to ensure that all elements of the SoS are in a safe state.

Like the Office of Road and Rail Regulation, the Defence Safety Authority and its Safety Regulators have made the requirements for safety Duty Holding clear, requiring them to have safety cases for activities. The MOD *"requires the appointment of Duty Holders where it has been assessed that there is credible and reasonably foreseeable Risk to Life (RtL) from a Defence activity"*. It also requires the use of a safety case as it *"provides the ability to understand the cumulative or interrelated risks from the use of the complex system and for this to be captured in a body of evidence"* (Defence Safety Authority 2016).

The Duty Holding construct applied in the UK is useful as it clarifies who is responsible for the safety components of SoS even for something as complex as a nuclear submarine with multiple Duty Holders which are regulated by several different regulator, including:

- Office for Nuclear Regulation;
- Defence Nuclear Safety Regulator;
- Defence Fire Safety Regulator;
- Defence Maritime Regulator; and
- Defence Ordnance, Munitions and Explosives Safety Regulator.

Understanding who the Duty Holders are is key to negotiating risk ownership and socio-technical controls for an evolutionary SoS – although this maybe complex and require careful negotiation. If Duty Holding is unclear, it may be necessary to talk to the regulators.

4 Summing the product safety cases

A variety of recent SoS Safety Cases have attempted to argue that as the constituent system components are safe, the overall SoS is safe. Unfortunately, these safety cases have not taken into account/explored the following:

- The safe activities the component systems and the SoS were required to undertake;
- The states that the system elements may be in (including maintenance);
- The potential SoS functional failures including the loss of safety features;
- Emergent properties; and
- The socio-technical safety management system.

Although equipment safety assessments are an important element in arguing that an activity is safe to undertake, it can never provide a complete *"body of evidence that provides a compelling, comprehensible and valid case that a"* SoS *"is safe for a given application in a given operating environment"* (00-056 Issue 6).

5 Safe and safely managed "SoS"

Recent SoS projects have highlighted the value of particular tools and techniques as these can aid the development of risk control systems.

5.1 Levers for change

The introduction of revolutionary SoS requires careful management. The MOD acquires new capability that is required to provide benefits across all the Defence Lines of Development (DLoD) of the Defence Enterprise. Table 1 below provide a definition of the DLoDs. These DLoDs provide a useful aid in the understanding all the elements of that need to be considered when conducting a SoS safety assessment.

Table 1 - Defence Lines of Development

Training	The provision of the means to practise, develop and validate the practical application of a common doctrine to deliver a capability.

Equipment	The provision of platforms, systems and weapons needed to outfit/equip an individual, group or organisation.
Personnel	The timely provision of sufficient, capable and motivated people to deliver outputs now and in the future.
Information	The provision of a coherent development of data, information and knowledge requirements for capabilities and all processes designed to gather and handle data, information and knowledge.
Concepts and Doctrine	A Concept is an expression of the capabilities that are likely to be used to accomplish an activity in the future. Doctrine is an expression of the principles by which military forces guide their actions and is a codification of how activity is conducted today. It is authoritative, but requires judgement in application.
Organisation	Relates to the operational and non-operational organisational relationships of people and organisation including contractors.
Infrastructure	The acquisition, development, management and disposal of all fixed, permanent buildings and structures, land, utilities and facility management services (both Hard and Soft facility management (FM)) in support of capabilities.
Logistics	Planning and carrying out the operational movement and maintenance of forces. In its most comprehensive sense, it relates to the aspects of operations, which deal with: the design and development, acquisition, storage, transport, distribution, maintenance, evacuation and disposition of materiel; the transport of personnel; the acquisition, construction, maintenance, operation, and disposition of facilities; the acquisition or furnishing of services, medical and health service support.
Interoperability (Overarching Theme)	In addition to the DLoD, Interoperability is included as an overarching theme that must be considered when any DLoD is being addressed. It is the ability to train, exercise and operate effectively together in the execution of assigned missions and tasks. It also covers interaction between Services, capabilities, other Government Departments and the civil aspects of interoperability, including compatibility with Civil Regulations.

The Digital Railways project developed the equivalent to the DLoDs, which were defined as "levers for change", again supporting a comprehensive understanding of how today's railway can be successful transformed through the proposed revolution. The levers for change used by Digital Railways (Digital Railway 2016) are:

- Government Policy and Legislation;
- Industry Standards and Business Policy;
- Processes;
- Business information;
- People (Skills);

- Technology;
- Infrastructure and Assets;
- Ways of working; and
- Organisational design.

Both sets of "levers for change" have been found to provide valuable prompts to safety engineers as these provide assurance that the SoS safety assessment is complete and the appropriate risk control systems can be put in place.

5.2 Soft System Methodology

Soft Systems Methodology (SSM) is an approach to organisational process modelling and leads itself to SoS understanding (Peter Checkland and Jim Scholes 2009). It has been successfully used over the past three decades. It provides for a seven-step process starting with a definition of the problem situation leading to the modelling of several "human" activity systems relevant to the problem situation. These models are used with stakeholders to explore: (1) what kind of changes can be feasibly imagined, are desirable and could be valuable in defining how safety aspects of the problem situation can be addressed; and (2) how safe systems of operation developed and tested. Nevertheless, as already discussed, the machine and human activities are being blurred as machines take on more management roles and the assessment needs to consider these changes.

5.3 Systems-Theoretic Accident Model and Processes (STAMP)

Accident and Hazard based assessment techniques tend to be based around equipment or platform failures and are poor at representing shared SoS risks, organisation and management deficiencies, and safety culture issues. Bowtie representations of hazards can address some of these deficiencies, if well constructed but as with all reductionist techniques do not easily allow for the sum of the organisational, management and safety culture issues to provide an overall understanding.

When considering all the tools and techniques available to a safety engineer then one stands out when considering the challenges offered by a revolutionary SoS, this being the Systems-Theoretic Accident Model and Processes (STAMP) (Leveson 2004). STAMP is predicated on the ability to define the accidents that may occur within the System of Systems environment. Potential accidents are relatively easy to define within a domain with some forward thinking regulators providing accident taxonomies. STAMP views safety as an accident control problem, and safety is managed by a socio-technical safety management system – even one that contains automation of management functions. The goal of the socio-technical safety management system is to enforce the safe design constraints on SoS development and the resulting System-of-System instantiation that in turn provides for the safe systems of operation and risk control systems.

5.4 Hybrid Development Lifecycles

As previously stated a SoS is a system that contains two or more independently managed elements with different parts of the system being subject to different management, investment and control policies. This results in components that have different lifecycles and changes in capabilities. These hybrid life cycles are more complex and therefore more difficult to manage.

The ability to create infrastructure, mechanical and structural engineering elements of systems continue to improve at a slow pace. The digitisation of systems brings its own problems as the technologies evolve at speeds that can mean technology is obsolete before the infrastructure, mechanical and structural engineering design is complete with disruptive technologies changing the businesses operations. Mike Wilkinson at Niteworks argues, *"A single lifecycle or acquisition, with gates and decision points fixed in time for all aspects of the system, is incapable of accommodating fast and slow spin technologies at one and the same time. One could go further and observe that the traditional engineering approach itself is incapable of dealing with such hybrid systems. This is the hybrid fast-spin/slow-spin system lifecycle incompatibility problem."* (Wilkinson 2014). The paper goes on to recommend that continuous lifecycle phases are matched to fast-spin technologies and low risk acquisition of well-defined incremental capabilities is planned.

Therefore the System-of Systems safety activities, including the production of an overall safety case, given safety documents produced by differing processes and standards need to match this fast-spin lifecycle and incremental capability changes. This means that the System-of-System safety argument needs to grow at the same pace as the fast spin systems, and therefore allow emergent unsafe properties to be identified and mitigated in the same incremental period.

6 Conclusion

Technological changes to our socio-technical safety management systems means that we cannot continue to focus on equipment/platform safety particularly for Systems-of-Systems. In consequence, further research is urgently needed in this area.

From our experience with the rise of the machine, it would be beneficial to develop a SoS safety methodology (code of practice) for the professional safety engineer. This methodology could be based on the following:

1. Establishing if the SoS change is revolutionary and whether existing safe systems of operation may no longer be effective and hence supporting a proportional approach to safety investment.
2. Understanding that the revolutionary SoS change may be due to the socio-technical safety management monitoring and enforcement system and not to the SoS under control.
3. The identification of all activity Duty Holders – ensuring they are engaged in the SoS development and safety assessment.
4. Using Soft Systems Methodology to aid the understanding of all stakeholders in defining safe design constraints, anticipating the safe systems of operation and risk control systems.
5. That the levers for change (e.g. Defence Lines of Development) aid the understanding of the complete socio-technical safety management system that should be in place to enforce safe design constraints and continued safe systems of operation.
6. Using STAMP is a valuable tried and tested, systems theory based, safety methodology that in conjunction with equipment and platform safety assessments provide for a complete SoS safety argument.
7. Ensuring that SoS Safety Management planning provides evidence that matches the fast-spin technologies life cycle.
8. How to utilize and combine safety artefacts and risk control systems from different development practices/standards.
9. Analyses techniques to identify emergent behaviours and properties.
10. Safety and evidence management at the fast-spin technologies cycle.

References

Boardman and Sauser (2006), Dr John Boardman and Dr Brian Sauser, 2006, System of Systems - the meaning of of; Los Angeles, CA, USA: IEEE, Proceedings of the 2006 IEEE/SMC International Conference on System of Systems Engineering.

Checkland and Scholes (2009), Peter Checkland and Jim Scholes, Soft Systems Methodology: a 30-year retrospection, 15 Dec 2009

Brown (2015), Daniel Brown, Director of Strategy and Policy, March 2015, ORR's approach to innovation.

Defence Safety Authority (2016), Defence Safety Authority, August 2016, Defence Policy for Health, Safety and Environmental Protection, DSA01.1 (https://www.gov.uk/government/uploads/system/uploads/attachment_data/file/548060/DSA01_Defence_Policy_for_Health_Safety_and_Environmental_Protection-20160804.pdf accessed September 2016)

Def Stan 00-056 (2015), Defence Standard 00-056, Safety Management Requirements for Defence Systems, Part 1: Requirements and Guidance. Issue 6, Date: 02 April 2015

DE&S ESAS (2016), Digital Railway: Approach to realising the future vision for the Railway, June 2016 presentation, provided to DE&S ESAS 2016.

Leveson (2004), Nancy Leveson, A New Accident Model for Engineering Safer Systems, Safety Science, Vol. 42, No. 4, April 2004, pp. 237-270

Chui, Manyika, et al (2015), Miremadi Michael Chui, James Manyika, and Mehdi Miremadi, Article McKinsey Quarterly November 2015 (http://www.mckinsey.com/business-functions/business-technology/our-insights/four-fundamentals-of-workplace-automation accessed September 2016)

Dunlop (2016), Ben Dunlop Director of Digital Railway at Atkins, Time for the digital railway revolution Atkins, 22 Mar 2016 (http://www.atkinsglobal.co.uk/en-GB/angles/all-angles/digital-railway-revolution accessed September 2016)

Wilkinson (2014), Mike Wilkinson, Continuous Capability Evolution – A Practical Approach to the Acquisition Strategy of Modern Defence Capabilities, Niteworks White Paper, June 2014

Progress Towards the Assurance of Non-Traditional Software

Rob Ashmore, Elizabeth Lennon

Defence Science and Technology Laboratory (Dstl)

Abstract *Traditional software development follows a hierarchical process, with system-level requirements allocated to software being progressively refined through high-level requirements into source code, coverage of which is a key measure of test completeness. This approach establishes a direct link between system-level requirements and the software implementation; it also assumes that executable behaviour is governed by the structure of the source code. Although it is still widely applicable, there are cases where the suitability of this traditional model is less apparent. Consider an algorithm implemented via a neural network. In this case the structure of the source code has much less effect on the software's behaviour than in the traditional case: the same generic neural network software could be trained to perform two very different functions. The link between the implementation and the high-level requirements is also harder to trace than in the traditional case. More generally, some of the approaches used to assure software may not be appropriate for new types of algorithm; nevertheless, such algorithms are becoming increasingly common. In response, this paper considers how current methods, notably the 'four plus one' software safety assurance principles, might be enhanced to support the assurance of non-traditional software.*

1   Introduction

This paper is an initial step in determining whether the approaches used to provide assurance of traditional software could be applied to non-traditional software. Note that, for brevity, we use the term 'ML software' to refer to software that is implemented using ML, or non-traditional, techniques.

We consider the most significant examples of 'non-traditional software' to be items developed using Machine Learning (ML) techniques. Examples include neural networks, support vector machines, random forests and Gaussian processes. Typical applications for this software include: classification, where unseen inputs are allocated to predetermined classes; clustering, where a natural method of grouping inputs has to be determined; and regression, where predictions have to be made for a continuous variable.

ML has shown great promise in many applications, including large-scale image recognition (Simonyan and Zisserman, 2014) and deeply tactical games like Go (Silver et al. 2016). This success is likely to lead to its use in safety-related applications, as is the drive towards ever larger and more complex safety-related systems.

We assume there is a need to provide direct assurance of the ML software's behaviour. An alternative approach would be to add a guardian, or monitor, function to watch over the ML software. Whilst the use of guardian functions may be a practical way of introducing ML software in safety-related systems, this may overly inhibit system performance and potentially undermine safety in unforeseen scenarios. We also note that providing confidence in ML software would provide additional confidence in the overall system, even if a guardian function is used.

The remainder of this paper is structured as follows: Section 2 discusses software safety assurance principles in the context of ML software; Section 3 considers the topic of verification; and Section 4 contains conclusions.

2   Software Safety Assurance Principles

This section considers the 'four plus one' software safety assurance principles (Hawkins et al. 2013) from the perspective of ML software.

2.1   Principle One

> Software safety requirements shall be defined to address the software contribution to system hazards.

According to (Hawkins et al. 2013), 'at this stage of the development, software is treated as a black box, used for enabling certain functions, and with little visibility into the way in which these functions are implemented'. That is, the nature of the software, whether ML-based or traditional, is irrelevant; this principle applies regardless.

As an example, an aircraft may have a system safety requirement that 'thrust reversers may only be engaged when the aircraft is on the ground', with this requirement being passed to the Full Authority Digital Engine Control (FADEC) software.

2.2   Principle Two

The intent of the software safety requirements shall be maintained throughout requirements decomposition.

The mention of 'decomposition' in Principle 2 implies a traditional hierarchical approach whereby high-level software requirements are traced through architectural design to low-level requirements.

For example, a key part of the system safety requirement identified in the preceding sub-section is an ability to determine if the aircraft is 'on the ground'. In a traditional approach, the system developer would consider what information can be used to make this determination: Weight-On-Wheels (WOW) sensors are one typical example; sensors that measure wheel-rotation speed are another. Further design considerations would introduce additional refinements until a set of low-level requirements had been established.

During traditional high-integrity software development, a separate argument can be provided to justify each stage of the design process. In particular, the strict hierarchical nature of the decomposition, combined with explicit traceability at each level, is a significant aid to constructing the link between the high-level requirement and the planned implementation, as demanded by Principle 2.

Conversely, neither the strict hierarchy nor the explicit traceability is apparent in the context of ML software. The following eight steps summarise how ML software is often developed in Academia (Wagstaff, 2012):

1. Phrase the problem as a machine learning task;
2. Collect data;
3. Select or generate features;
4. Choose or develop algorithm;
5. Choose metrics, conduct experiments;
6. Interpret results;
7. Publicise results to relevant user community;
8. Persuade users to adopt technique.

For real-world systems a precursor step may involve gathering data and determining whether ML was an appropriate solution. If so, the problem would be rephrased as an ML task (step 1); additional data may then be collected (step 2); and so on. For example, if the intent was to train an algorithm to determine whether an aircraft was on the ground then a training data set could conceivably include parameters from any relevant aircraft system: engine revolutions; indicated air speed; air temperature; air pressure; throttle position; values of WOW sensors; and so on. In a supervised learning approach examples of these parameters, together with the observed state (ie, 'on ground' or 'not on ground') would be provided and the algorithm would learn which combination of features best described the 'on ground' state. Generally speaking, there would be no deductive argument explaining why these features matched the high-level requirement.

It has been suggested (Varshney, 2016) that ML software should only be used in safety-related contexts if the logic behind the learnt algorithm can be explained and understood. Whilst understandable, this is a rather restrictive constraint; it limits the ML techniques to cases where they mimic a human-derived, step-by-step argument and, as such, removes a significant amount of their power. Furthermore, there is a danger that any *post hoc* explanation of a ML algorithm could be plausible but incorrect (Lipton, 2016), leading to an inappropriate level of confidence.

In summary, the principle that there should be a demonstrable link between (system-level) software safety requirements and the planned software implementation is applicable to ML software. However, the lack of hierarchical decomposition means the specific wording of Principle 2 may need adjustment; this suggests a Principle 2' (read two-primed).

To recap, Principle 2 states:

> The intent of the software safety requirements shall be maintained throughout requirements decomposition.

Conversely, Principle 2' could say something like:

> The software detailed design shall embody the intent of the software safety requirements.

Note that here the term 'detailed design' covers training and test data sets as well as source code (since both data sets contribute to the deployed system).

2.3   Principle Three

> Software safety requirements shall be satisfied.

The third software safety assurance principle relates to verifying the software implementation. Generally speaking, it is infeasible to completely test software (Butler and Finelli, 1993). As such, testing tends to be about providing appropriate confidence by conducting a suitable breadth and depth of software testing. From a traditional perspective, there are two main aspects of measuring test completeness: covering each requirement; and achieving a specified level of code coverage (in both nominal and robustness contexts).

The first aspect of measuring test completeness (ie, covering each requirement) is largely agnostic of the type of software. This is not, however, the case for the second aspect. For example, DO-178C (RTCA, 2011a) adopts statement, branch and modified condition / decision coverage (MC/DC) as a graduated scale of increasingly complete verifications of the implemented logic. Put another way, from the perspective of traditional software these measures provide confidence that a suitable amount of the software's potential behaviour has been checked. This, in turn, provides confidence that we understand what the software does and, also, what it does not do.

The measures adopted in DO-178C are typical of those used in many software-related standards. However, they assume that code structure has a dominating effect on software behaviour, which need not be the case for ML software. This suggests that additional measures of test completeness and thresholds of acceptability may need to be developed.

Formal methods, which can provide mathematical proofs of certain properties, offer an alternative to using code coverage as a measure of test completeness. Use of this alternative is, for example, defined in a supplement to DO-178C, specifically DO-333 (RTCA, 2011b).

It should also be noted that, at least from the perspective of safety-related software, verification is about more than testing. It is also concerned with other parts of the software development process including, for example: process documentation; change management; configuration control; and independent quality assurance. These considerations also apply to ML software.

Regardless of the way that it is achieved, demonstrating that the implemented software satisfies the software safety requirements (ie, Principle 3) is directly applicable to ML software.

2.4   Principle Four

Hazardous behaviour of the software shall be identified and mitigated.

The first three principles provide assurance that the software meets the system-level requirements; the fourth principle is about showing that the software has not introduced any new hazards. According to (Hawkins et al. 2013), hazardous implications may be identified by a *'systematic and thorough consideration of potential software failure modes and their effects (both on the software and other systems)'*. This approach is echoed in UK Defence Standard 00-055 (MOD, 2016), which mentions both Hazard and Operability Study (HAZOP) and Functional Hazard Assessment in this context.

ML software can introduce hazards in (at least) two different ways:

1.   Firstly, there could be traditional errors in the software that implements the algorithm; for example, a memory leak that gradually erodes system resources until a crash occurs.
2.   Secondly, there could be errors in the behaviour of the trained system; for example, it could mis-classify a sample, or it could give a classification an inappropriately high degree of confidence.

Since they are traditional in style, the first type of failure can be analysed using existing techniques. The second type of failure may require additional work before it can be used in an ML context: a new way of interpreting the standard set of HAZOP guide words may be required. Nevertheless, Principle 4 is directly applicable to ML software.

2.5   Principle 'Four Plus One'

The confidence established in addressing the software safety principles shall be commensurate to the contribution of the software to system risk.

Dissimilar pieces of software can pose different levels of system risk. These different pre-mitigation risk levels require varying levels of rigor to be applied during the software development to reduce the final risk to an acceptable level. Principle 4 + 1 thus helps developers distribute their resources across different parts of the system in an appropriate manner.

More specifically, from the perspective of this work, if the other principles can be mapped into the context of ML software then this principle poses no additional challenges.

2.6   Principle 'Four Plus Two'

For traditional software, the principles discussed previously are generally achieved while the software is 'in the factory'. This is important as it gives confidence before the software is used operationally. However, ML software poses some challenges to this 'prove then use' philosophy:

- There may be emergent effects, especially when unexpected environmental factors are encountered. We note there are many definitions of the term 'emergent' (Johnson, 2006), here we are referring to unexpected properties exhibited by a complex system.
- The software could continue to learn and adapt during operational use. This could mean that a different response is made to the same situation, depending on how long the system has been operating.

- The software may be used in a way not envisaged during its design. The potential adaptability of ML software means this challenge may be faced more frequently than is the case for traditional software.

It could also be argued that ML software requires a new concept, to adequately capture the importance of maintaining assurance throughout system adaptations. Perhaps something like:

> Software required to produce behaviour not predictable at design time should consider the consequence of behavioural adaptations on its environment.

The relationship between this proposed principle and the normal 'four plus one' principles is illustrated in Figure 1.

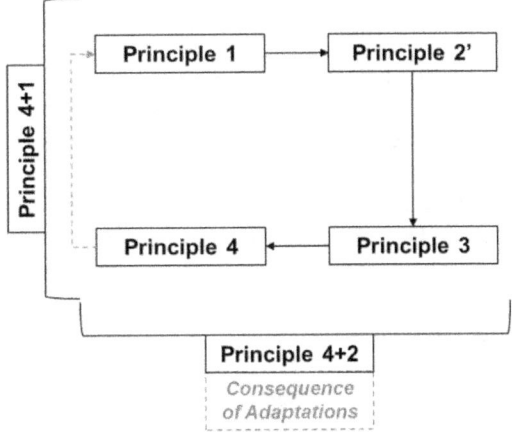

Fig. 1. Relationship between Principles

At the risk of oversimplification, Principle 1 establishes requirements, Principle 2 (or 2') traces these into the detailed design, Principle 3 confirms the implementation and Principle 4 ensures there are no new issues (which could lead to new requirements, hence the dashed line to Principle 1). Principle 4+1, which is about balancing resources and risk, cuts across the other four principles.

Similarly, the proposed new principle cuts across the first four principles, in that we would like to be assured that any adaptations do not undermine any part of the safety argument. It could be argued that this is most relevant to Principle 4 and, possibly, Principle 3 (if the adaptations change aspects of the implementation). For that reason, the proposed new principle is illustrated at the bottom of Figure 1. However, we leave open the possibility that an adaptation may change some of the original requirements captured under Principle 1 and the tracing under Principle 2'. Hence, we adopt the term Principle 4+2.

3   Software Verification

The previous section has identified a number of areas where the 'four plus one' principles may need enhancing for ML software. We have arbitrarily selected one area for more detailed investigation. In particular, this section considers Principle 3 and, more particularly, the issue of software verification. It is based on the philosophy adopted by DO-178C (RTCA, 2011a). This was selected due to its widespread use in the air domain: it is also referenced from UK Defence Standard 00-055 (MOD, 2016).

3.1   Philosophy

It is important to note that DO-178C is targeted at the traditional situation where software is verified before it is used. As noted previously, this situation does not capture all aspects of ML software. This limits the extent to which DO-178C's verification philosophy can address all issues associated with ML software. Nevertheless, it remains a potentially useful framework.

The five aspects of DO-178C's verification philosophy are considered below.

3.1.1   Requirements-based

Software test cases should be based on requirements. Once all requirements have been tested, the level of coverage achieved by the tests should be determined. If this is insufficient then further requirements-based tests should be implemented. This approach, supported by traceability information, ensures there is a tight link between the requirements (which can be interpreted by domain experts and systems engineers) and the software's behaviour.

Although the nature of a requirement may change, the notion of requirements-based testing does not depend on the specific nature of the software under test. Hence, this part of DO-178C's verification philosophy can be readily applied to ML software.

3.1.2   Independence

Depending on the criticality of the software under test, some of the objectives listed in DO-178C may need to be satisfied with independence. This means that the same individual cannot both develop and test a piece of software; often separate teams are responsible for development and for testing.

It is relatively easy to map this notion of independence into the context of non-traditional software. When developing ML software the available data is often split into a training set (which is used to develop an algorithm) and a test set (which is used to determine the quality of the learnt behaviour). The development process is often iterative and items of data can move between the training and test data sets on different iterations. Both training and testing are usually performed by the same team, which would be analogous to the software development team in a traditional approach.

At least in theory, an independent verification team could use a separate (or overlapping) set of data in order to provide confidence in the algorithm's behaviour. Whilst there may be some challenges in partitioning available data between training, test and verification sets, these should not be unsurmountable. For example, since the verification set is intended to represent actual use of the algorithm it may be feasible to draw some parts of this set from trials and / or simulated activities. That said, care will be needed to prevent independence being inadvertently undermined by, for example, the verification team sourcing data in the same way, and from the same sources, as the development team.

In summary, it should be possible to interpret the independence part of DO-178C's verification philosophy in the context of ML software.

3.1.3   Accurate

From the perspective of DO-178C, verification involves the production of: test cases, which describe how a particular test will exercise parts of the software to demonstrate that a particular requirement has been satisfied; and test procedures, which detail how a particular test case is implemented.

The question of accuracy is concerned with whether a test procedure has correctly implemented the associated test case to an appropriate degree of accuracy. This is often achieved by a review of the test script. This concept, and the associated review procedure, can be directly mapped to ML software.

3.1.4   Robust

There is a long history of using traditional software in safety-related systems. This has provided the opportunity to identify a number of common errors. Robustness testing is largely about ensuring these errors have been avoided.

Although the field of ML is less mature than that of traditional software development, there has been sufficient work to allow some common problems to be identified, for example (Hamner, 2014) and (Kaufman et al., 2012). Examples include:

- Overfitting, where the algorithm learns the specific data rather than the generic relationship.
- Data leakage, where information that should not legitimately be available is included, either explicitly or implicitly, in training data.

Further work may be required to produce a suitably authoritative list, along with ways of avoiding or overcoming common problems, but there appears to be no fundamental difficulty in extending the concept of robustness tests to ML software.

3.1.5   Sufficiently-complete

As discussed in Section 2, DO-178C (RTCA, 2011a) uses measures based on structural code coverage to determine whether the requirements-based testing has been sufficiently complete. This assumes a close connection between code structure and software behaviour, which is not evident in ML software. Hence, DO-178C's notion of test completeness cannot be extended to ML software.

3.2   Test Completeness

The following sub-sections briefly discuss other measures of test completeness, from both the software engineering and the machine learning communities.

3.2.1   Alternate Measures from the Software Engineering Community

Although DO-178C's measures of verification (or test) completeness are not appropriate for ML algorithms, this need not necessarily be the case for all measures developed by the software engineering community. Whilst an exhaustive survey has not been conducted, a small number of possibilities are listed below. (This list does not include items that are not explicitly used in DO-178C but are implicitly based on code structure; for example Linear Code Sequence And Jump and Observable MC/DC.)

- The number of fault-free hours of representative testing could conceivably be used. However, despite recent growth in processing power and the parallel nature of testing, it may be impossible to conduct enough hours of testing, especially for safety-critical software (Butler and Finelli, 1993).

- Combinatorial Interaction Testing (Arcuri and Briand, 2012) in which, for example, all three-way combinations of inputs are tested is at least superficially attractive. In particular, this measure is focused on coverage of the input domain. However, it is not clear whether it would make sense to test, for example, every three-way combination of inputs for ML software as not all of these may be feasible.
- Some test measures have been proposed for Service-Oriented Architectures (Sneed and Verhoef, 2015). Examples include: *parameter* coverage, where each input is given at least one value; *value* coverage, where each input takes each possible value at least once, and *state* coverage, where every possible combination of inputs is used. These measures make no assumptions about internal structure, which makes them attractive for applying to ML software. However, there is apparently little evidence as to the efficacy of the less-strict levels and it may not be feasible (or appropriate) to achieve state coverage.

We suggest that, whilst measures from the Software Engineering community may provide a useful starting point, in their own right they are not suitable as measures of test completeness for ML software.

3.2.2 Alternate Measures from the Machine Learning Community

There does not appear to be a clear distinction between *testing* an ML algorithm and *measuring* its performance. Furthermore, different measures are appropriate depending on the nature of the ML algorithm: measures that are suitable for a regression algorithm are unlikely to be suitable for a classification algorithm.

Even if a specific type of ML algorithm is considered, there are further difficulties. Consider, for example, a binary classifier (ie, a classifier with only two categories) that is intended to identify pictures of cats. Three commonly used performance measures are:

- *Precision*, which, in this case, is the fraction of pictures the classifier identifies as cats that actually are cats - that is, precision is measuring the fraction of selected items that are correct;
- *Recall*, which is the fraction of actual cat pictures that the classifier identifies as cats - that is, recall is measuring the fraction of correct items that are selected;
- *Accuracy*, which is the fraction of pictures that are correctly identified.

Put another way, using cat identification as a 'positive' result:

$$Precision \; = \; \frac{TruePositives}{TruePositives + FalsePositives} \; = \; \frac{TP}{TP + FP}$$

$$Recall \; = \; \frac{TruePositives}{TruePositives + FalseNegatives} \; = \; \frac{TP}{TP + FN}$$

$$Accuracy \; = \; \frac{TP + TrueNegatives}{TP + FP + TrueNegatives + FN}$$

One difficulty with using these measures as a generic way of establishing test completeness is that desirable values for them are heavily dependent on the context. For example, (Wagstaff, 2012) notes that whilst an 80% accuracy might be good for recognising species of iris, it probably would not be appropriate for deciding whether a particular mushroom was poisonous.

The importance of identifying 'unknown unknowns' (ie, those inputs where ML software displays a high degree of confidence, but it is wrong) is another potential problem (Attenberg et al., 2011).

Furthermore, work focused on the performance of Deep Neural Networks (DNNs) in image classification tasks highlights difficulties with the performance measures used in the ML community. Even with a well-performing DNN it is relatively easy to find cases where small changes to an image, which would be almost imperceptible to a human, cause the DNN to mis-classify an image (Szegedy et al., 2013). Likewise, finding cases where the DNN confidently attributes a class to a 'nonsense image' is also relatively straightforward (Nguyen et al., 2015).

Given these issues, we suggest that the specific measures currently used by the ML community are better suited to being requirements than to being measures of verification completeness.

3.2.3   Sampling Density

At an abstract level, ML software can be viewed as mapping from an input domain to an output range. From that perspective, test completeness is related to the number of inputs that have been considered. More specifically, given the importance of context, the density of the inputs that have been processed is potentially a good measure of how well an ML algorithm has been tested. Note that this measure does not capture the mapping from the real-world to the ML software's input domain; this mapping could be quite complex and difficult to reason about. Nevertheless, sampling density is judged to be worthy of further consideration.

In order to calculate sampling density some form of distance metric needs to be defined across the input domain. As well as *numerical* inputs, a metric may also have to handle *ordinal* inputs (ie, things that can be put in order, like:

small; medium; large) and *categorical* inputs (ie, things that have no natural order, like: apple, banana, orange). That said, it seems plausible to assume that a meaningful distance metric can be defined.

Since the training and test data are used to develop the ML software, it could be argued that only the verification data set's density is relevant to measuring verification completeness. However, the key thing about this data set is that it facilitates independence in the verification process. The way that ML software is developed means that both the training and the test data set have contributed to its performance. Hence, it is appropriate to include these data sets in calculations of verification completeness. More specifically, the combined density of the training, test and verification data sets is a potential measure of verification completeness.

There are a wide variety of potential applications of ML software, with very different input domains. Furthermore, any distance metric used across the input domain will, by definition, be context specific. For these reasons it is inappropriate to define an absolute data set density that is required. Instead, some form of *relative* measure seems more appropriate.

For simplicity, the term 'sample point' is used to refer to an item in any of the training, test or verification data sets. A simple, three-step algorithm for determining a relative measure of the density of sample points is:

1. For each sample point calculate the minimum distance between it and any other sample point.
2. Calculate d, the smallest, non-zero element from this set of 'minimum distances'.
3. Normalise the set of minimum distances by dividing each element by d.

The resulting set of values shows how the distance between neighbouring sample points varies (as a multiple of the smallest distance between any two sample points). This concept is essentially the same as the gap ratio (Bishnu et al., 2015).

It is assumed that the smallest distance between two sample points is a reasonable measure of the density that is required to understand algorithm behavior in the most challenging part of the input domain. That density is not necessarily required across the entire input domain, but it seems reasonable to argue that the least sampled regions should not have sampling densities many orders of magnitude lower than the most sampled regions.

To reiterate an earlier point, the sampling density includes the training, test and (crucially) verification data sets. This means that, in principle, a required sampling density can be achieved, for example, by adding extra points to the verification data set without any effect on the risk of overfitting.

4   Conclusions

A graphical summary of this paper is shown in Figure 2, below:

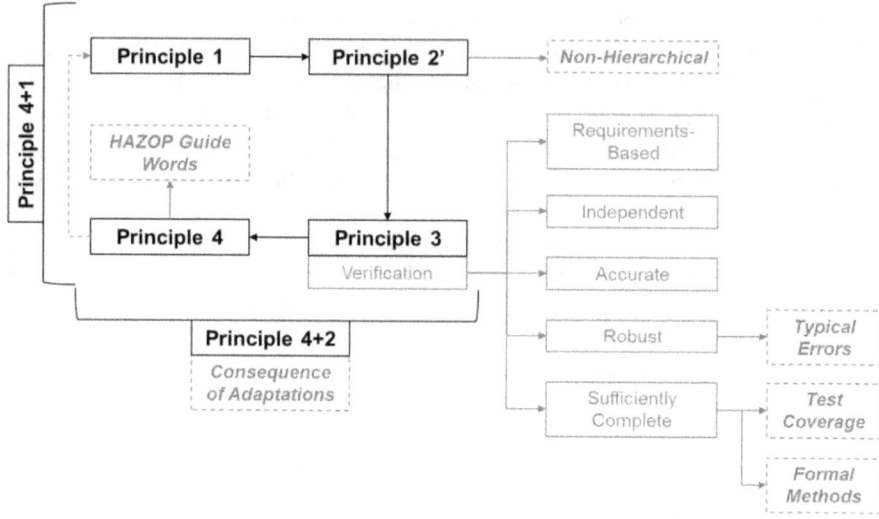

Fig. 2. Paper summary

We conclude that the 'four plus one' software safety principles provide a good structure against which to measure the evidence required to support the operational use of ML software in safety-related systems. There are two places where, in our opinion, this structure can be enhanced to provide more compelling assurance:

1. To account for the lack of hierarchical decomposition in the generation of ML software, Principle 2 could be amended to Principle 2' which reads, '*the software detailed design shall embody the intent of the software safety requirements*'.
2. To account for the potential impact of system adaptations it may be appropriate to create Principle 4+2: '*software required to produce behaviour not predictable at design time should consider the consequence of behavioural adaptations on its environment*'.

In addition, there are a number of areas where further work is required. These include:

- Demonstrating, without relying on a hierarchical structure, that intent has been maintained when moving from system-level software requirements to planned software implementation (Principle 2 / 2').
- New interpretations of HAZOP guide words tailored for ML software (Principle 4).

- We also conclude that the software verification philosophy embodied in DO-178C is a helpful, if incomplete, framework for considering the verification of ML software. Additional work is required in a number or areas, including:

- Understanding typical errors and how to avoid them, so as to inform robustness testing.
- Ways of demonstrating that testing activities have been sufficiently complete.

Overall, we hope that we have provided a structure to help guide and inform work on the assurance of non-traditional software.

Acknowledgments This work has benefitted from numerous discussions with our colleagues at Dstl. We would particularly like to thank Alec Banks and Paul Caseley for their insightful contributions.

References

Arcuri, A, Briand, L (2012), Formal Analysis of the Probability of Interaction Fault Detection Using Random Testing, IEEE Transactions on Software Engineering, 2012, 38(5), 1088-1099.

Attenberg J, Ipeirotis, P G, Provost, F J (2011) Beat the Machine: Challenging Workers to Find the Unknown Unknowns, Human Computation, 2011, 11, 11.

Bishnu, A, Desai, S, Ghosh, A, Goswami, M, Paul, S (2015) Uniformity of point samples in metric spaces using gap ratio. In International Conference on Theory and Applications of Models of Computation (pp. 347-358). Springer International Publishing.

Butler, R W, Finelli, G B (1993) The infeasibility of quantifying the reliability of life-critical real-time software, IEEE Transactions on Software Engineering, 1993, 19(1), 3 –12.

Hamner, B (2014) Machine Learning Gremlins, in Strata Conference, O'Reilly, 11 Feb 2014.

Hawkins, R, Habli, I, Kelly, T (2013) The principles of software safety assurance. 31st International System Safety Conference, Boston, Massachusetts USA.

Johnson, C W (2006). What are emergent properties and how do they affect the engineering of complex systems?. Reliability Engineering & System Safety, 91(12), 1475-1481.

Kaufman, S, Rosset, S, Perlich, C, Stitelman, O, Leakage in data mining: Formulation, detection, and avoidance, ACM Transactions on Knowledge Discovery from Data (TKDD), 2012, 6(4), 15.

Lipton, Z C (2016). The Mythos of Model Interpretability. arXiv preprint arXiv:1606.03490.

MOD (2016), Requirements for Safety of Programmable Elements (PE) in Defence Systems - Part 1: Requirements and Guidance, Def Stan 00-055, Issue 4, Ministry of Defence.

Nguyen, A, Yosinski, J, Clune, J (2015) Deep neural networks are easily fooled: High confidence predictions for unrecognizable images, In: 2015 IEEE Conference on Computer Vision and Pattern Recognition (CVPR), IEEE, 2015, 427–436.

RTCA (2011a) Software Considerations in Airborne Systems and Equipment Certification, DO-178C, RTCA, 2011.

RTCA (2011b) Formal Methods Supplement to DO-178C and DO-278A, DO-333. RTCA, 2011.

Silver, D, *et al.* (2016). Mastering the game of Go with deep neural networks and tree search. Nature, 529(7587), 484-489.

Simonyan, K, Zisserman, A (2014). Very deep convolutional networks for large-scale image recognition. arXiv preprint arXiv:1409.1556.

Sneed, H M, Verhoef C (2015) Measuring test coverage of SoA services, In: 2015 IEEE 9th International Symposium on the Maintenance and Evolution of Service-Oriented and Cloud-Based Environments (MESOCA), IEEE, 2015, 59–66.

Szegedy, C, Zaremba, W, Sutskever, I, Bruna, J, Erhan, D, Goodfellow, I, Fergus, R (2013) Intriguing properties of neural networks, arXiv preprint arXiv:1312.6199, 2013.

Varshney, K, R (2016) Engineering Safety in Machine Learning. arXiv preprint arXiv:1601.04126.

Wagstaff, K (2012) Machine Learning that Matters. In: Proceedings of the 29th International Conference on Machine Learning (ICML-12), 2012, 529–536.

Disclaimer This article is an overview of UK MOD sponsored research and is released for informational purposes only. The contents of this article should not be interpreted as representing the views of the UK MOD, nor should it be assumed that they reflect any current or future UK MOD policy. The information contained in this article cannot supersede any statutory or contractual requirements or liabilities and is offered without prejudice or commitment.

Functional Safety:
Where have we come from?
Where are we going?

Audrey Canning

Virkonnen Ltd.

Abstract *This paper provides a personal perspective of the history and evolution of the discipline of Functional Safety. It starts with the concerns that provided the early drivers, the initiatives that moved the discipline from a diverse and adhoc collection of different control measures, to an internationally recognised, systematic framework with a broad spectrum of risk reduction principles, to sector customised approaches for the consequences, risk reduction customs and system behaviours found in different industry sectors. It also poses the question of where future advancements will lie, in further refinement into sector and industry specific practices or through development of a more comprehensive framework able to adapt to a wide variety of different threats in a uniform manner.*

1 Introduction

This paper provides a personal perspective of the history of Functional Safety from the mid 1980s through to the current day.

Although I cannot claim to be aware of (or even remember) all twists and turns of the developing discipline, perhaps one day, when put together with the memoires of other practitioners, a broader history of the discipline will emerge and its relevance not just to complex computer-controlled applications, but to the whole range of engineering disciplines and applications might be more widely appreciated. If in doing so this helps to improve safety, or at least prevent expensive rework, my time will not be wasted.

2 The Early Years

2.1 Pre-history

My story in functional safety begins some ten years after the first stirrings of concern over the potential for complex control systems to lead to heavy loss of human life (see Figure 1). Although significant loss in individual incidents was not uncommon in coal mining and rail transport, the explosion at the Flixborough chemical plant in 1974, attributed to a failure to recognise the severity of the consequence and the complexity of the systems, led to a drive for improved process plant safety.

The Moorgate incident in 1975 had a similar wake-up call on the railway industry, leading to the introduction of 'End of Terminal Station' protection to reduce the speed of any overrun collision and eventually leading to the introduction of automatic train protection throughout the UK rail network. As I write, the enquiry into the September 2016 New Jersey train crash is commencing, a testament to the ongoing importance of 'End-of-terminal-station' protection.

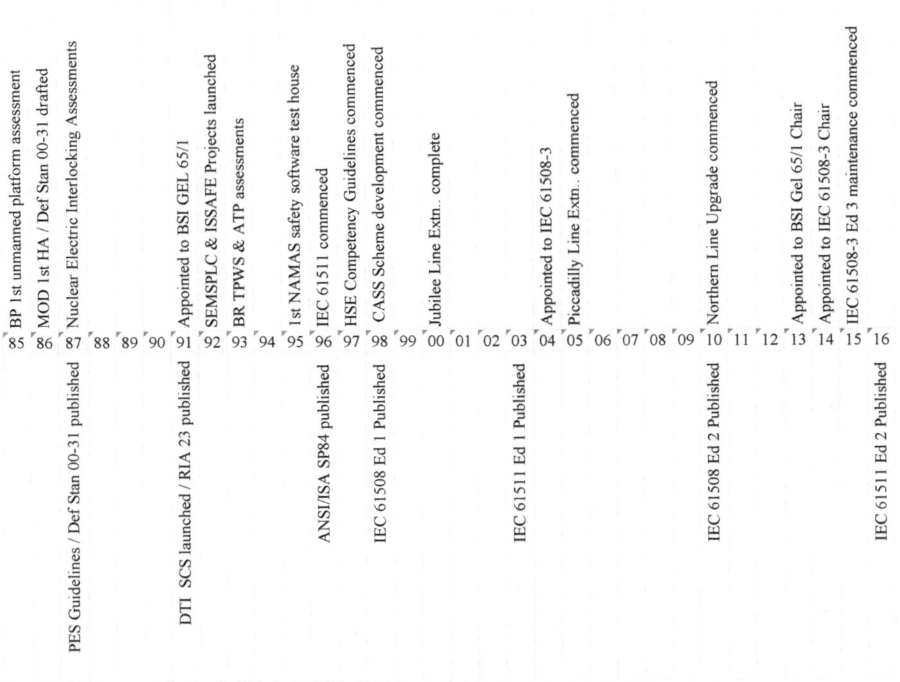

BP 1st unmanned platform assessment
MOD 1st HA / Def Stan 00-31 drafted
Nuclear Electric Interlocking Assessments
Appointed to BSI GEL 65/1
SEMSPLC & ISSAFE Projects launched
BR TPWS & ATP assessments
1st NAMAS safety software test house
IEC 61511 commenced
HSE Competency Guidelines commenced
CASS Scheme development commenced
Jubilee Line Extn. complete
Appointed to IEC 61508-3
Piccadilly Line Extn. commenced
Northern Line Upgrade commenced
Appointed to BSI Gel 65/1 Chair
Appointed to IEC 61508-3 Chair
IEC 61508-3 Ed 3 maintenance commenced

85 86 87 88 89 90 91 92 93 94 95 96 97 98 99 00 01 02 03 04 05 06 07 08 09 10 11 12 13 14 15 16

PES Guidelines / Def Stan 00-31 published
DTI SCS launched / RIA 23 published
ANSI/ISA SP84 published
IEC 61508 Ed 1 Published
IEC 61511 Ed 1 Published
IEC 61508 Ed 2 Published
IEC 61511 Ed 2 Published

Fig 1. Approximate Timeline of Key Events

2.2 The Process Control Industry

My own significant involvement in the functional safety industry began in 1985, with a request to undertake an Independent Safety Assessment for the first unmanned BP offshore platform. A unique aspect of this Platform was that it was to have a computer based Emergency Shutdown System, the first ever to be used on a BP Platform. With work already commenced in the HSE on the development of guidelines for Programmable Electronic Systems (PES) used in Safety Related Applications (HMSO 1987) it was considered pragmatic to obtain an independent view of the architecture and implementation of the system. With experience in electronic design, firmware and software development (including the use of finite state machines and languages with safe sub-sets) I was selected to carry out this first independent assessment.

With no formal guidance in existence at that time, but with a need to have objective criteria against which to carry out the assessment, my approach was to collate 'good practice' guidelines from organisations such as the IEEE, as well as to obtain a copy of the draft PES Guidelines. The PES Guidelines in turn introduced me to the concept of hazard analysis which, at that time, was largely applied only in the chemical industry. With earlier experience in reliability modelling for communications (derived largely from an early edition of David Smith's excellent book (Smith 1985) I had sufficient knowledge to model the reliability of the 1 out of 2 to shut down voting architecture. Recognising that it was unlikely that software based self-test functions would completely reveal all faults, the model predicted a decaying 'saw-tooth' estimate of the instantaneous reliability. This first independent assessment, 'though perhaps not as thorough as would be advocated by today's authorities, never-the-less was effective in providing BP management with early visibility of the suppliers progress, and led to two further commissions, expanded to address the process control system and Fire and Gas protection system, as well as the Emergency Shutdown (ESD) system.

2.2 The Defence Industry

Meanwhile, my employer's leading role in standardisation of token passing ring communications protocols for the Ministry of Defence (MOD) procurement led to a new enquiry – to assist in the update of (Def Stan 00-31 1987) for software used in Avionics Systems. Based on the success of the ESD system assessment, I was also asked to undertake a pilot application of hazard analysis on a computer controlled weapon range system, one of two initial pilot applications carried out at that time by the MOD. Adapting the recommendations of the Chemical Industries Association for Hazard and Operability (HAZOP) Studies (CIA 1977), I replaced the chemical process flow model with a functional model of the control system and then applied the HAZOP guidewords to identify the ways in which the functions could fail. However, being aware that timeliness as well as functionality is a critical parameter in software engineering, I added the guidewords 'too early' and 'too late'. To the best of my knowledge, this was the first use of the extension of the HAZOP guidewords to address timeliness.

My work on Defence Standard 00-31 was quickly overtaken by a Joint Forces initiative to develop a Cross-Services standard for safety-critical software. With a contract already in place, my work moved to assist Aquila in drafting a new standard. Several years later I did discover the very same wording in an Annex to Defence Standard 00-56 that I had written in the range assessment proposal, describing the approach to be undertaken for audit as distinct from assessment.

2.3 The Nuclear Industry

The BP and MOD experience had brought together concepts of software engineering, functional hazard analysis and reliability, but it was the nuclear industry, and in particular Mr Antony Greenway, who initiated me into logic proving and 'defence in depth'. In 1987 I was asked to undertake the analysis of nuclear reactor interlocks. The systems were implemented in solid state relays with a preferential failure mode and the task was to prove that the logic could not fail into an unsafe condition – to assume that it could, would result in tautology. The approach was conclusive for simple logical relationships between the output and its inputs and which comprised the majority of the interlocking. However, the approach was laborious when applied to even the simple feedback loops within this system since every potential sequence of operation, covering each of the internal states needed to be evaluated. Furthermore, formal proof alone was not sufficient. For critical circuits not only was the interlock logic based on triple redundant channels, but in addition three further channels based on distinctly different physical input data were used to provide diversity and thus eliminate the possibility of common cause failure in the physical sensing circuits.

3 The Harmonisation Years

3.1 Harmonisation through Assessment

By the late 1980s/early 1990s my work had expanded into independent safety assessment of software based protection systems for the railway industry, carried out to British Rail Standards and, in particular, Railway Industry Association (RIA) 23 (RIA 23 1991). By this stage my assessment activities had developed a formalised approach, comprising of:

- Functional failure analysis
- Traceability between each level of software and hardware specification level (with the aim of identifying all required functions were implemented and no additional functions were present)
- Static analysis and dynamic testing of the source code back against the requirements whilst demonstrating coverage and reachability from source code level to the top level specifications

- Conduct of audits for good software and hardware engineering practices.

A few years later the processes and evidence produced through these early activities was submitted for accreditation under the UK's National Measurement Accreditation Service (NAMAS - now United Kingdom Accreditation Service - UKAS), with the result that my group became the first organisation to obtain national accreditation for safety-critical software assessment.

The experience of running an accredited test house, and in particular the difference between process assessment and product assessment, was also brought to bear when Stuart Nunns undertook an initiative to establish an open assessment framework throughout the European Union. For some time it had been apparent that whilst the UK had some excellent safety assessment activities, it was losing ground in that the Germany TUVs had a more dominant international presence. Stuart's contribution was to lead the development of a scheme known as 'the CASS Scheme Ltd' – since transformed into the '61508 Association'. Based on my experience in developing the NAMAS accredited test house, my contribution (besides undertaking requirements capture) was to propose that there was a difference between assessing the processes used in development of safety-critical systems and assessing the actual behaviour of the product. The distinction led to the development of the CASS Targets of Evaluation (TOEs) separating audit of an organisation's Functional Safety Management (FSM) system from assessment of the attributes of the Product. Schedules developed along this principle by a range of practitioners brought together by Stuart are still maintained on the 61508 Association web site (CASS TOES 2015).

3.2 Harmonisation through Standards

The practical work I was undertaking in the process sector throughout the late 1980's came to the attention of Ron Bell. By this time it was becoming apparent that, unless similar standards to the PES Guidelines were adopted internationally, the UK, whilst a leader in technical aspects of Computer System Safety, stood a good chance of becoming non-competitive on the international stage. Ron, together with Professor Phil Bennett (also working in the process control and railway industries), had the foresight to take the emerging UK guidance into the international standardisation arena, starting the initiative which led to the development of IEC61508 (IEC 1998 & 2010) and eventually to its promotion to a Basic Safety Standard under the IEC Advisory Committee on Safety (ACOS). My contribution to the early development of IEC61508 was limited to review as part of the British Standards Institute (BSI) Shadow Committee responsible for IEC61508, but later, when work started on Edition 2, I was invited to represent the UK on Part 3 (Software Requirements) as a member of the International Committee.

3.3 Harmonisation through Research

Meanwhile, the UK Department of Trade and Industry (DTI), under the guidance of Professor Bob Malcolm, had established a collaborative research programme, 'The Safety-critical Systems Programme', aimed at achieving a co-ordinated approach to safety related systems across UK industry. With initiatives evolving largely independently in the different high hazard industries, there was a risk that the UK led initiative would become segmented, with different industry sectors adopting widely differing approaches, thus confusing the regulatory situation and further reducing the UK influence internationally.

From my work in the process sector, I was aware that conventional software engineering practice, as represented through the IEEE and the emerging drafts of IEC61508, did not sit well with the typical computer architecture in common use in safety systems – namely the Programmable Logic Controller (PLC). In this type of architecture the functionality of an application is captured in a database and interpreted by a hardware/software platform which can be reused in many different applications. Further, these architectures are susceptible to the mis-use of the database, since faults are not well controlled by traditional software engineering practices. This 'problem' became the focus of the collaborative research project 'Software Engineering Methods for Programmable Logic Controllers' (SEMSPLC) and had the effect of bringing together a diverse range of industries, including a computer manufacturer, software tool supplier, process operator, control system integrator and the University of York under Professor John McDermid. The result were later published in the SEMSPLC Guidelines (IEE 1996). A second, more speculative, collaborative research project exploring the synergies between inductive software engineering and requirements capture (ISSAFE) was formulated at the same time, bringing together practitioners from the safety world and from advanced software research in artificial intelligence, (also one of my early research experiences).

One further initiative, in which I was privileged to be given a role, was Ron Bell's initiative to provide competency guidelines (IEE 1999). Working for the IEE (now IET) in collaboration with the BCS and funded by the HSE, the majority of the technical work was carried out by Rod May with a sound knowledge of competency schemes such as National Vocational Qualifications (NVQs) and Institute of Railway Signalling Engineers (IRSE) licencing arrangements. However, the contribution of which I am proud, and which was required in response to professional review, was to propose a means to simplify the complexity of multi-level competency schemes, by proposing that there are three critical distinctions in expertise; ability to work autonomously within a well understood field, ability to work autonomously 'outside the box' i.e. in a novel situation and those practitioners not yet judged to be capable of working without supervision. Thus was born the simplification of competencies into just three levels; Practitioner, Expert and Supervised Practitioner, intended to reflect technical competency in a specific application as distinct from position within an organisation. The practical result, as far as the published document itself was concerned, was a significant reduction in complexity of an already complex subject.

4 Diversification

4.1 The Seeds of Divergence

By the early 2000's the UK had established a good cross-sectoral sharing of safety engineering practices (ably sustained through the Safety-critical Systems Club under the auspices of Professor Tom Anderson and Felix Redmill) and was also leading in standardisation internationally. Particularly in the software industry, IEC61508-3 and the railway sector specific standard CENELEC 50128 presented very similar approaches (owing much to RIA 23), whilst much of the guidance in RTCA DO-178B presents similar recommendations for the avionics sector.   When the requirements of the normative Annex A of IEC61508-3 are addressed, then the Defence Standard 00-55 focus on formal methods to implement safety systems of the highest integrity is also consistent with the recommendations of IEC61508-3.

The evidence of harmonisation in systems engineering was less obvious. Whilst IEC61508 was based on a 'whole lifecycle framework', the separation between hardware and software engineering in traditional engineering applications was such that standards tended to be 'ring fenced' to one or the other discipline. Furthermore, the original aspirations of encompassing not just electrical engineering expertise, but all engineering disciplines, was not well received by the non-electrical engineering institutions. Never-the-less, the importance of the emerging concepts of safety management, hazard and risk assessment, the lifecycle framework, reliability modelling and the importance of independent review can be seen in most of the standards developed for the high hazard industries, and their adoption has continued as the standards have evolved.

By the mid 1990s, the US was keen to make its contribution to international standardisation, specifically with the practices proposed by the American National Standards Institute (ANSI) in its SP84 publication (SP84 1996). Accordingly a new IEC work item proposal was tabled by the US, under the chairmanship of Vic Maggioli who, in retrospect, can be credited with much of the uptake of functional safety in the US process sector. With experience in the process sector, I was appointed by the UK, together with four other experts, to work on the new standard – later to become known as IEC61511 (IEC 2011& 2016). After considerable discussion, it was agreed that the starting point would be IEC61508 rather than SP84, but that this would be adapted to address the US process sector practices and terminology.

4.2 Limitations of IEC61508 Ed 1

One of the most significant limitations of the first edition of IEC61508 when applied to the process sector (other than the understanding of the specific vocabulary which had grown up around the new discipline of functional safety) was that observed in the SEMSPLC project – i.e. that by the mid 1990s an industrial control system was *never* developed from scratch. Thus the waterfall lifecycle and methods and techniques used to manage the development of early computer systems were too idealistic. Instead, systems were 'configured' from existing hardware and software components, customised to undertake application-specific functionality. Furthermore, the underlying platforms constrained the complexity of the implementation and controlled configuration errors, but with techniques that required considerable knowledge of software engineering to be able to trace to the requirements of IEC61508. As a result, significant interpretation of IEC61508 was required, for which the control engineers implementing these system were not well prepared.

The solutions to these 'gaps' adopted by IEC61511 varied from modifications in terminology to make the requirements more understandable to control engineers, to proposing alternative routes to compliance to address pre-existing systems (albeit with stricter limits on the level of safety that could be claimed). In most cases, the safety integrity claim was limited to Safety Integrity Level (SIL) 2, but subject to some additional assessment activities, SIL3 could be claimed. The stricter claim limits were not considered a limitation in the process sector, since it reflected the common practice at that time.

An innovation proposed by myself (based on experience from the railway industry) was to 'encapsulate' underlying components built to IEC61508, with a set of instructions for how they were to be applied to ensure that their configuration also complied with IEC61508. This has the benefit of 'delegating' to the platform supplier responsibility for ensuring that his platform constrained the user in such a way as to comply with IEC61508 when used in accordance with the Manual. This suggestion made its way into IEC61511 as the 'Safety Manual' and was later adopted into IEC61508 and other standards.

Another limitation of the 1st Edition of IEC61508 was its focus on shut-down systems to bring a process to a safe state. Although the potential for continuous control was recognised, little guidance was provided on how to handle systems where the control system itself must maintain safety (for example in flight control) or where there may be no safe state (for example radiotherapy; autoclave sterilisation). Neither did it well address the concept of a continuous demand as encountered in railway signalling, or make clear the importance of the process safety time.

Yet further concerns relate to whether a consequence can be *so* severe that the *frequency* of the hazardous situation should not be taken into account, thus negating the concept of 'risk' in selecting the appropriate set of implementation techniques. In order to address this concern IEC61511 formalised the concept of 'layers of protection' requiring diversity between the different layers. Standards in other sectors (e.g. marine) have opted for a 'consequence driven' approach, relating the implementation techniques purely to the severity of the potential consequence.

4.3 A Multiplicity of Functional Safety Standards

Given the limitations of the first edition of IEC61508 and the example of the process sector in developing its own sector specific standard, it is hardly surprising that over the last 10 years the number of application specific standards has blossomed. In addition to the railway, avionic and process sector standards, over the last ten years functional safety standards have been published for electrical drives (needing continuous control), machinery and explosive atmospheres (needing to comply with the EU Directives), automotive (with high human intervention), nuclear (potential high consequence), medical devices (needing to comply with EU as well as US regulations), water management, and even farm vehicles.

Furthermore, IEC61508 itself has developed, feeding back the experience from IEC61511 and other standards into Edition 2. One of the main criticisms of Part 3 was the prescriptive nature of the methods and techniques proposed in Annexes A and B. Not only can they be difficult to apply, for example where application languages were used (process control, engine management), but also they place an embargo on the use of any system that had been built prior to wide adoption of the standard. Working with Thuy Nguyen (of EDF France), I proposed that if it were possible to extrapolate to the properties of the software that the mandated techniques could deliver (for example ''correctness', ''understandable' or 'testable'), alternative combinations to achieve the same properties, with the same level of rigour, could be used. Although an approach would be to move to requiring achievement of the software properties alone, this was felt by members of the International Committee to be open to too wide an interpretation of the rigour required. On this basis, the need to relate the properties (in Annex C) to the examples of how they could be achieved dictated the structure of Annex C. On the third iteration of the structure a consensus of the International Committee was achieved, based on a normative, prescriptive requirement in Annex A, supported by informative guidance in Annex C. Whilst criticised as 'putting the cart before the horse', there is anecdotal evidence that Annex C has value in structuring an argument as to why alternative combinations of methods to those specified in Annex A are of equal validity.

We also recognised the need to handle pre-existing software in any realistic development programme. The first step was to require that the software be supported by a Safety Manual, such that it could be safely applied to another application. More controversial was the decision to introduce clauses which enable 're-engineering' of pre-existing software that has not been developed to IEC61508. Based on the principle that if 're-engineered' software can meet the equivalent properties to software developed in full compliance to IEC61508, the alternative route to compliance known as 'Route 3s' was conceived.

Other concerns that Edition 2 of IEC61508 Part 3 attempted to address included requirements on tools (which often gave rise to queries as to whether the tools needed to comply with the same rigour as the product) and some efforts to address data driven software.

5 Where Next?

5.1 Overhead to Industry of Multiple Requirements

Should the emerging multiplicity of diverging standards be of concern? Many, from their individual sector standpoints, would say 'no' – at least within their own sector. However, it is increasingly likely that a specific system will need to comply with several of the different functional safety standards – for example process applications will likely need to address both process control and machinery, making both IEC62061 and IEC61511 relevant. The railway industry although governed by the CENELEC standards, often must demonstrate compliance with IEC61508. A particular issue in which I am currently engaged in my current role as Chairman of IEC61508 Part 3 is that of diverging definitions – where in a bid to make the definition more understandable to certain sectors (or possibly to make the requirement more attractive to a particular sub-set of standards users) variations on a definition are developed. Unfortunately this can also subtly change the meaning of a requirement between different standards – for example the definition of 'fault' and the definition of 'safe failure'. Where multiple standards apply it can be difficult to reconcile, manage, or even identify, that a requirement is interpreted differently in different standards.

5.2 Technology Innovations and Concerns

New expectations are emerging both from standards makers and from industrial developers which will influence all functional safety standards. Will they be able to resolve them in a harmonised manner or will further variations arise as each sector grapples with its own industry drivers? The IEC61508 Part 3 maintenance committee has already identified several areas where such new expectations will need to be addressed.

Cybersecurity

Not only is there an increasing expectation from regulators that cyber-security will be addressed (and from industry an increasing perception of the threat), there are several (un-coordinated) standards initiatives under development that recognise that a system cannot be safe if not also secure. However, the converse does not automatically hold and there is an open question as to how the two sets of requirements should be integrated – for example would it be possible for standards such as IEC61508 to simply refer out to security standards, or is interpretation and guidance required specifically for the safety application?

Agile Lifecycles

There is a growing body of evidence that agile development processes, if executed correctly, can significantly improve productivity. It is inevitable that there will be a growing pressure from industry to apply these techniques for competitive safety solutions. Whilst not strictly 'outlawed' by functional safety standards, this type of lifecycle represents a significant departure from the traditional waterfall lifecycle model and the requirement that a preceding lifecycle phase should be fully complete before moving to the next phase. There will be a challenge to maintain integrity in terms of management of change, impact analysis, demonstration of completeness and coverage of requirements and regression testing.

Tool Qualification

Edition 2 of IEC61508 Part 3 addressed some requirements on tools, but there is a body of opinion that further guidance on tool qualification is required. Such guidance should explicitly recognise the impact of the use of a tool on integrity, as well as setting out recommendations for tool development. There is a view within the International Community that there is much to learn from the avionics sector in adopting a risk based approach to the impact of tool error. Work is ongoing to provide guidance on tool characterisation, integration and validation, and configuration management, as well as criteria for the development of tools themselves.

Proven-In-Use

Since the publication of the first edition of IEC61511, the concept of 'proven-in-use' has generated much discussion. It is closely linked to how to handle products or systems that have been accepted into operation under earlier norms, the expectation of different national regulators and to the viability of reliability modelling of systems with predominantly systematic modes of failure. Some guidance on qualification through Proven-in-use is available in IEC61511 and this has been further formalised in the IEC61508-2 routes to compliance 2_H and 2_S. However in an effort to ensure the properties mandated through the normative Annexes of Part 3 are not undermined, further guidance on the interpretation of route 2_S for software has recently been published under IEC Technical Specification 61508-3-1:2016. Currently the guidance deals with the application of proven-in-use software components up to SIL 2 (and places a claim limit at that level), but there is an ambition to address higher SILs in future work.

Concurrent and Multi-core architectures

Concurrent and multi-core architectures are now the norm in many commercial platforms and their deployment, at least for the lower safety integrity requirements, is inevitable. For information-based safety systems e.g., surveillance and communication systems, the scale of data processing and the response times required necessitates their use. Extension of the guidance in IEC61508 to better address multi-core and concurrency is ongoing, with a shift in emphasis from avoiding dangerous interference to designing safe communication between cores.

Integrity of Diagnostics

The question of the integrity of software based self-test functions has been of concern since my earliest work on the BP systems, yet in many systems there is an implicit assumption that once the self-test cycles have been completed the system can be assumed to be restored to perfect health. As any health monitoring practitioner is aware, the adequacy of the diagnostics is only as good as the level of instrumentation and the ability to foresee the critical failure modes of the monitored system; yet to date the requirements on the diagnostics (of both hardware and software) have barely been addressed.

Cloud

Given commercial pressures, it is inevitable that proposals to use cloud hosting will start to enter safety applications. Arguably the earliest uses might be the hosting of virtual development environments and information management, but their extension to storage of health monitoring data for critical systems is easily foreseen. Beyond that could come plans for use during maintenance and even for control and monitoring. Not only will this emphasise the importance of the topics covered above, but also the sustainability, responsiveness, ownership, jurisdiction and governance of the control and data pertinent to safety.

Autonomy

As systems become more complex, capable of self navigation and complex decision making, the approach to safety will need to deal with the emergency of novel hazardous situations, as well as the more frequent application of continuous safety control systems. For example, the use of exoskeletons for mobility impairment brings new possibilities as to how they could lead to, or fail to prevent, hazardous situations for their wearers. The use of autonomous vehicles requires complex decision-making regarding 'sense and avoid' functionality. Future safety systems are likely to need increased guidance on decision making in the presence of uncertainty and lack of complete information.

Refresh of Current Guidance

Of course, although standards represent the consensus opinion of experts, it is often the case that changing industrial norms bring in the need to revisit previous consensus. For example, the rise to interconnectedness of sensors/controls via internet technologies has changed the security risks of safety systems far beyond that anticipated in the early years. Further, systems of systems pose new challenges in the complexity of analysis and argument. With IEC61508 having been published for more than 17 years, techniques such as object orientation have moved into the mainstream. Therefore work is also ongoing to update current guidance, particularly in the areas of formal methods, object orientation, software reliability, metrics and data driven systems[1] – provided of course that the practical evidence and critical consensus of the gathered experts deems that it is right to do so, without erosion of the safety principles.

5.3 Compliance to a 'Gold' Standard

Judging by the history of the last ten years, as well as the number of new concerns that can be foreseen and the refinement of opinion on old concerns, it would at first glance seem that the divergence in safety practices between industries and applications is set to grow significantly. This would act as a barrier, both to trade and to movement of expertise between different sectors. And it would further escalate the effort and cost to industry to ensure compliance with the full range of applicable regulations, standards and good practice guidelines. The question to ask – is this in the best interests of industry and, if not, what can we do to mitigate?

One initiative that is currently underway is to undertake a trial of a pilot process to assess conformance to the IEC Basic Safety Standards. Under the IEC governance structure, there is a requirement that all IEC standards comply with the relevant Basic Safety Standards to ensure consistency. The currently foreseen components of the pilot call for *training* to understand the responsibilities of members of IEC standards committees, *identification* of the relevant Basic Safety Standards at the point of submitting a proposal for a new standard or to initiate maintenance to an existing standard, and *development of a compliance statement* at the point of submitting a document for review by National Committees. The compliance statement will show traceability from a Basic Safety Standard requirement to the relevant clause in the proposed document and list any deviations with an explanation of their significance. It will then be a matter for the National Committees comprising the membership of the IEC to decide whether they will vote positively for adoption of the document. The process is expected to be trialled initially on the maintenance of IEC61511 against the IEC61508 standard, as well as to be applied to proposed changes to IEC61508 itself.

One of the objections to the pilot is the complexity of showing adherence to a standard such as IEC61508. In order to simplify this task, an item of work

---

[1] Based on the recommendations of the SCSC Data Safety Initiative Working Group see http://www.scsc.uk/groups.html?group=gd through the membership of Fan Ye

being taken forward within IEC61508-3 is whether it would be possible to 'tag' those requirements that relate to application agnostic principles to differentiate them from prescriptive implementation requirements. However, this work is at an early stage and will require substantial effort to complete.

6 Concluding Remarks

This paper has aimed to show from a personal viewpoint some of the history of reaching a consensus on the emerging discipline of functional safety over the last 30 years, the drivers that have developed a harmonised, comprehensive approach, but also the factors that are causing fragmentation. The IEC is putting steps in place to provide a better level of coherence for the future, but with many different bodies and regulators future coherence cannot be guaranteed. To maintain a sufficient presence in the face of commercial pressures I advocate that the time is coming to strive toward future harmonisation, rather than increasing divergence.

Acknowledgments I am grateful for the support, guidance and encouragement of the many friends and colleagues over the years, a small number of whom are mentioned above. With regard to the future, I would like particularly to acknowledge the ongoing work of the IEC61508-3 Maintenance Committee for their diligence in identifying and addressing potential future functional safety concerns.

References

Application of Safety Instrumented Systems for the Process Industries ANSI/ISA–84.01–1996

CASS TOES for Functional Safety Management Assessment to IEC61508-1:2010; 61508 Association. CASS web site http://www.61508.org/downloads/index.php

CASS TOES for Software Assessment to IEC61508-3:2010 v2 (Full Requirements); 61508 Association. CASS web site http://www.61508.org/downloads/index.php

Guide to Hazard and Operability Studies. Chemical Industries Association 1977

Defense Standard 00-31, The Development of Safety Critical Software for Airborne Systems 3 July 1987, Obsolescent 1 July 1996

D.J.Smith Reliability Maintainability and Risk. Practical Methods for Engineers 2nd Edition. Macmillan 1985

Functional Safety of Electrical/Electronic/Programmable Electronic Safety-related Systems (All Parts) IEC 61508 Edition 1 1998 and Edition 2 2010

Functional Safety - Safety Instrumented Systems for the Process Industry Sector (All Parts) IEC 61511 Edition 1 2003 and Edition 2 2016

PES Programmable Electronic Systems in Safety Related Applications. Health and Safety. Executive HMSO 1987

Safety Related Software for Railway Signaling, RIA Technical Specification No. 23, British Railways Board, London Underground Limited 1991

Safety, Competency and Commitment, Competency, Guidelines for Safety-Related Practitioners. The Institution of Electrical Engineers 1999

SEMSPLC Guidelines. Safety-Related Application Software for Programmable Logic Controllers. The Institution of Electrical Engineers 1996

Going 'Back to the Future': Developing safety-critical embedded systems using modern Time-Triggered software architectures

Michael J. Pont

SafeTTy Systems Ltd

UK

Abstract   *This paper is concerned with the development of software for real-time, safety-related embedded systems. The particular focus of the paper is on 'Time-Triggered' (TT) systems. TT design can be viewed as a subset of more generic 'Event Triggered' (ET) designs. When compared with ET alternatives, TT designs have a simple software architecture and – once constructed – are generally accepted as being easier to test. As a consequence, forms of TT design have been used for many years in industries such as aerospace, because they have been found to provide the basis for safe and reliable systems. Despite the growing demand for safety-related embedded systems in sectors such as industrial control, automotive and household goods, use of TT architectures is less common than ET architectures in these areas. This paper explores some of the benefits of modern TT designs, and considers some of the reasons why this approach is less commonly used than ET architectures in current safety-related designs.*

1 Introduction

Prof. Tony Hoare provided some advice about software design back in 1980:

> There are two ways of constructing a software design: one way is to make it so simple that there are obviously no deficiencies, and the other way is to make it so complicated

that there are no obvious deficiencies. The first method is far more difficult.
(Hoare, 1981)

This paper is concerned with the development of safety-related embedded systems using 'Time Triggered' (TT) software architectures. In keeping with the sentiments of the quotation from Prof. Hoare, such TT systems are undoubtedly simple. For example, implementation of software for a TT System will typically start with a single interrupt that is linked to the periodic overflow of a timer. This interrupt will drive a task scheduler (a very basic form of 'operating system'). The scheduler will – in turn – release the system tasks at predetermined points in time.

Such a TT architecture can be viewed as a subset of a more general event-triggered (ET) architecture. Implementation of a system with an ET architecture will typically involve use of multiple interrupts, each associated with specific periodic events (such as timer overflows) or aperiodic events (such as the arrival of messages over a communication bus at unknown points in time).

The 'sting in the tail' of the quotation from Prof. Hoare is the suggestion that the creation of simple designs can be challenging. Such a claim might reasonably be levelled at TT designs. The creation of a TT design requires planning and modelling during the development process: these processes may be a little more challenging than those required to construct an equivalent ET design, but – in the author's experience – most development teams can acquire the skills needed to create effective TT designs very quickly, once they understand the potential benefits of this approach.

The goal of this paper is to encourage developers of reliable embedded systems to consider use of some form of TT design in future safety-related projects. To this end, the paper explores some of the characteristics of modern TT designs. The paper goes on to consider why this approach to software development is currently less widely employed than might be expected.

2 Engineering modern TT designs

This section provides an overview of some of the key characteristics of modern TT Systems.

2.1 Tick Lists

In a TT system, each processor releases tasks in accordance with a predetermined task schedule. For example, Figure 1 shows a set of tasks (in this case Task A, Task B, Task C and Task D) that might be executed by a TT System.

In Figure 1, the release of each sub-group of tasks (for example, Task A and Task B) is preceded by what is usually called a timer 'tick'. In most designs, the tick is implemented by means of a timer interrupt. These timer ticks are usually periodic. In an aerospace application, the 'tick interval' (that is, the time interval between timer ticks) of 25 ms might be used, but shorter tick intervals (e.g. 1 ms or 100 μs) are more common in other systems.

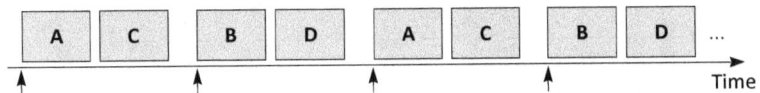

Fig. 1. A set of tasks being released according to a pre-determined schedule.

In Figure 1, the task sequence executed by the processor is as follows: Task A, Task C, Task B, Task D. In many designs, such a task sequence will be determined at design time (to meet the system requirements) and will be repeated 'forever' when the system runs, unless: [i] the system changes mode; [ii] the system is powered down; or [iii] a system failure occurs.

During the design process, it can be useful to think of this task sequence as a 'Tick List': for example, the Tick List corresponding to the task set shown in Figure 1 could be represented as follows:

```
[Tick 0]
Task A
Task C
[Tick 1]
Task B
Task D
```

Once the system reaches the end of the Tick List (that is, the end of the hyperperiod), it starts again at the beginning.

2.2 A simple modelling example

Consider a set of three periodic tasks, with worst-case execution time (WCET) values as shown in Table 1. In this example, the total (average) CPU load is 95%: as this figure is less than 100%, we may be able to schedule the task set.

Such a basic analysis is useful, as an initial 'sanity check' during the early phases of a system development, but it does not provide any detailed information about the system CPU load.

To provide detailed information, the CPU load for every tick over the hyperperiod can be considered (Figure 2).

Table 1: A set of 3 periodic tasks

Task	Period (ms)	WCET (ms)	CPU load (%)
A	5	2	40%
B	10	4	40%
C	20	3	15%
			95%

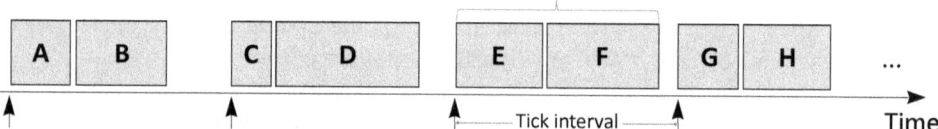

Fig. 2. Determining the maximum CPU load for a system.

For each tick:
(CPU load) = (Sum of the WCETs for all executing tasks) / (tick interval)

We need to repeat this calculation for all ticks in the hyperperiod in order to determine what the maximum CPU load will be during the program execution.

This process is straightforward. We can also determine other key system characteristics through analysis of a Tick List, such as response times (e.g. how long the system will take to respond to a switch press), and the level of 'jitter' (timing variation) in the task releases.

2.3 Working with TTC Schedulers

Many TT designs are implemented using co-operative tasks and a 'TTC' Scheduler. Figure 3 shows a schematic representation of a key components in such a scheduler. First, there is function SCH_Update(): in this example, this is linked to a timer that is assumed to generate periodic 'ticks' – that is, timer interrupts – every millisecond.

The SCH_Update() function is responsible for keeping track of elapsed time.

Within the function PROCESSOR_Init() there will be function calls that will initialise the scheduler, initialise the tasks and then add the tasks to the schedule.

In Figure 3, function main(), the process of releasing the tasks is carried out in the function SCH_Dispatch_Tasks(). The dispatcher releases any tasks that are due to run in a particular Tick. The dispatcher then moves the processor into a power-saving mode, ready to be wakened by the next Tick.

```
uint32_t main(void)                                  ┌──────────────┐
    {                                                │  1 ms timer  │
    PROCESSOR_Init();                                └──────────────┘

    SCH_Start();

    while(1)                         void SysTick_Handler(void)
        {                                {
        SCH_Dispatch_Tasks();            Tick_count++;
        }                                }

    return 1;
    }
```

Fig. 3. A schematic representation of a key components in a simple TTC Scheduler.

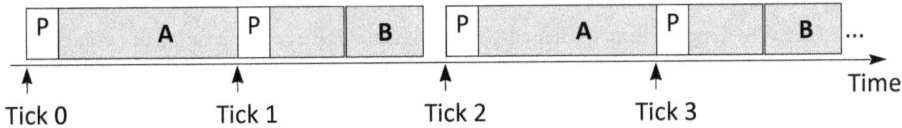

Fig. 4. Executing tasks using a TTH scheduler. See text for details.

2.4 Supporting task pre-emption

The designs discussed so far in this paper involve co-operative tasks: this means that each task 'runs to completion' after it has been released. In many TT designs, higher-priority tasks can interrupt (pre-empt) lower-priority tasks during the system execution.

For example, Figure 4 shows a set of three tasks: Task A (low-priority), Task B (low-priority), and Task P (high-priority). In this example, the low-priority tasks may be pre-empted periodically by the high-priority task. More generally, this kind of 'time triggered hybrid' (TTH) design may involve multiple co-operative tasks (all with an equal priority) and one or more pre-empting tasks (of higher priority).

We can also create 'time-triggered pre-emptive' (TTP) SCHEDULERs: these support multiple levels of task priority.

We can – of course – record the Tick List for TTH and TTP designs. For example, the task sequence for Figure 4 could be listed as follows: Task P, Task A, Task P, Task B, Task P, Task A, Task P, Task B.

It should be noted that correctly-designed TT systems do not suffer from problems of 'priority inversion', because the timing of task releases is always known in advance (allowing us to ensure that lower-priority tasks always make way for tasks of higher priority).

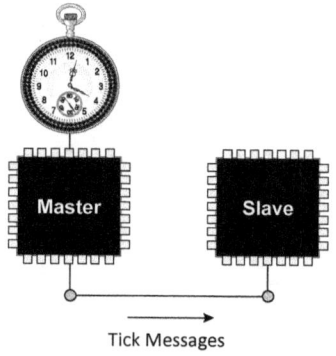

Tick Messages

Fig. 5. The simplest form of Shared-Clock Scheduler involves the sending of periodic 'Tick Messages' from the Master to the Slave.

2.5. Supporting multiple processors

Two broad classes of multi-processor design can be identified:

1. In <u>distributed</u> designs, we have two or more (often many more) processors that are physically separated in different parts of a vehicle, factory or build-ing (for example). A passenger car might contain some 50+ processors controlling brakes, door windows and mirrors, steering, air bags, and so forth. Similarly, an industrial fire detection system might typically have 200 or more processors, associated – for example – with a range of differ-ent sensors and actuators.

2. In <u>local</u> designs, we will have at least two processors, often located on the same PCB. Such designs will usually involve cross-checking between the two processors in order to meet particular safety requirements.

Where our design is distributed or local, the same challenge arises in a TT design: how can the activity on the multiple processors be synchonised?

Consider a simple example with two processors. A single clock can be 'shared' between the two processors, as illustrated schematically in Figure 5.

In Figure 5, we have one clock source on the Master processor in the network. This clock is used to drive the scheduler in the Master processor in exactly the manner discussed previously in this paper.

The Slave processors also have schedulers: however, the interrupts used to drive these schedulers are derived from 'Tick messages' generated by the Master (Figure 5). For example, in a design based on Controller Area Network (CAN), the Slave processor will have a scheduler driven by the 'receive' interrupts generated through the receipt of a CAN message sent by the Master.

In this manner, the activity on a network of processors can be synchronised very easily.

2.6. Changing the Processor Software Mode

Almost all practical embedded systems, each processor will support at least two Processor Software Modes (PSMs), called something like 'NORMAL mode' and 'FAIL-SAFE mode'. However many processors support additional PSMs. For example, Figure 6 shows a schematic representation of a software architecture for an aircraft system with PSMs corresponding to the different flight stages (preparing for take off, climbing to cruising height, etc).

We consider that the PSM is changed if the task set is changed. It should therefore be clear that there is likely to be a different Tick List for each PSM.

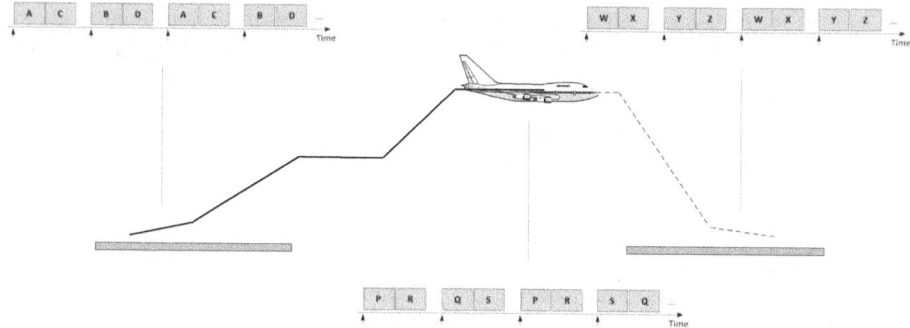

Fig. 6. An example of a system with multiple PSMs.

There are two particular features of these PSM changes that should be noted:

- within a given PSM, the tasks are **always** released according to a schedule that can be validated and verified when the system is designed;
- the timing of the transition between PSMs need **not** generally be known in advance.

2.7. The need for run-time monitoring

One key reason for choosing a TT architecture is that the resulting system will have deterministic behaviour: as we have noted, key characteristics such as the maximum CPU load, the response time and levels of task jitter can therefore be modelled precisely at design time.

Such models are necessarily based on various assumptions, including the following:

- we have operational processors in our system;
- we are running the correct program on each processor;
- we have operational peripherals on each processor;
- we have an operational scheduler on each processor;
- we can transfer data between tasks on the same processor without corruption;
- we can transfer data between processors without corruption;
- we know the WCETs (and probably the BCET (best-case execution time)) of all tasks on each processor;
- we know the execution sequence of the tasks in each PSM on each processor.

Should any of these assumptions prove to be invalid, this may invalidate the system models, with the consequence that the system may not behave as expected in the field.

Some of the threats that may need to be considered are as follows:

- Hardware failures that may result (for example) from electromagnetic interference, or from physical damage;
- Residual software 'bugs' that may remain in the system even after test and verification processes are complete;
- Deliberate software change that may be introduced into the system, by means of 'computer viruses' and similar security-related attacks.

Safety-related designs need to incorporate mechanisms that can test the design assumptions at run time (and the system designers need to consider what action will be taken if these tests are not passed).

As an example, a schematic representation of a predictive monitoring mechanism for use in a TT design is shown in Figure 7. This mechanism is designed to monitor the release of tasks by the scheduler at run time (to ensure that that are released in the intended sequence).

The approach involves performing the usual scheduler calculation (in the Dispatcher function), in order to determine which task to run next. However, before we release this task, we check with a second (independent) representation of the required task schedule in order to be sure that we are about to release the correct task.

This is a very simple mechanism to implement, but it is also very powerful, because it allows us to prevent incorrect task releases (not simply react to such releases after they have taken place).

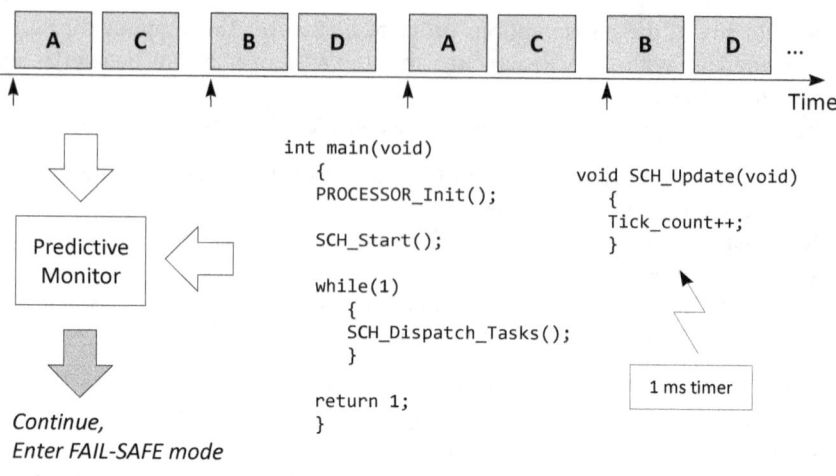

Fig. 7. A schematic representation of the operation of a predictive monitoring unit.

2.8. A summary of a typical TT development approach

In summary, the development process for typical TT designs has three stages:

1. modelling the system using one or more Tick Lists;
2. building the system (in compliance with the models);
3. adding support for run-time monitoring (to ensure that the system continues to comply with the models in the field).

3 Why don't more people use TT designs?

The technique for developing TT systems that we summarised in Section 2.8 are not particularly complicated to implement, but they can provide development teams, organisations and certification authorities with a high level of confidence that a given system will operate as expected in the field.

In light of this, Section 3 will consider some of the reasons why ET systems are still rather more widely used than TT systems.

(Please note that no attempt is made to pretend that what follows is any form of balanced review - it simply reflects the views and experience of the author of this paper).

3.1 Different understandings of the phrase 'real time'

The author has had many experienced engineers inform him that a particular TT design is 'not real time' in nature. This concern usually amounts to an observation that response to an interrupt on an ET system will usually be faster than the 'polled' response that can be obtained in a TT system.

At one level, this is perfectly fair criticism. Where there are a small number of interrupts enabled in an ET system, and the processor is lightly loaded, then a rapid (and reasonably deterministic) response to interrupts can be expected.

However, as we add additional interrupts to such an ET system and / or increase the processor load in other ways, it becomes much more difficult to understand what state the system hardware and software will be in when Interrupt X arrives and – therefore – how long the system will take to respond to this event. To be on the safe side, we might try to keep the processor fairly lightly loaded: this may then allow us to estimate with reasonable accuracy what the key timing behaviour (such as jitter and response time) will be.

In a TT system, the response time can be modelled precisely (to provide a worst-case figure). The modelled figure will not vary even if further tasks are added to the system (and maximum CPU loads of 90% or more can be safely achieved in many cases).

If the design has a requirement for 'fastest average response time to Event X', then an ET solution may be more appropriate. If the design has a requirement for 'deterministic response time to Event X', then a TT design should be considered.

3.2 Perceived lack of flexibility

Where people take the trouble to learn a little about TT designs, another concern raised is often that the resulting system will be 'too static'. Given the name of this architecture, such an assumption is not unreasonable. However, in practice, TT designs can be very flexible.

For example, please consider Figure 6 again. In this figure, the plane can switch between PSMs at times that are required by the flying conditions: the timing of such PSM transitions may vary based, for example, on the prevailing weather and / or on the density of the air traffic during the flight. The timing of the transitions between PSMs will not be determined in advance: in effect, the transition between modes is this TT design is ET in nature.

By implementing an appropriate set of PSMs in a TT design, we can address most concerns about flexibility (without sacrificing the deterministic nature of the behaviour within each PSM). This type of flexible design does not only apply in an aircraft: it can also be useful in a washing machine and many other systems.

3.3 The challenge of configuring the task sets

One key challenge with any form of TT architecture is what can be called the 'fragility' of the design. By this we mean that changes to even a single task in a working design (for example, a slight increase in execution time) may require revisions to the schedule parameters.

In fact, this is just one aspect of a larger problem. If (for example) we implement a TTC or TTH design, a number of key scheduler parameters must be determined at design time (including the tick interval, task order, and initial delay – or phase – of each task). Inappropriate choices may mean that a given task set cannot be scheduled (at all). Where the parameter set does ensure that all tasks are scheduled, inappropriate decisions may still lead to unnecessarily high levels of task jitter and / or to increased system power consumption. It has been demonstrated in various previous studies that the problem of determining these parameters is NP-hard (Brucker et al. 1977; Baker and Shaw 1989; Tindell et al. 1992; Ekelin and Jonsson 2001; Cucu and Sorel 2004; Baruah 2006).

Effective and practical solutions now exist for such problems. For example, Gendy and Pont (2008) have described techniques for configuring a task set that can ensure that: (i) task constraints are met; (ii) power consumption is 'as low as possible'; (iii) a fully co-operative scheduler architecture is employed whenever possible.

Please note that this approach does not attempt to perform an exhaustive search, but it provides results which are close to those obtained in a branch and bounds search, in a fraction of the time (typically a few seconds on a laptop computer, or a mobile phone).

3.5 The ET / TT threshold

Suppose that we have a design with 1000 tasks. Assume that we create a TT version of this design and an ET version of this design. Most people would probably agree that it would be easier to ensure that the TT version of this system had deterministic behaviour.

Suppose that we have only one task in the system. In this case, it may well be very easy to create a deterministic ET design (subject to a small number of constraints). It may well also be easier to create the ET design than the TT design.

Most practical systems will run somewhere between 1 task and 1000 tasks on a single processor. Where does the 'ET / TT' threshold lie? In other words, how large must the system be before it may become a good idea to consider use of a TT (rather than an ET) system?

Providing a precise answer to this question is (of course) impossible: we need to define what we mean by a task, and we need to consider the particular system characteristics. However, a 'rule of thumb' is possible: for the author, this threshold is around 10 tasks on a processor. Below this level, an ET design may be easy to model with reasonable accuracy; above this, use of a TT design may start to offer greater benefits.

If we accept – for the purposes of this argument – that [i] an ET / TT threshold exists; and [ii] it arises at a level of around 10 tasks, then we can begin to understand another challenge for those who wish to encourage use of TT designs.

Most people begin to explore embedded systems (on YouTube® or in university classes) with small systems. For such systems, an ET design can be very effective. Available code examples (and modern processors) further encourage the use of an ET approach. By the time an engineering or computer-science graduate has begun to develop her first large-scale automotive control system (and begins to experience the challenges of developing larger ET designs) the damage has already been done …

To address this, we need to have more universities and colleges consider in greater detail the creation of larger (and more representative) real-time systems.

4 Conclusions

There are no simple solutions to the challenges involved in developing complex real-time embedded systems, and it is not intended to suggest that the form of TT architecture summarised in this paper will meet the needs of every application. However, it is the author's experience that many organisations and development teams do not even consider the use of a TT approach, and that this results in the creation of some systems that are less reliable (and less safe) than they could be.

Will the world ever get to the point where TT (rather than ET) becomes the default architecture for real-time embedded systems? This seems unlikely in the short term, but there are perhaps some positive signs (from the author's perspective) in the latest edition of the influential international standard IEC 61508. In this standard (IEC 2010):

- TT architectures are "Highly Recommended" for systems of Safety Integrity Level (SIL) 2 or above. [IEC 61508-3 (2010), Table A.2]

- Use of a TT architecture "Greatly reduces the effort required for testing and certifying the system" [IEC 61508-3 (2010), Table C.1]
- Static synchronisation of access to shared resources – a key characteristic of all TT designs — is "Recommended" (SIL3) / "Highly Recommended" (SIL4) [IEC 61508-3 (2010), Table A.2]
- Limited use of interrupts – a defining characteristic of TT designs – is "Recommended" for SIL1 and SIL2 systems and "Highly Recommended" for SIL3 and SIL4 systems. [IEC 61508-3 (2010), Table B.1]

References

Baker, T. P. and Shaw, A. (1989) "The cyclic executive model and Ada," *Real-Time Systems*, Vol. 1, No. 1, pp. 7-25.

Baruah, S.K. (2006) "The non-preemptive scheduling of periodic tasks upon multiprocessors," *Real-Time Systems*, Vol. 32, No.1-2, Feb. 2006, pp.9-20,

Brucker, P., Garey, M.R. and Johnson, D.S. (1977) "Scheduling equal-length tasks under treelike precedence constraints to minimize maximum lateness, "*Mathematics of Operations Research*, Vol. 2, No. 3, pp. 275-284.

Buttazzo, G.C. (2005) "Rate monotonic vs. EDF: Judgement day," *Real-Time Systems*, Vol. 29, No. 1, 2005, pp. 5-26.

Cucu. L. and Sorel, Y. (2004) "Non-preemptive multiprocessor scheduling for strict periodic systems with precedence constraints," *In Proc. 23rd Annual Workshop of the UK Planning and Scheduling Special Interest Group*, PLANSIG'04, Cork, Ireland, Dec. 2004.

Ekelin, C. and Jonsson, J. (2001) "Evaluation of search heuristics for embedded system scheduling problems," *in Proc. Int. Conf. Principles and Practice of Constraint Programming, Paphos*, Cyprus, 2001, pp. 640 – 654.

Gendy, A.K. and Pont, M.J. (2008) "Automatically configuring time-triggered schedulers for use with resource-constrained, single-processor embedded systems", *IEEE Transactions on Industrial Informatics*, 4(1): 37-46.

Hoare, C.A.R. (1981) "The emperor's old clothes", Turing Award Lecture, CACM Vol. 24, No. 2, February 1981, pp.75-83.

IEC (2010). IEC 61508: 2010. Functional safety of electrical / electronic / programmable electronic safety related systems

Pont, M.J. (2001) *'Patterns for Time-Triggered Embedded Systems: Building Reliable Applications with the 8051 Family of Microcontrollers'*, Addison-Wesley / ACM Press. ISBN: 0-201-331381.

Pont, M.J. (2016) *"The Engineering of Reliable Embedded Systems: Developing software for 'SIL 0' to 'SIL 3' designs using Time-Triggered architectures"*, (Second Edition) SafeTTy Systems. ISBN: 978-0-9930355-3-1.

Tindell, K., Burns, A. and Wellings, A. (1992) "Allocating hard real-time tasks: An NP-hard problem made easy," *Real-Time Systems*, Vol. 4, No. 2, pp. 145–165.

Product Integrity Assurance Argument Framework for Vehicle Autonomy

John Birch, Mark Cousen, David Ward

HORIBA MIRA Ltd.

Abstract *Increasing autonomy of vehicle control features requires product integrity considerations that push the boundaries, and lie beyond the scope, of the present edition of the automotive functional safety standard ISO 26262. They include reliability, availability and cyber security as well as the safety-related aspects of Safety of the Intended Functionality (SOTIF), fail-operational functionality and the human-machine interaction. The purpose of this paper is to suggest a framework for the construction of an explicit product integrity assurance argument that caters for such considerations. The proposed framework builds on work previously developed by MISRA for arguing the achievement of functional safety within the scope of ISO 26262. The paper asserts that there is particular value in the creation of an explicit, holistic, product assurance argument for vehicles that feature autonomy as the considerations are inherently interrelated and extend beyond the scope of any one particular industry standard.*

1 Introduction

There is an increasing trend towards the development of autonomous functionality within the automotive industry, the ultimate expression of which would perhaps be the release of a fully autonomous vehicle that would ferry its occupants to their desired location in an unconstrained environment without the need for manual controls (Sedgwick 2016), (O'Brien 2016). Whilst the advent of autonomous vehicles promises a multitude of benefits, it also brings a plethora of new challenges to the industry related to the assurance of product integrity (the term 'product integrity' is expanded in section 2).

Recent high-profile incidents that serve as examples of the need to consider a broader view of product integrity include the fatal accident involving a Tesla vehicle whilst being driven with the use of the 'Autopilot' feature (Golson 2016) and the hacking and subsequent recall of a Jeep vehicle (Greenberg 2015). These instances not only highlight the technical nature of the development challenges of such functionality, but also the public interest in such events.

There is a real need, therefore, for the developers of such systems to assure themselves, the general public and safety regulators such at the National Highway Traffic Safety Administration (NHTSA) that such systems have been developed to an adequate level of product integrity. For example, arguing that the use of such systems and functions does not introduce an unreasonable safety risk to individuals that come into contact with them.

The purpose of this paper is to propose an argument framework for those developing autonomous systems and functions as an aid towards successfully claiming adequate product integrity.

2 Product Integrity

The term 'product integrity' is proposed as an overarching term comprising the following four individual facets, which, for the purposes of this paper, have been defined thus:

1. Reliability: *The probability of a system to perform its required function(s) for a desired period of time without failure*
2. Availability: *The probability that a system is operating correctly when it is requested for use*
3. Functional Safety: *The absence of unreasonable risk due to the behaviour of an electrical or electronic system*
4. Cyber Security: *The absence of unreasonable risk due to unauthorized access or attack*

Traditionally, these facets have been addressed individually, and with the possible exception of functional safety, there is typically no attempt made within the industry at developing an explicit argument to demonstrate assurance that all facets have been achieved satisfactorily.

This paper asserts that, due to the complex interrelationships between these facets for autonomous systems and functions, there is real value in developing an assurance argument to address them collectively.

3 Assurance Cases and their Value

The international standard ISO 15026-1:2013 (ISO/IEC 2013) defines an assurance case thus:

> A reasoned, auditable artefact created that supports the contention that its top-level claim (or set of claims) is satisfied, including systematic argumentation and its underlying evidence and explicit assumptions that support the claim(s).

An example of an assurance case would be a safety case where the top-level claim is one related to safety. The term 'safety case' has a number of definitions, one of which is cited below (ISO 2011):

> An argument that the safety requirements for an item are complete and satisfied by evidence compiled from work products of the safety activities during development.

For any automotive system (or 'Item') there is value in developing an explicit safety argument, rather than an implicit argument, appealing to the collation of a set of work products required by a particular standard (Birch et al. 2013). The claim that "*The system is free from unreasonable risk because it was developed according to ISO 26262*", even if it were of value for traditional automotive systems, is not valid with regards to autonomous functionality as the respective safety considerations lie beyond the scope of ISO 26262, or indeed its successor (ISO 2016). The scope of ISO 26262 is limited to considering hazards caused by the malfunctioning behaviour of electrical or electronic systems, whereas for an autonomous system many key safety considerations relate to whether the intended functionality itself presents an unreasonable level of risk.

There are ongoing efforts to standardise the treatment of the safety of the intended functionality (SOTIF) in the forthcoming Publically Available Specification (PAS) (ISO 2016). However, the requirements of the PAS are more amenable to being applied to discrete functions treated in isolation from one another than they are to the possibly non-deterministic behaviour of a highly complex set of interacting functions as may be found in an autonomous vehicle.

The recently released Federal Automated Vehicles Policy document from NHTSA (NHTSA 2016) states:

> The Department of Transport (DOT) anticipates that manufacturers and other entities planning to test and deploy Highly Autonomous Vehicles (HAVs) will use this Guidance, industry standards and best practices to ensure that their systems will be reasonably safe under real-world conditions.

In doing so, NHTSA is acknowledging that there is no 'silver bullet' solution to arguing autonomous vehicle safety by appealing to the compliance with any particular standard. The corresponding requirements relating to safety, cyber security and other facets of product integrity within the policy document are necessarily goal-based, rather than prescriptive, placing the burden of arguing the achievement of the absence of unreasonable risk, for example, firmly with the system developer.

Therefore the assertion is made that there is value in developing an explicit assurance argument for the safety of autonomous functionality that sits above claiming compliance with any particular standard, (even if the argument makes use of claims regarding compliance with certain aspects of them).

As all product integrity facets are closely coupled, and may need to be traded with one another, this assertion is extended to the development of a wider assurance argument for product integrity as a whole. This is illustrated in the example argument framework provided in the following sections of the paper.

4 Proposed Assurance Argument Framework

In this section of the paper, a proposed framework is provided for structuring a holistic assurance argument. There is no intention for it to be rigidly adopted but rather that it may serve as a starting point for an argument that is tailored according to the needs of the application. The framework is presented in the form of a partial argument for a theoretical 'Autonomous Driver' (AD), introduced below.

4.1 'Autonomous Driver' Definition

The notion of an AD is that of a substitute for a human driver within a manually-driven vehicle by an autonomous system, depicted in the 'black-box' diagram below in Figure 1.

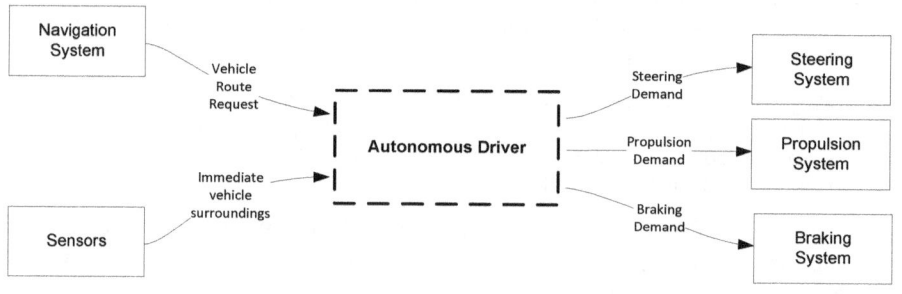

Fig. 1. Simplified 'Autonomous Driver' Black-Box View

The AD is informed of the intended route the vehicle should follow by the vehicle navigation system, and of the immediate vehicle surroundings by the vehicle sensor set. Based on this information the AD decides what the forthcoming vehicle movement should be, from which it issues a set of demands to the principal vehicle actuators (steering, propulsion and braking). In reality, of course, a human driver performs a host of other functions such as communication to the drivers of other vehicles and controlling the illumination of exterior lights etc. but such functions have been omitted for the sake of simplicity.

Building on the 'black box' view of the AD shown in Figure 1, suppose that the system designers have developed the AD according to the example set of High Level Requirements (HLRs) provided in Table 1, yielding the functionality shown in Figure 2. Note that the function steps in Figure 2 are labelled according to the corresponding HLRs from Table 1.

Table 1. Autonomous Driver High Level Requirements

Requirement Identifier	Requirement
HLR1	The Autonomous Driver shall calculate safety risk at time step n based on knowledge of the immediate vehicle surroundings
HLR 2	The Autonomous Driver shall determine all options for actuator demands for time step $n+1$ that exist within the physical limitations of the vehicle in its environment
HLR 3	The Autonomous Driver shall predict what the safety risk at time step $n+1$ would be for each actuator demand option
HLR 4	The Autonomous Driver shall determine if there are any actuator demand options that would yield safety risk at time step $n+1$ that would be less than the Acceptable Risk Threshold
HLR 5	If the Autonomous Driver determines that there are actuator demand options that would yield safety risk at time step $n+1$ that would be less than the maximum acceptable risk threshold, then it shall enact the option from this set that yields the optimum route following
HLR 6	If the Autonomous Driver determines that no actuator demand options exist that would yield safety risk at time step $n+1$ that would be less than the maximum acceptable risk threshold, then it shall enact the option that yields the minimum safety risk

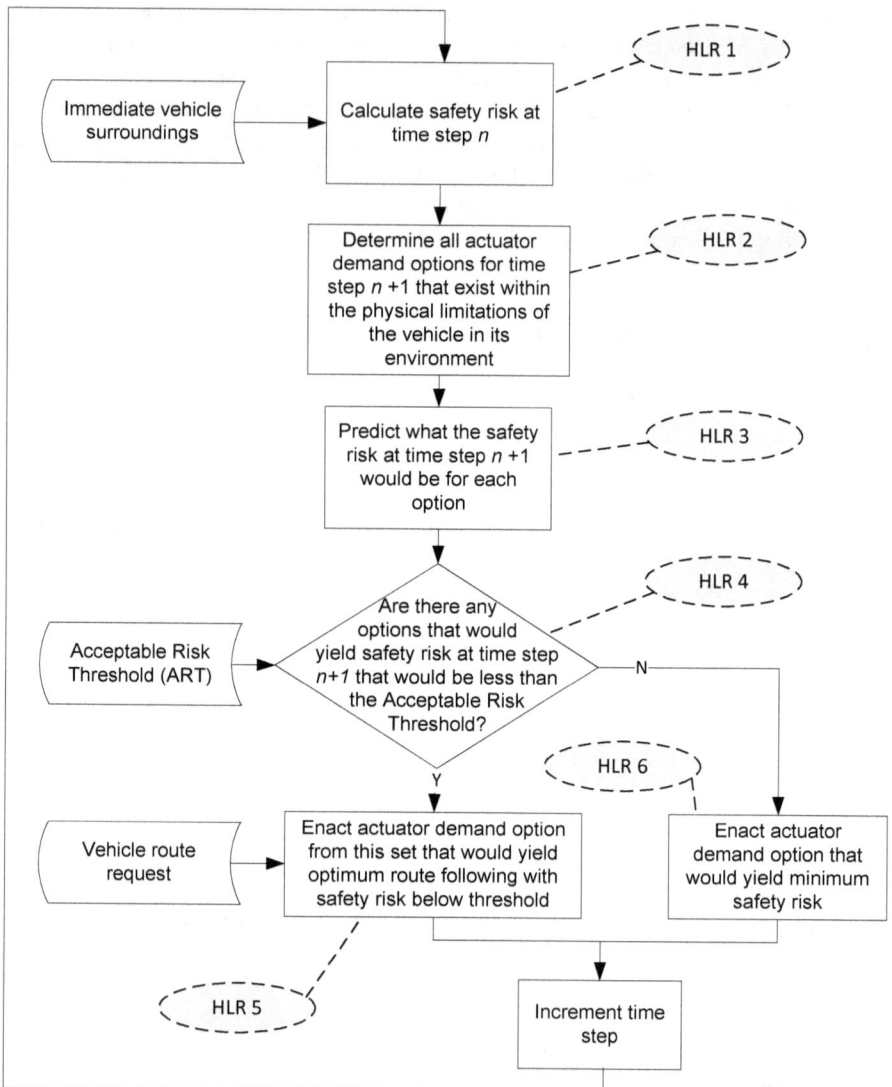

Fig. 2. Autonomous Driver High-Level Functionality

Given the black box view in Figure 1, the requirements in Table 1 and the proposed high-level functionality in Figure 2, the next section of the paper provides the corresponding assurance argument that might be made for the AD.

4.2 Top-Level Argument

Shown below in Figure 3 is the top-level assurance argument for the AD. The argument is shown using Goal Structuring Notation (GSN), for which a definition and guidance on its use is provided in the corresponding GSN standard (Origin Consulting (York) Limited 2011). The argument references a claim for each of the 'downstream' argument structures corresponding to the individual facets of product integrity (as described in section 2). For brevity only the Functional Safety argument structure has been subsequently developed.

Each product integrity facet argument structure inherits a common context, which is a single 'Item Definition' (Item Definition being a term borrowed from ISO 26262 (ISO/IEC 2011). The Item Definition would contain the detail required by ISO 26262 such as the functionality of the AD (e.g. as described in Figure 2), the intended operating environment, the boundary of the Item and its interfaces etc. The fact that this is a shared context ensures that if, for example, the Item Definition is updated due to a safety-related development, there is assurance provided that a reasonable degree of availability has been maintained. The Functional Safety argument structure is developed further in the following sections of this paper.

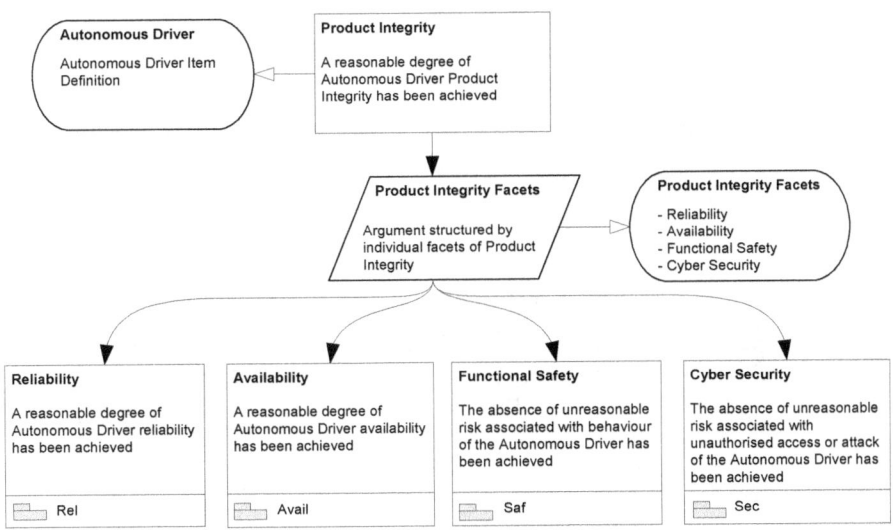

Fig. 3. Autonomous Driver Top-Level Assurance Argument

4.3 Functional Safety Argument Structure

Shown below in Figure 4 is the Functional Safety argument structure referenced from Figure 3.

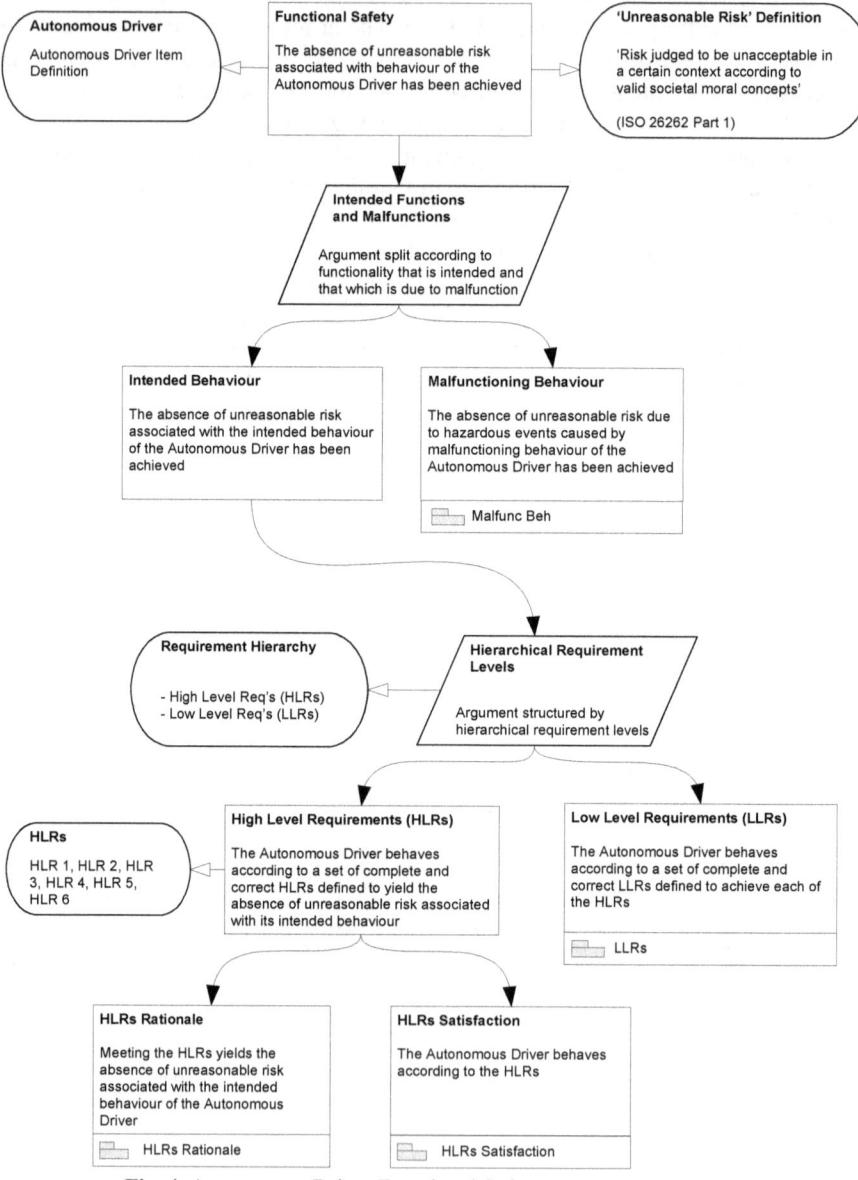

Fig. 4. Autonomous Driver Functional Safety Argument Structure

Note that the term 'Functional Safety' used in the top-level claim in Figure 4 has a wider scope than that defined by ISO 26262 (provided in section 2).

Note also that in Figure 4, and subsequent figures, the AD Item Definition is shown as explicit context to the top-level claim to serve as an aid to understanding, although in reality this is unnecessary as the context is inherited from the parent argument in Figure 3.

Coming from a relevant international standard the ISO 26262 definition of 'unreasonable risk' is used here.

4.3.1 Malfunctioning Behaviour

The argument in Figure 4 is split according to risk associated with the intended behaviour of the AD and that associated with its malfunctioning behaviour. The latter argument structure is one that corresponds to the scope of ISO 26262 and is developed to the next level below in Figure 5. This structure is proposed by the Motor Industry Software Reliability Associated (MISRA) (Birch 2016) in the context of an activity to publish a set of forthcoming industry guidelines for developing automotive safety case arguments. As such, the argument structure in Figure 5 is not developed further within this paper, rather the reader is referred to the example argument for an alternative Item presented by Birch (Birch 2016).

The argument structure presented in Figure 5 makes the use of an Assurance Claim Point (ACP) between the 'Malfunctioning Behaviour' claim and the 'Hazardous Events' context, indicated by a solid black square. As ACPs are not described in the GSN standard there is a brief note provided in 4.3.2 of this paper to explain their value and use.

One particular consideration that should be given to the development of the structure in Figure 5 for an AD, however, is how it is argued that the AD would successfully transition to a 'safe state' (ISO/IEC 2011) or a 'minimal risk condition' (NHTSA 2016) in the event of a malfunction without the aid of human driver intervention, and what requirements (and assurance argument burden) this would place on AD availability.

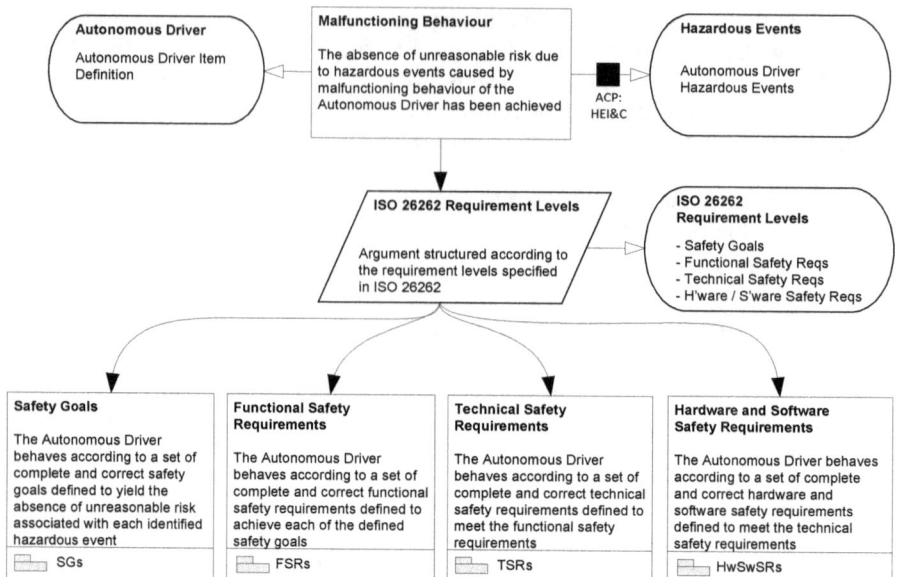

Fig. 5. Autonomous Driver Malfunctioning Behaviour Argument Structure

4.3.2 Intended Behaviour

The 'intended behaviour' structure in Figure 4 is shown as being divided according to two hierarchical levels of requirement, termed 'High Level Requirements' (HLRs) and 'Low Level Requirements' (LLRs). This is to reflect the adoption of good Systems Engineering practice in the use of levels of requirement abstraction. In practice there may be several more requirement abstraction levels used. For the brevity the LLRs argument structure is not developed further within this paper.

The HLRs argument structure is made in the context of the HLRs provided in Table 1 and is further sub-divided according to the 'rationale' for the HLRs and their 'satisfaction'; a technique previously proposed by MISRA (Birch 2016).

HLRs Rationale. Shown in Figure 6 is the 'rationale' argument for the HLRs from Table 1. The product assurance argument up until this point could look very similar for different systems, but the rationale behind a set of requirements is of course unique to those requirements. The argument shown in Figure 6 (and that subsequently provided in Figure 7) should therefore be considered as an example instantiation of the framework, rather than the framework itself.

The argument is structured according an 'Acceptable Risk Threshold' (ART) strategy that is referred to in the HLRs and the functionality in Figure 2. This strategy is presented in the context of two pieces of information related to the ART: the specification of the risk threshold and its corresponding risk matrix. At this point the risk threshold is presented 'as is'; its justification is provided lower in the argument structure.

To reiterate, the three claims below this strategy are:

1. The ART is specified such that if it is respected then the resulting risk will not be unreasonable
2. Meeting the HLRs will ensure that the vehicle behaviour will only fail to respect the ART when prohibited by the physical limitations of the vehicle in its environment
3. Failing to respect the ART due to the physical limitations of the vehicle in its environment does not yield unreasonable risk

Of the above claims, the second can be supported simply by inspecting the requirements themselves.

The third claim is shown as being supported by a further sub-claim and associated solution appealing to the public recognition that there will be instances in which the AD will not be able to avoid a high-risk scenario due to its physical limitations, as would be the case with a human driver. Such instances might include, for example, a person running out in front of the vehicle from behind a parked truck, or the vehicle being side-struck by another vehicle that ignored a red light at a crossroads. Going back to the definition of 'unreasonable' risk at the top of Figure 4, if it can be shown that such risks are accepted by 'society' then they could be argued to be reasonable. Of course it may be that there are greater societal expectations for a vehicle with an AD than a human driver, hence the need in the argument for a solution appealing to the results of a survey that gauges public expectations of autonomous vehicle safety.

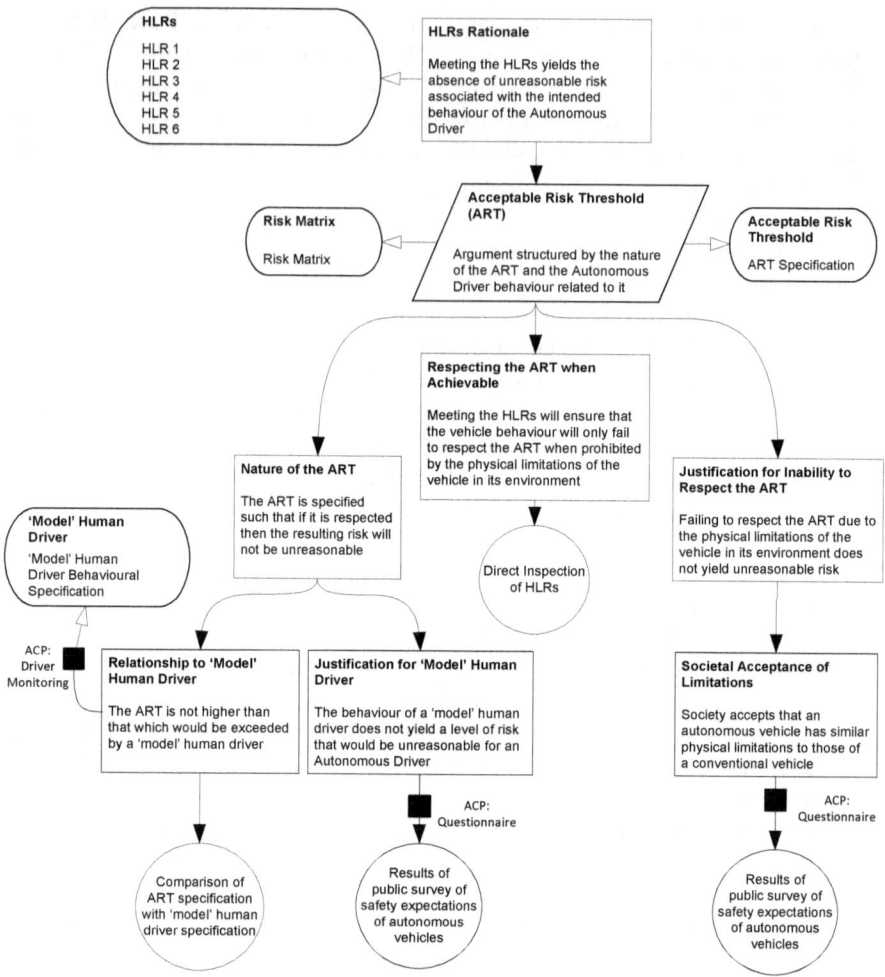

Fig. 6. Autonomous Driver High Level Requirements 'Rationale' Argument Structure

ACPs have been used in the argument structure in Figure 6 between some of the claims and their corresponding solutions. The use of ACPs is proposed as a means of claiming confidence, or assurance, in the relationship between two argument entities (Hawkins et. al). The 'Questionnaire' ACP in Figure 6 could be used to reference a separate argument for the way in which the public survey was conducted, how many people were consulted, their age range, their socio-economic background etc. without 'cluttering' the main argument. Being able to make such confidence claims is of value if there is a non-trivial relationship between the claim and its solution, or if a claim is particularly critical in its role or supporting a top-level claim, for example.

The first of the three claims listed above is more challenging to support. It has been done with two sub-claims; firstly that the ART is not higher than that which would be exceeded by a 'model' human driver and secondly that it is not unreasonable to specify the ART with regards to a model human driver. A model human driver is a notion that is proposed as a means of encapsulating the best aspects of human driver behaviour, i.e. the ability to perceive and act upon risk quickly and correctly, without making mistakes or being distracted, whilst still making good progress towards reaching the intended destination. The specification for a model human driver shown as context in Figure 6 might therefore include, for example, a reaction time (that is perhaps lower than 99.9% of human drivers) and some kind of measure of risk appetite. Arriving at a justifiable specification of course would not be no trivial task and the means by which this was achieved would need to be detailed in an argument referenced by the 'Driver Monitoring' ACP, which may, for example, appeal to the extensive monitoring of the behaviour of a wide range of human drivers. Again, the results of a public survey of the safety expectations of autonomous vehicles might serve as suitable evidence for supporting the claim that if the AD behaves in a manner similar to the model human driver then the resulting risk would not be unreasonable.

HLRs Satisfaction. The way in which it is argued that the HLRs have been satisfied by the AD is shown in Figure 7. Again the argument is structured according to an ART strategy, this time contextualised by a specification of both the virtual and physical test scenarios employed.

The key claim in the argument is that the diversity and number of test scenarios yields sufficient confidence in having met the HLRs. This claim is contextualised by a set of statistics that describe the testing that has been completed (e.g. number of miles of urban driving, number of miles of motorway driving, number of hours of night time driving, number of hours of driving in fog etc.) and a recognised industry measure for required successful test completion. Such a measure might be derived from an international standard (e.g. from International Organization for Standardization), a regulator (e.g. NHTSA) or a consortium of companies (e.g. as is the case in the 'Pegasus' project (VRA 2016)).

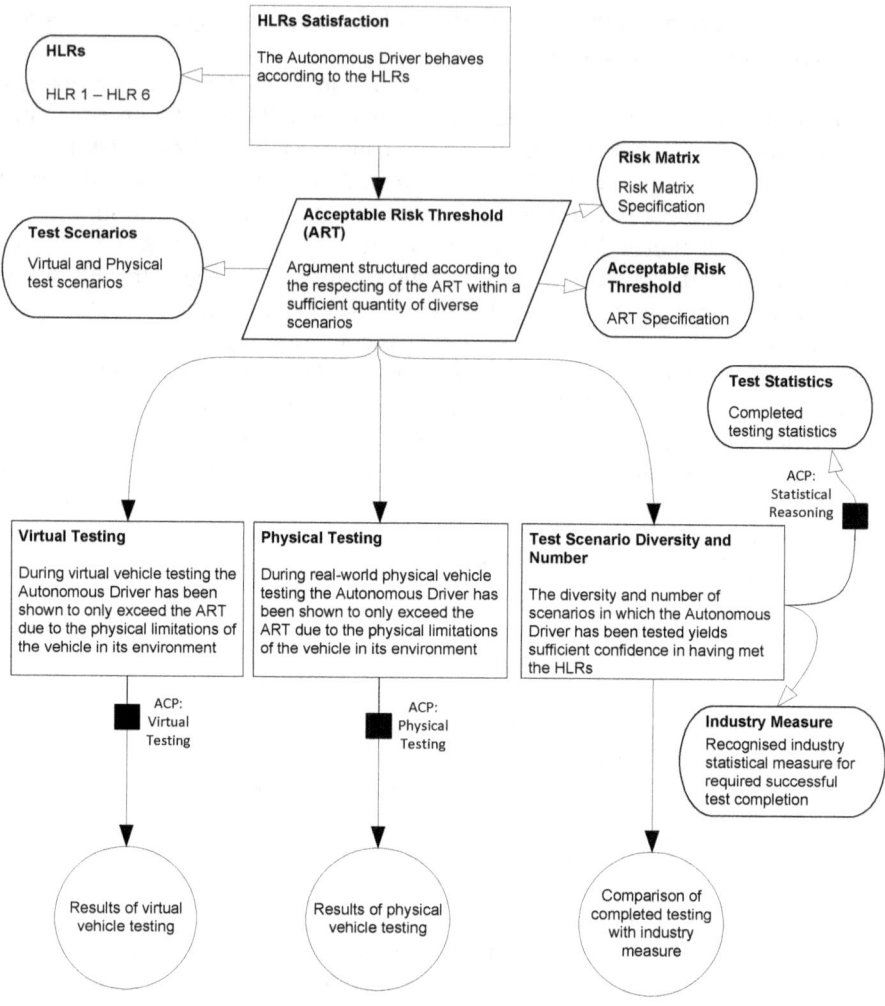

Fig. 7. Autonomous Driver High Level Requirements 'Satisfaction' Argument Structure

5 Concluding Remarks

This paper has outlined the product integrity challenges faced by developers of autonomous functionality and the value in creating an explicit, holistic product assurance argument to demonstrate that a sufficient level of integrity has been achieved. It has sought to provide an outline argument framework, which has been partially instantiated to show how an argument for an autonomous driver might look. It has shown the argument claims that need to be made in addition to those for demonstrating that freedom from unreasonable residual risk has been achieved due to malfunctioning behaviour (as per the scope of ISO 26262).

Whilst it is recognised that the challenge for developers lies in the detail of being able to provide and justify the solutions to the types of claim made in the argument, it is hoped that at least being able to see the claims that need to be made and their relationship to one another will provide a valuable stepping stone for the industry in the development of high integrity autonomous functionality.

References

Birch J (2016) Automotive Safety Arguments: The Good, the Bad and the Ugly. Available at: http://scsc.uk/file/404/05---John-Birch---Automotive-Safety-Arguments---MISRA---09-Jun-16---Final.pptx

Birch J et al (2013) Safety Cases and their role in ISO 26262 Functional Safety Assessment. Available at: https://www-users.cs.york.ac.uk/~ihabli/Papers/2013Habli_Safecomp.pdf

Golson (2016) Tesla driver killed in crash with Autopilot active, NHTSA investigating. Website accessed 06-Oct-2016:
http://www.theverge.com/2016/6/30/12072408/tesla-autopilot-car-crash-death-autonomous-model-s

Greenberg (2015) After Jeep Hack, Chrysler Recalls 1.4M Vehicles for Bug Fix. Website accessed 06-Oct-2016:
https://www.wired.com/2015/07/jeep-hack-chrysler-recalls-1-4m-vehicles-bug-fix/

Hawkins R et al. A New Approach to Creating Clear Safety Arguments. Available at: https://www-users.cs.york.ac.uk/rhawkins/papers/HawkinsSSS11.pdf

ISO (2011) ISO 26262-1:2011 Road vehicles -- Functional safety -- Part 1: Vocabulary. Available at: http://www.iso.org/iso/catalogue_detail?csnumber=43464

ISO (2016) ISO/DIS 26262-1: Road vehicles -- Functional safety -- Part 1: Vocabulary. Available   at:
http://www.iso.org/iso/home/store/catalogue_tc/catalogue_detail.htm?csnumber=68383

ISO (2016) ISO/AWI PAS 21448 Road vehicles -- Safety of the intended functionality. Available at: http://www.iso.org/iso/catalogue_detail.htm?csnumber=70939

ISO/IEC (2013) ISO/IEC 15026-1:2013 Systems and software engineering -- Systems and software assurance -- Part 1: Concepts and vocabulary. Available at:
http://www.iso.org/iso/home/store/catalogue_tc/catalogue_detail.htm?csnumber=62526

NHTSA (2016) Federal Automated Vehicles Policy. Available at:
https://www.transportation.gov/AV

O-Brien (2016) One-Third of all Long Haul Trucks to be Semi-Autonomous by 2025. Website accessed 06-Oct-2016:
https://www.trucks.com/2016/09/12/one-third-trucks-autonomous-2025/?utm_content=bufferbeacb&utm_medium=social&utm_source=linkedin.com&utm_campaign=buffer

Origin Consulting (York) Limited (2011) GSN Community Standard Version 1. Available at: http://www.goalstructuringnotation.info/

Sedgwick (2016) Delphi to expand driverless-taxi test. Website accessed 06-Oct-2016:

http://www.autonews.com/article/20161001/OEM10/310039973/delphi-to-expand-driverless-taxi-test

VRA (2016) Pegasus. Website accessed 06-Oct-2016:
http://vra-net.eu/wiki/index.php?title=PEGASUS

Experiences of avionics safety certification of an ARINC 653 RTOS on multi-core processor architecture

Paul J. Parkinson

Wind River

Swindon, UK

Abstract   *The avionics industry is currently undergoing a transition to multi-core processor architectures. This transition provides the potential for increased functionality on a common computing platform, while reducing size, weight and power (SWaP). The advent of multi-core processor architectures also presents new challenges in the design, implementation and safety-certification of real-time operating systems (RTOS). In this paper, the impact of multi-core architecture on RTOS design, the ARINC 653 software standard, and safety-certification under RTCA DO-178C and EUROCAE ED-12C will be considered. Additionally Wind River's experiences of the DO-178C DAL A safety certification of VxWorks 653 Multi-core Edition on the NXP QorIQ T2080 multi-core processor will be presented.*

1 Processor Selection Factors for Avionics Applications

During the last two decades, there has been dramatic growth in the number and complexity of aviation electronics (or avionics) systems in civil and military aircraft, driven by the need to implement new capabilities to meet operational requirements. The selection of microprocessors for avionics systems during this period has been influenced by a number of factors:

1.   US DOD directive on use of COTS

The directive by US Defense Secretary William J. Perry (Perry 1994) paved the way for the use of commercial-off-the-shelf (COTS) technologies on US Department of Defense (DOD) programmes. This also led to the decline in the availability of military-grade processors which provided better support for harsh environments (including temperature and shock), and also provided better data for single-event upsets (SEU). More recently this has also resulted in programme challenges in terms of obsolescence management due to the shorter lifecycles of commercial microprocessors which are incompatible with the long lifecycles of aerospace programmes.

2.   Adoption of Integrated Modular Avionics architectures
The increasingly widespread adoption of integrated modular avionics (IMA) architectures has provided benefits in terms of reducing Size, Weight and Power (SWaP) requirements for airborne platforms. This has been achieved through the use of common computing platforms with high-performance processors to host multiple applications concurrently. This in turn has resulted in a reduction in the number and type of line replaceable units (LRUs, as shown in Figure 1), and as a consequence fewer processors (of potentially different architectures) when compared to federated systems.

3.   Advent of Multi-core processors
The advent of multi-core processor (MCP) architectures and proliferation of numerous multi-core architecture variants, combined with the diminishing availability of single-core processor architectures, has led to a fragmentation of processor selection in the avionics market segment.

Fig. 1. Republic F-105B with avionics layout (Source: US Air Force – public domain)

1.1 Processor Selection History

Aerospace suppliers historically have selected microprocessors for avionics programmes based on criteria which have included (but not limited to) performance, power dissipation, availability of extended temperature range, and longevity. Safety-critical avionics programmes have also had additional selection criteria related to potential for access to proprietary information on processor design, manufacture and testing to support DO-254 and EUROCAE ED-80 objectives for hardware safety certification (DO-254 2002, ED-80 2002).

Many avionics projects have selected for COTS and custom board designs which have used PowerPC™ 750 and PowerPC MPC7410, MPC7448 and MPC7457 processors with external devices for I/O (NXP 2016). In addition, PowerPC architectures with integrated on-chip peripherals (such as MPC8349E, MPC8548E, MPC8572 and MPC8641D) need fewer external devices, and have been used in applications with more stringent SWaP requirements.

A number of these avionics applications have successfully completed avionics hardware certification under DO-254 and ED-80, and avionics software safety certification under DO-178B and EUROCAE ED-12B (DO-178B 1992, ED-12B 1992) and the later revisions, DO-178C and ED-12C (DO-178C 2011, ED-12C 2011). Their successful deployments on safety-critical programmes has resulted in these microprocessors being regarded as a low technical risk for subsequent safety-critical programmes, creating a "virtuous circle".

1.2 Safety Certification of the VxWorks RTOS

Wind River has ported the VxWorks® real-time operating system (RTOS, Wind River 2016) to a broad range of processor architectures, including ARM®, MIPS™, PowerPC, SPARC® and x86. This has been undertaken due to the differing processor requirements of multiple vertical market segments (including aerospace & defence, automotive, industrial automation, medical, railway and telecommunications).

However, undertaking DO-178B and ED-12B Level A software certification of an RTOS is extremely expensive, costing millions of dollars and is specific to an underlying processor architecture. Hence, it would have been cost-prohibitive for Wind River to undertake DO-178B and ED-12B safety certification on many different processor architectures, with no guarantee of being able to recoup the non-recurring engineering (NRE) costs.

For these reasons, Wind River decided to develop the first DO-178B and ED-12B Level A safety COTS certification evidence package for VxWorks 5.4 Cert on the most widely-used processor in avionics, which at the time in 2000, was the PowerPC 750.

In subsequent years, Wind River has developed DO-178 Level A COTS certification packages for additional processor architectures for VxWorks Cert for federated avionics systems and VxWorks 653 for integrated modular avionics systems (Parkinson and Kinnan 2015) based on market trends and specific customer requirements. This has included additional PowerPC variants (MPC74xx, MPC8245, MPC82xx, MPC8349E, MPC8548E, MPC8560 and MPC8641D) and Intel processors variants (Pentium III, Core and Atom) as shown in Figure 2. Although this has included several multicore Intel and PowerPC processors, these have historically been configured for single-core operation in civil safety-critical applications, due to civil certification authorities not having published guidance on MCP architectures at that time. Military avionics programmes which have the ability to perform self-certification have had the option of running VxWorks in an Asymmetric Multi-Processing (AMP) configuration on multiple cores.

This COTS evidence approach has enabled the significant DO-178 and ED-12 certification NRE costs to be amortised across multiple customers and programmes using the same processor architecture, reducing the cost of certification on each programme. This has also resulted in a virtuous circle, as these processors have provided the lowest cost options for follow-on certification projects, due to the ability to reuse existing DO-178 and ED-12 certification evidence, rather than having to develop it for a new processor architecture and associated incremental costs. At the time of writing (September 2016), VxWorks and VxWorks 653 have been used by over 300 Wind River customers on 450 projects on over 85 aircraft.

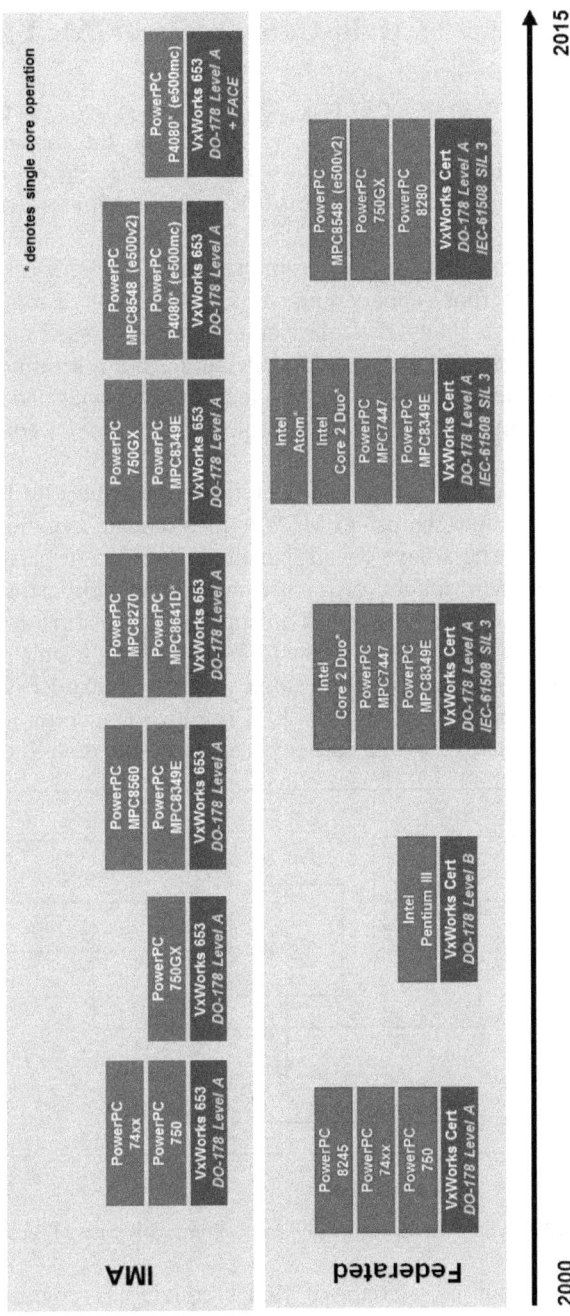

Fig. 2. VxWorks DO-178 COTS certification package timeline

2   The Challenges of Multi-Core Processor (MCP) Selection

Over the last decade, in order to meet the demands of ever increasing performance from the commercial market, and faced with the fundamental performance limit of single-core processors due to clock speed ceiling, semiconductor manufacturers transitioned to MCP architectures to achieve performance gains.

The introduction of MCP architectures has provided performance gains for enterprise general purpose applications; it has also presented some unique challenges for their use in safety-critical avionics systems. This is because avionics applications have specific requirements, including (but not limited to) application isolation and determinism, and these are not the primary considerations of semiconductor manufacturers when designing MCPs for the commercial market.

The avionics industry, academia and certification authorities have undertaken research projects into the use of MCP architectures in avionics applications. A number of researchers have found that there is variation between MCP designs in terms of their suitability for use in avionics applications, due to the impact of architectural design features on application isolation and determinism (Kinnan 2009, FAA 2011). These relate to factors arising from shared resources on the device, which include use of a single memory controller or shared bus is used by multiple cores (providing a risk of resource contention), and similarly use of separate or shared Level 2 caches per core, as shown in Figure 3.

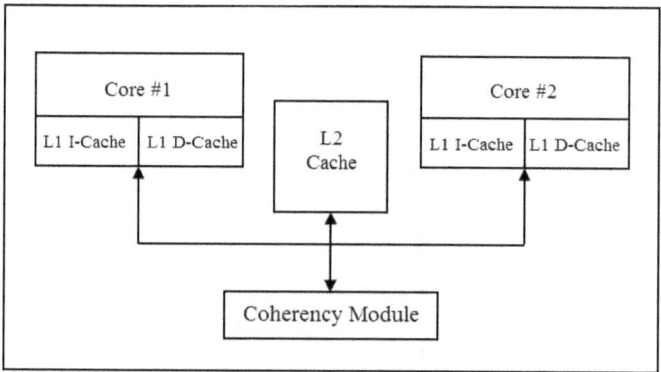

Fig. 3. Notional multi-core cache architecture with shared L2 cache

This uncertainty about the selection of MCPs for avionics programmes has been compounded by the following factors:

1. Although the avionics safety certification agencies EASA and FAA have published the MULCORS research report (Jean et al. 2012) and the CAST-32 position paper (FAA 2014) respectively, on the use of MCPs in avionics, this does not constitute formal policy or guidance.

2. Single-core processors which have been used in safety-critical avionics applications are now nearing the end of silicon availability or are no longer available (NXP 2014).

3. The historical dominance of PowerPC architecture in the embedded market appears to be somewhat in decline, and the long term future appears to be uncertain with NXP (formerly Freescale) developing ARM-based processors as well as PowerPC. In addition, the large number of PowerPC QorIQ® multicore processor architecture variants makes it unclear if there will be a de facto choice for avionics.

4. The increasing performance of ARM-based processors means that they may be considered as a viable option for some types of avionics application where PowerPC processors had been used previously.

5. Intel processors which historically were not widely considered for use in avionics applications due in part to their power dissipation requirements are now being considered due to Intel's high-performance, low-power 14nm processor devices (Intel 2014).

These market dynamics have resulted in fragmentation of processor selection for avionics, resulting in a lack of an obvious, single successor for widely-deployed PowerPC single core processors. We are now facing a wide range of contenders in terms of ARM multi-core, PowerPC QorIQ architecture families and Intel Core and Atom architectures. These selection issues were also discussed in more detail in section 9 of (Jean et al. 2012) based on information publicly available in 2012; section 12.2 of (Jean et al. 2012) also provides a processor selection guide checklist.

3   The Challenge of Multi-core Certification

The route to multi-core certification currently presents a challenge to avionics programmes due to lack of formal policy / guidance published by FAA and EASA. However, the EASA MULCORS research report (Jean et al. 2012), and FAA CAST-32 position paper (FAA 2014), should be taken into consideration when planning a safety-critical multi-core avionics project.

Programmes might wish to consider the selection of a multi-core processor in their next hardware platform even if their current processing requirements do not exceed that provided by a single core, in order to provide adequate processing capacity to meet future processing requirements. The selection of an MCP may also become a necessity due to the lack of availability of single core processors as mentioned earlier. Similarly, some programmes might wish to use MCPs which have more than two cores, as 4-core and 8-core devices are now relatively common. Although CAST-32 does not consider MCPs with more than two active cores, certification authorities are understood to be considering the use of MCP processors with more than two active cores.

3.1 Core Deactivation

In both of the above scenarios, programmes will need to be able to utilise certain processor cores and deactivate the unused cores. To meet the multi-core determinism objectives of CAST-32, programmes will need to demonstrate that a deactivated core cannot unexpectedly become active and interfere with the operation of the processor's other cores. This could either use an approach of regularly reading control registers which are critical to safe operation and resetting the register value in the event of a change of state being detected; or by regularly overwriting the control registers to ensure that the desired state is maintained. Some processors may also provide performance monitoring units which enable the state of an individual core to be determined independently.

The software implementation of core deactivation is processor-specific, and depends on whether individual processor architecture provides the ability for a core to be able to write to a control register to deactivate another core or not. For example, on the PowerPC QorIQ T2080™ processor, deactivation of an individual core can be achieved by setting the relevant bit field in the Core Disable Register during Pre-Boot Initialisation or when the core is in boot hold off mode, and once a core has been deactivated it can only be re-enabled via power-on, hard reset or core reset (NXP 2015).

The ability of safety-critical avionics programmes to be able to deactivate individual cores and develop a safety-case which includes robust arguments for the deterministic operation of the processor may depend on the ability to obtain detailed technical information on the design and operation of the processor from the semiconductor manufacturer. Some semiconductor companies may make this information publicly available, while others may only provide certain levels of information under non-disclosure agreement. For programmes undertaking DO-254 certification of airborne electronic hardware (AEH) this is an important requirement to achieve certification (FAA 2005), and they will need to ensure that the selected semiconductor manufacturer will provide access to the required information, even if they do not formally support DO-254 certification in the same way as companies such as Altera (Altera 2016).

3.2 Multi-Core Interference

Programmes intending to use MCPs in safety-critical avionics applications will need to manage contention between cores for shared resources. In particular, consideration should be given as to whether potential interference paths will result in actual interference channels (Jean et al. 2012, FAA 2016).

Wind River has undertaken research to measure inter-core perturbation on the QorIQ P4080 processor due to shared cache(s) and share memory controller(s) (Kinnan 2013). This included the development of a benchmark suite which continuously passed through a code loop of configurable size, to access consecutive data in a data set of configurable size. The intent was to generate continuous cache misses on both the instruction and data caches. The benchmark was run on a Wind River SBC P4080 reference board which contained a QorIQ P4080 processor with 64KB L1 cache per core (32KB for instructions, 32KB for data); 128K L2 cache per core (64KB for instructions, 64KB for data); a 1024KB L3 cache per memory controller, and two memory controllers. When the benchmark was run in a partition in virtualized environment (known as a *virtual board* or *VB*, or virtual machine), on a single core of the P4080, the performance degraded in predictable manner as the data size increased (as shown in Figure 4). The vertical axis indicates time in ticks measured by the PowerPC 64-bit Time Base Register (TBR), and the horizontal axis indicates the number of iterations of the benchmark that were performed. The lines of the graph are shown in order of increasing size of benchmark, and show that when the benchmark is small, the code and data are contained within the on-processor L1 instruction and data cache. As the benchmark size is increased, the benchmark overflows into shared L2 cache and eventually into the L3 platform cache with predictable degradation in performance.

One VB on Core 0

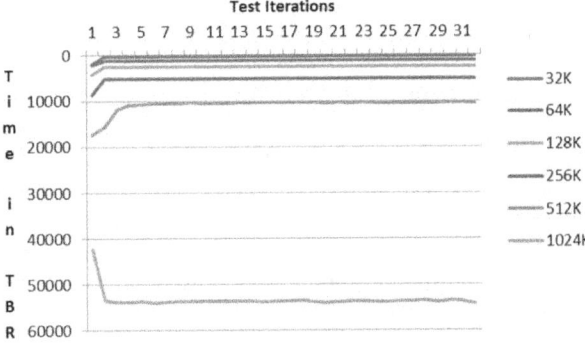

Fig. 4. Benchmarks results for P4080 single core operation

However, when the benchmark was run simultaneously on two cores on the P4080 using the same memory controller, the results became unpredictable once the data size had overflowed into the L3 cache, shown by the wavy lines for 512KB and 1024KB benchmark sizes in figure 5.

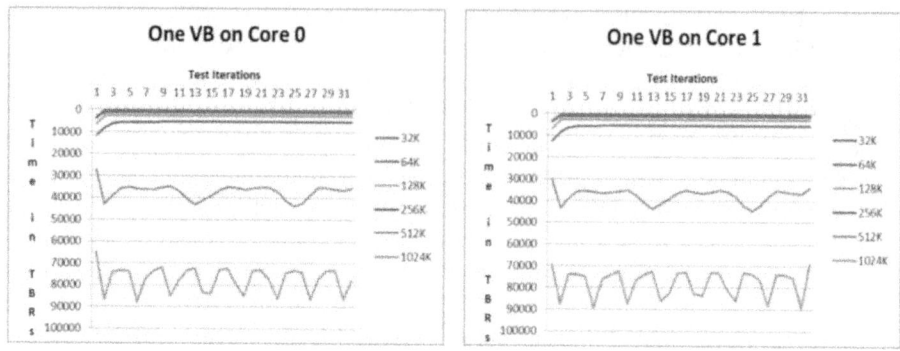

Fig. 5. Benchmarks results for P4080 dual core operation with single memory controller

When the test was re-run simultaneously on two cores on the P4080 but this time using different memory controllers for each core, this resulted in predictable results once more, shown by the 512KB lines in figure 6.

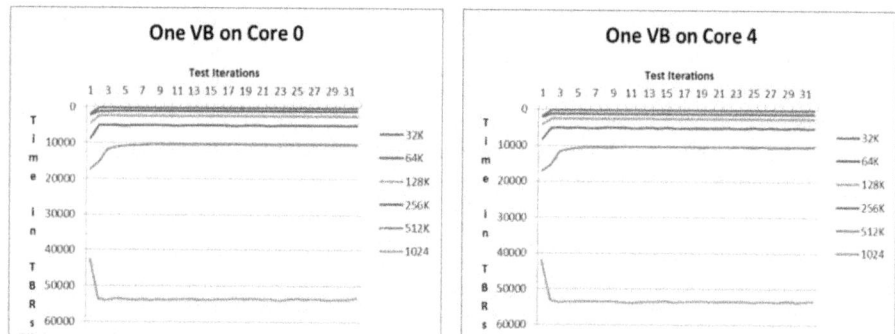

Fig. 6. Benchmarks results for P4080 dual core operation with dual memory controllers

These benchmark results illustrate the potential interference paths for a specific processor architecture, but this does not necessarily indicate the actual interference channels which will occur for an avionics system, as this is dependent on the characteristics of the applications. Therefore multicore interference analysis cannot be performed on the underlying operating system in isolation, but needs to be undertaken at the system-level including the application.

4 Certification of an ARINC 653 RTOS on multi-core processor architecture

We will now consider the challenges of avionics software safety certification and multicore, which were previously discussed in sections 1.2 and 3 respectively in the context of an ARINC 653 RTOS.

4.1 ARINC 653 update for multicore

ARINC 653 is the leading industry open standard for avionics software architecture in an integrated modular avionics environment, with ARINC 653-based systems having being widely deployed in civil and military aircraft. ARINC 653 Part 1 Supplement 3, Required Services (ARINC653P1-3 2010) which was published in 2010 did not address the use of ARINC 653 in multicore processor avionics systems. However, as there was strong market demand for support for multicore, the AEEC APEX Subcommittee (AEEC 2015) undertook the updating of ARINC 653P1-3 to support the use of MCPs. This industry effort involved Tier-1 suppliers, system integrators, and COTS software suppliers, including proactive participation and contribution by Wind River.

The evolution of the standard resulted in the publication of ARINC 653 Part 1 Supplement 4 (ARINC653P1-4 2015) in 2015 to support the use of MCPs, and contains an important provision, stating (page 5) that an application developed under ARINC 653P1-3 to run on a single-core processor, should also run on a single core on a multi-core platform under ARINC 653P1-4 with the same behaviour. This preserves the investment of previously-developed ARINC 653 applications when migrating to multi-core platforms.

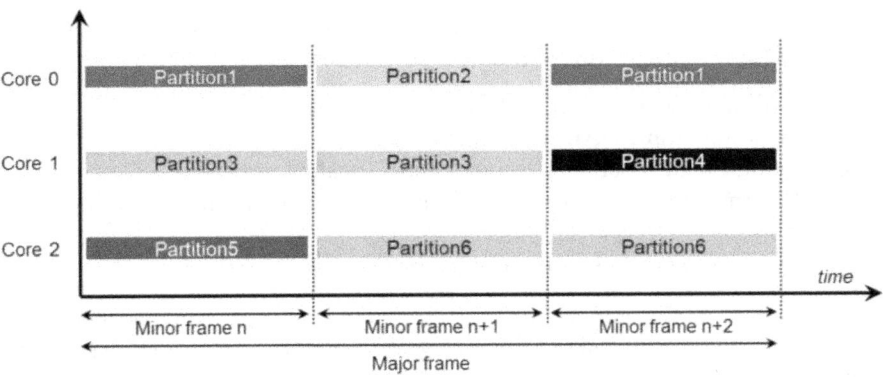

Fig. 7. ARINC 653P1-4 Multicore Scheduling Example

ARINC 653P1-4 also provides ability to support multiple partitions on each processor core using a timeslot scheduling approach where an individual partition will execute on a specific core within a timeslot of defined duration (known as a *minor frame*). An ARINC 653 schedule can be composed of multiple minor frames of similar or differing durations, and the time taken to execute all of the minor frames within the ARINC 653 schedule is known as the *major frame* period. After the execution of the last minor frame has been completed, the ARINC 653 scheduler will then schedule the first minor frame in the schedule again, in a cyclic manner. The ARINC 653P1-4 architecture provides the potential for many potential scheduling configurations (as illustrated by figure 7). However, as discussed in section 3.2, the System Integrator will need to ensure that the configuration of specific applications on a particular IMA platform will provide deterministic behaviour, and that potential interference paths are reduced to the minimum number of interference channels.

ARINC 653P1-4 does not include the ability to run an instance of a partition across multiple cores (known as a *multicore partition*), but states that this capability may be added in a future update of the standard.

4.2 VxWorks 653 RTOS Multi-core Requirements

For the earlier releases of the VxWorks 653 RTOS targeting single-core operation, the requirements were defined in the Software Requirements Specification (SRS) contained in the VxWorks 653 DO-178B Level A certification evidence package for the respective processor architecture (see figure 2). The software architecture of VxWorks 653 2.x RTOS is discussed in (Parkinson & Kinnan 2015).

For *VxWorks 653 Multi-core Edition*, the following high-level goals were defined for the product on multi-core architectures:

1. Certifiable to DO-178C DAL A
2. Support multiple DALs on multiple cores
3. Perform fault isolation and containment (health monitors)
4. Perform static configuration and enforcement in accordance with ARINC 653
5. Enable role-based development as per DO-297

These goals were addressed through the product requirements, design implementation and certification strategy using an agile development process in conjunction with DO-178C processes. This enabled the product definition to evolve and to track enhancements to the ARINC 653P1-4 standard.

4.3 VxWorks 653 Multi-core Edition RTOS Design Considerations

In order to achieve the high-level goals of support the safety certification of multiple applications at different DALs up to and including DAL A on multiple cores, the design of VxWorks 653 Multi-core Edition RTOS therefore needed to support isolation of applications running individual partitions through spatial partitioning, temporal partitioning, resource partitioning and multi-core partitioning. The RTOS design also needed to minimise the potential for multicore interference paths where possible.

4.3.1 Hypervisor & Hardware Virtualisation Support

Microprocessors have typically provided two privilege levels: *User mode*, for user application context, and *Supervisor Mode*, which is usually reserved for use by the operating system kernel. However, an increasing number of modern multicore processors (such as the QorIQ T2080 and Intel 64-bit processors) implement a third privilege level, known as *Hypervisor Mode* (NXP 2015). This enables an operating system to be run at hypervisor privilege level, utilising the processor's full hardware virtualisation support to run multiple virtual machines containing unmodified guest operating systems (GOS) and applications.

Wind River took the decision to utilise this capability within the VxWorks 653 Multi-core Edition by implementing the Module Operating System (MOS) at hypervisor level running across the processor's multiple cores, with an unmodified guest OS in virtual machines, as shown in figure 7. A MOS board support package (BSP, not shown in figure 8) is used to map the board-independent functionality of the MOS to the hardware architectures of a specific PowerPC QorIQ board. This approach enables the MOS to run on different boards by changing the underlying BSP.

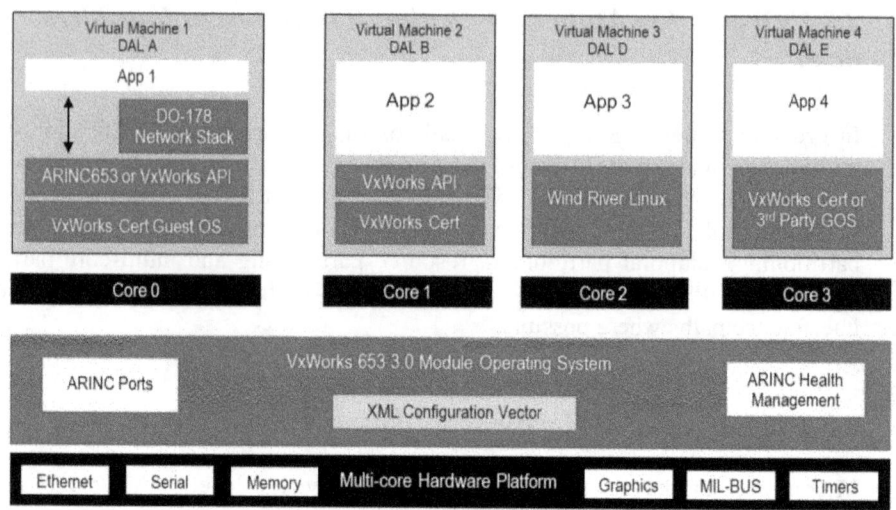

Fig. 8. VxWorks 653 3.0 RTOS Architecture

At the partition-level, Wind River decided to continue to use a partition operating system (POS) approach which had previously been used in VxWorks 653 2.x on single-core processor architectures, with runtime libraries supporting the deployment of applications which use ARINC 653 processors, POSIX APIs, or VxWorks native applications, any of which could be written in C or Ada programming languages. This architecture also enables a GOS running in a partition at *Supervisor level* to have memory protection from the application running at *User Level* within the virtual machine. Full hardware virtualisation enables a physical device to be allocated to specific virtual machines on a processor core, enabling a device driver within the Guest OS in the virtual machine (VM) to efficiently access the device in a controlled manner without significant performance penalty overheard.

VxWorks Cert 6.6.x was used for the POS to enable applications at different Development Assurance Levels (DAL) up to and including DO-178C DAL A. This also facilitates the migration of federated applications developed to run on VxWorks Cert 6.6.x to run in a VxWorks 653 3.x system. The ability to support unmodified GOS also enables consolidation of federated applications develop to run under Linux and third-party / legacy OS on an IMA platform running VxWorks 653.

4.3.2 Two-Level OS Scheduler

The two-level OS scheduler was also extended to support multi-core architectures with a scheduling model that uses the Partition OS to perform scheduling of contexts (ARINC 653 processes, POSIX threads or VxWorks 653 tasks) without having to perform a system call into the MOS and the associated overhead of processor context switch. This is efficient and very scalable because as the number of partitions increases, the scheduling overhead on the MOS remains constant. This is because the MOS is only responsible for scheduling of partitions at fixed timeslot durations as defined by the ARINC 653 schedule (as shown by the example in figure 8), rather than having to manage an increasing number of processes, threads and tasks across multiple partitions. This approach minimises the potential for performance degradation and partition jitter in a system with a large number of ARINC 653 partitions, and helps to reduce worse-case execution times (WCET). Figure 9 also illustrates how schedules may be defined on each core, and VxWorks 653's ability to support multiple ARINC 653 schedules.

```
<Cores>
    <Core Name ="0" InitialScheduleNameRef="init">
        <Schedule Name="init">
            <Window Duration="100000000"/>
        </Schedule>
    </Core>
    <Core Name ="1" InitialScheduleNameRef="init2">
        <Schedule Name="init2">
            <Window Duration="100000000" PartitionNameRef="vxworks1"/>
            <Window Duration="100000000"/>
        </Schedule>
        <Schedule Name="main2">
            <Window Duration="100000000" PartitionNameRef="vxworks1"/>
            <Window Duration="100000000" PartitionNameRef="vxworks2"/>
        </Schedule>
    </Core>
</Cores>
```

Fig. 9. VxWorks 653 3.0 MOS Schedule XML configuration example

4.3.3 Multi-core Communication

ARINC 653P1-4 defines inter-partition communications in terms of APplication/EXecutive (APEX) services via *ports*, which provide access at both ends of a communications channel between partitions. Applications access the ports via a pre-defined port name, and ports at both ends of the channel can either be both configured as sampling ports or queuing ports.

In single-core processor systems, the underlying transport mechanism can be implemented by performing memory copy of message buffers between POS and MOS address spaces, and then from MOS to POS for the destination partition. However, extending APEX ports to an MCP system presents some challenges. An application using APEX ports should not be aware of whether the application at the other end of the communications channel is also executing on the same processor core or on a different core, which is known as location transparency. This would enable the System Integrator to reconfigure a system, and migrate an application partition from one core to another core, but without impacting the APEX port configuration at the application level.

Wind River achieved this in the implementation of VxWorks 653 Multi-core Edition by maintaining location transparency of ARINC 653 ports at the partition-level, and mapping an APEX port to an underlying Safe IPC transport mechanism within the MOS. The mapping of APEX ports to Safe IPC channels is defined in XML, enabling the System Integrator to reconfigure the system without requiring individual applications to be modified and recompiled.

4.3.4 Fault Isolation and Containment

In VxWorks 653 2.x, the MOS ran in the processor's Supervisor Mode to provide isolation from POS applications running in User Mode, and also uses the processor's memory management unit (MMU) to prevent an application from making a programmed I/O access outside its allocated address space (e.g. through de-referencing an invalid pointer).

For the design of VxWorks 653 Multi-core Edition, Wind River decided extend the isolation capabilities to utilise the full hardware virtualisation capabilities available on the QorIQ T2080 processor to verify that DMA transfers to/from a partition are using only valid source and destination address ranges for that partition. This prevents illegal DMA transfers from occurring.

VxWorks 653 Multi-core Edition continues to provide support for ARINC 653 Health Monitoring (HM), by providing the ability to perform cold and warm starts of partitions, and cold restart of the entire module. This provides System Integrators with the ability to configure an ARINC 653 system via an HM framework, enabling a system to provide resilient operation.

4.3.5 System Configuration and DO-297 Role-based Development

In VxWorks 653 2.x, system configuration was defined using XML configuration and a process known as independent build link and load (IBLL) to configure and initialise an IMA platform using a single configuration vector (CV). This approach enables platform providers, application developers and system integrators to collaborate according to role separation (DO-297 2005) and also reduce the cost of change, as system and partition configurations can be changed without rebuilding the entire application or platform, which significant reduces the impact-analyses burden when upgrading and modifying an existing system. This is discussed further in (Parkinson and Kinnan 2015). This software architecture continues to be supported in VxWorks 653 Multi-core Edition due to the significant benefits which it has provided.

4.4 DO-178C DAL A Certification Strategy for VxWorks 653 on QorIQ T2080

Since 2000, Wind River has developed and released DO-178 Certification Packages as certifiable COTS components which organisations could use in their certification programmes. When COTS MCPs started to become widely available, customers started asking Wind River to provide DO-178C Certification Packages on MCP processors, utilising multiple cores. However, as EASA and FAA had not published formal policy on multi-core certification, Wind River regarded undertaking development of a safety-critical multi-core RTOS platform as presenting a significant technical risk with no guarantee of success.

For these reasons, Wind River decided not to start certification until after the publication of the EASA MULCORS research report (Jean et al 2012), and FAA CAST-32 position paper (FAA 2014). Although these documents do not constitute formal policy at this time, they provide insights into multicore considerations and best practices for multicore certification. Therefore, Wind River developed its Plan for Software Aspects of Certification (PSAC) for VxWorks 653 Multi-core Edition on QorIQ T2080 at DO-178C DAL A, with reference to the FAA CAST-32 paper's objectives in relation to MCP Determinism, MCP Software and MCP Error Handling. Although the CAST-32 paper only addresses two cores, it is largely applicable to more than two cores, with the caveat that the slow down due to MCP interference, known as the *interference penalty* can grow exponentially as the number of cores increases (FAA 2016). This means that system integrators will need to enforce restrictions, known as *interference mitigations*, to reduce interference.

Wind River is working with a lead customer and the FAA on an avionics programme in order to gain early feedback from DO-178C certification audits on the design and certification approach and guidance on application of CAST-32 from the certification authority through the four *Stage of Involvement* (SOI) audits. This approach was regarded as presenting lower technical risk, increasing the probability of successful completion of certification, and in shorter overall timescales.

4.5 Future Challenges

Although ARINC 653P1-4 does not currently support multi-core partitions, it indicates that this may be supported in a future update of the standard. This would enable more computationally-intensive applications to be hosted on ARINC653 systems, enabling further consolidation of avionics LRUs onto IMA common computing platforms. ARINC 653 support for multi-core partitions may also increase the potential of using *manycore*[2] processors such as the MPAA® (Kalray 2014) and Tilera (now part of Mellanox) in IMA applications.

At the time of writing, Wind River is in the process of porting the VxWorks 653 Multi-core Edition RTOS to Intel 64 bit multi-core processor architectures (Core, Xeon D). The release dates for the Early Access Release (EAR) and Generally Available (GA) release are published in the current official published Wind River Product Roadmap.

The DO-178C certification of an ARINC 653 RTOS, on other MCP architectures could present different requirements, as other architectures have different initialisation sequences. For example, Intel processors use a BIOS or Intel Firmware Support Package (Wong et al. 2014, Yao et al. 2015), which might require optimisation in order to meet the AC2511-B start-up time requirement for an avionics flight display (FAA 2014) and undergo DO-178C certification.

Finally, as ARM-based system on chip (SoC) devices increase in processing performance, these may become an attractive option for an IMA platform, especially if DO-254 certification artefacts are provided by the semiconductor manufacturer.

4   Conclusions

The avionics market is currently undergoing a significant transition from single-core to MCP architectures, driven by demands for greater system functionality and the semiconductor product lifecycles which primarily target the much larger commercial market segments. The advances made by semiconductor manufacturers now present a much broader range of viable processor choices for avionics applications than was available in the past.

Although there currently is some uncertainty about the best choice of processor for individual aerospace application use cases, it is likely that positive experiences gained by early adopters on multi-core programmes will result in a virtuous circle of support, further adoption and success, in a similar way to single-

---

[2] Manycore processors are specialist multi-core processors designed for a high degree of parallel processing.

core avionics programmes of previous decades generated a rich supplier ecosystem of COTS avionics certification solutions.

The evolution of the ARINC 653 standard to support MCP architectures, combined with the provision of ARINC 653 multi-core software support from COTS RTOS suppliers will enable previously-developed ARINC 653 applications to be re-hosted on MCP IMA platforms, facilitating software reuse and preserving investment. Experience gained from DO-178C certification on multi-core programmes should enable further adoption and proliferation to other MCP architectures.

Acknowledgments   The author wishes to thank the following Wind River colleagues for their input into this paper: A. Wilson, C. Downing, and S. Olsen.

References

AEEC (2015) Airlines Electronic Engineering Committee, APEX Subcommittee, http://www.aviation-ia.com/aeec/projects/apex. Accessed 20th September 2016

Altera (2016), DO-254 Safety Solutions web page, Altera website, https://www.altera.com/solutions/industry/military/applications/do-254/mil-do-254.html. Accessed 20th September 2016

ARINC653P1-3 (2010) Avionics Application Software Standard Interface, Part 1, Required Services, ARINC Specification 653 Part 1 Supplement 3, ARINC.

ARINC653P1-4 (2015) Avionics Application Software Standard Interface, Part 1, Required Services", ARINC Specification 653 Part 1 Supplement 4 (653P1-4), ARINC. http://www.aviation-ia.com/cf/store/catalog_detail.cfm?item_id=2510. Accessed 20th September 2016

DO-178B (1992) Software Considerations in Airborne Systems and Equipment Certification. RTCA Inc. and ED-12B (EUROCAE)

DO-178C (2011) Software Considerations in Airborne Systems and Equipment Certification. RTCA Inc. and ED-12C (EUROCAE)

DO-254 (2002) Design Assurance Guidance for Airborne Electronic Hardware, RTCA Inc. and ED-80 (EUROCAE). http://www.rtca.org/store_product.asp?prodid=752. Accessed 20th September 2016

DO-297 (2005) Integrated Modular Avionics (IMA) Development Guidance and Certification Considerations, RTCA. http://www.rtca.org/store_product.asp?prodid=617. Accessed 20th September 2016

FAA (2005) Advisory Circular AC20-152, US Federal Aviation Administration. http://www.faa.gov/documentLibrary/media/Advisory_Circular/AC_20-152.pdf. Accessed 20th September 2016

FAA (2011) Microprocessor Evaluations for Safety-Critical, Real-Time Applications: Authority for Expenditure No. 43 Phase 5 Report", DOT/FAA/AR-11/5, US Federal Aviation Administration. https://www.faa.gov/aircraft/air_cert/design_approvals/air_software/media/11-5.pdf. Accessed 20th September 2016

FAA (2014) Multi-core Processors, Position Paper, CAST-32, Certification Authorities Software Team, US Federal Aviation Administration. https://www.faa.gov/aircraft/air_cert/design_approvals/air_software/cast/cast_papers/media/cast-32.pdf. Accessed 20th September 2016

FAA (2014) Electronic Flight Displays, Advisory Circular AC25-11B, US Federal Aviation Administration.   http://www.faa.gov/documentLibrary/media/Advisory_Circular/AC_25-11B.pdf. Accessed 20th September 2016

FAA (2016) White Paper on Issues Associated with Interference Applied to Multicore processors,   29   January   2016,   US   Federal   Aviation   Administration. https://www.faa.gov/aircraft/air_cert/design_approvals/air_software/media/SDS_DO005_White_Paper.pdf. Accessed 20th September 2016

Intel (2014) Advancing Moore's Law – The Road to 14nm, presentation, Intel website, 11th August 2014.   http://www.intel.com/content/www/us/en/silicon-innovations/advancing-moores-law-in-2014-presentation.html. Accessed 20th September 2016

Jean X, Gatti M, Berthon G, Fumey M (2012) MULCORS - Use of MULticore proCessORS in   airborne   systems,   Research   Project   EASA.2011/6,   EASA. http://easa.europa.eu/system/files/dfu/CCC_12_006898-REV07%20-%20MULCORS%20Final%20Report.pdf. Accessed 20th September 2016

Kalray   (2014)   MPPA®   MANYCORE   processor,   product   data   sheet,   Kalray. http://www.kalrayinc.com/IMG/pdf/FLYER_MPPA_MANYCORE.pdf.   Accessed   20th September 2016

Kinnan, LM (2009) Use of multicore processors in avionics systems and its potential impact on implementation and certification, Digital Avionics Systems Conference

Kinnan LM (2013) Multicore in Avionics – Current Practices, Trends and Outlook for Certification, 32nd Digital Aviation Systems Conference, Syracuse, New York.

NXP   (2014)   Product   Longevity   –   Archived   (September   2014),   NXP   website. http://www.nxp.com/pages/product-longevity-archived-september-2014:LONGEVITY-ARCHIVED. Accessed 20th September 2016

NXP (2015) QorIQ T2080 Family Reference Manual, T2080RM Rev 1, NXP, May 2015. https://www.nxp.com/webapp/Download?colCode=T2080RM. Accessed 20th September 2016

NXP (2016) Host and Integrated Host Processors (8xxx, 7xxx, 7xx, 6xx) web page, NXP website.   http://www.nxp.com/products/microcontrollers-and-processors/power-architecture-processors/integrated-host-processors:MPC8XXX7XXX. Accessed 20th September 2016

Parkinson, Paul J. (2011) Safety, Security and Multicore. Proceedings of the Nineteenth Safety-Critical   Systems   Symposium. http://www.springerlink.com/content/w2751nx7l28mj35r. Accessed 20th September 2016

Parkinson PJ, Kinnan LM (2015) Safety Critical Software Development for Integrated Modular   Avionics,   technical   white   paper,   Wind   River http://www.windriver.com/whitepapers/safety-critical-software-development-for-integrated-modular-avionics/. Accessed 20th September 2016

Perry, William J. (1994) US Secretary of State for Defense, Specifications and Standards: – A New Way of Doing Business. U.S. Department of Defense

Wind   River   (2016)   VxWorks   RTOS   product   page,   Wind   River   website. http://www.windriver.com/products/vxworks/. Accessed 20th September 2016

Wong SH, Sun J, Mahesh D (2014) Intel® Firmware Support Package for Intel Architecture, white paper, Intel. http://www.intel.my/content/dam/www/public/us/en/documents/white-papers/fsp-iot-royalty-free-firmware-solution-paper.pdf. Accessed 20th September 2016

Yao J, Zimmer VJ. Rangarajan R, Ma M, Estrada D, Mudusuru G (2015) A Tour Beyond BIOS Using the Intel® Firmware Support Package Version 1.1 with the EFI Developer Kit   II,   Intel,   April   2015.   https://firmware.intel.com/sites/default/files/resources A_Tour_Beyond_BIOS_Creating_the_Intel_Firmware_Support_Package_Version_1_1_with_the_EFI_Developer_Kit_II.pdf. Accessed 20th September 2016

Confidence in a connected world: safe, secure, resilient and autonomous

Robin E Bloomfield, Kate Netkachova, Peter Bishop

Adelard LLP and City, University of London

Abstract *Since establishment of the SCSC in 1992 the world of safety-related computing and assurance has changed enormously, but the fundamental principles of the approaches articulated in the first decade of the club remain valid. However, since 2000 the dramatic growth in the Internet and the changes to security threats exemplified by the attacks of 9/11 have changed the safety engineering world. The need for change is further illustrated by the impact of climate change, the growing importance of interdependencies and lessons from Fukushima accident emphasising the need for resilience. Innovation in technology, in particular the increasing autonomy of systems, also provides another driver for change. The paper therefore discusses these three challenges: security-informed safety, resilience and autonomy. But in addition there is a significant challenge in that we must not forget "normal business": the challenge of continuing to apply what we know in a rigorous and competent manner in the light of organisational change, project pressures and resource limitations.*

1 Introduction

This volume of proceedings marks the UK Safety Critical Club's 25th anniversary. As someone who was involved with the club from its inception to the present one of us (Bloomfield) has been asked to provide some reflection on the past 25 years of safety engineering. This historical aspect is interesting as it is the SCSC anniversary but it also important as how we view our history shapes our safety culture. Our focus is on work done at Adelard and City, University of London. It is therefore a very personal perspective: we hope that brings with it the benefit of being rooted in experience rather than being a narrow and self-serving view. Many people and organisations have made very significant contributions to safety engineering in the past 25 years and by its nature this review does not provide the overall balanced history that the subject deserves. Also, the emphasis is on publicly accessible information which means the focus is on policy, technical themes and research: the myriad of interesting projects have to remain confidential.

The paper first sets the scene around the time of the Club's formation in 1992 and then continues to describe some of the themes of the 1990s. It then considers how since 2000 the dramatic growth in the Internet and the increase in security threats have changed the safety-engineering world. The need for change is further illustrated by the impact of climate change, the growing importance of infrastructure interdependencies, lessons learnt from the Fukushima accident, and the need for resilience. Changes in technology, and in particular the increasing autonomy of systems, also provide another driver for change. This paper therefore discusses these three challenges: security-informed safety, resilience and autonomy. But in addition there is a significant challenge we must not forget: although achieving and assuring safety is a relatively mature undertaking we must not be complacent particularly in the light of organisational change, project pressures and resource limitations. So the fourth and final challenge discussed is that of "normal business": the challenge of continuing to apply what we know in a rigorous and competent manner.

2 Safety and computer based systems in the 1990s

2.1 The early 1990s and the formation of the club

This Section reviews the world of safety and computer based systems just prior to the formation of the Club and a few years afterwards. It draws heavily on (Bloomfield 2010). In the UK there were a number of policy initiatives that pre-dated the Club and led to its formation. The ACARD report (ACARD 1986) and subsequent IEE/BCS and HSE studies (IEE 1989, HSE 1987) set the scene and in 1988 the Interdepartmental Committee on Software Engineering (ICSE) established its Safety-Related Software (SRS) Working Group to coordinate the Government's approach to this important issue. Members were drawn from a wide range of departments and agencies: CAA, CEGB, DES, DoE, DTI, DoH, DoT, HSE, MoD, RSRE and SERC[1].

---

[1] Civil Aviation Authority, Central Electricity Generating Board, Department of Education and Science. Department of the Environment, Department of Transport, Department of Health, Department of Trade and Industry, Health and Safety Executive, Ministry of Defence, Royal Signals and Radar Establishment, Royal Signals and Radar Establishment, Science and Engineering Research Council

The UK Health and Safety Executive (HSE) was active in taking the lead in ICSE and this, with support from DTI, led to a consultation document known as SafeIT (Bloomfield 1990) and an associated standards framework (Bloomfield and Brazendale 1990). SafeIT identified four main areas of activity requiring a coordinated approach: standards and certification; research and development; technology transfer; education and training. The implementation of the technology transfer and awareness led to the founding of the Safety Critical Systems Club when Newcastle University CSR responded to a DTI ITT.

According to ICSE, the SafeIT work was motivated 'not by a recognition of particular present dangers; rather by a desire to anticipate and forestall hazards which may arise with the very rapid pace of technical change' (Bloomfield 1990). However, this was a somewhat overly optimistic and reassuring message: Peter Neumann's interest in the computer-related risks was already well established and he published in 1985 a list of 'Some computer-related disasters and other egregious horrors' (See Table 1).

Table 1. The 1985 perspective: Computer-related disasters and egregious horrors. Compiled by Peter G. Neumann 1985) mostly from back issues of ACM SIGSOFT Software Engineering Notes Vol 10, 1985, Quoted from (Bloomfield 1989)

System
Microwave arthritis therapy reprogrammed pacemaker, killed patient (SEN 5 1)
Three Mile Island (SEN 5 3)
SAC: 50 false alerts in 1979 (SEN 5 3); Simulated attack triggered a live scrambles [9 Nov 79] (SEN 5 3); WWMCCS false alarms triggered scrambles [3-6 Jun 80] (SEN 5 3)
F14 off aircraft carrier into North Sea (SEN 8 3)
NORAD alert based on BMEWS report of the moon as incoming missiles (SEN 8 3)
Mercury astronauts forced into manual re-entry (SEN 8 3)
Frigate George Philip fired missile in opposite direction (SEN 8 5) Credit/debit card copying easy despite encryption (DC Metro, DF BART, etc.)
Remote (portable) phones (free calls)
Software
First Space Shuttle backup launch computer synchronisation (SEN 6 5 [Garman])
Second Space Shuttle operational simulation: tight loop upon cancellation of an attempted abort; required manual override (SEN 7 1)
Second Shuttle simulation: bug found in jettisoning and SRB (SEN 8 3)
F16 simulation: plane flipped over whenever it crossed equator (SEN 5 2)
Mariner 18: abort due to missing NOT (SEN 5 2)
F18: plane crashed due to missing exception condition (SEN 6 2)
Gemini V 100 mi landing error, program ignored orbital motion around sun (SEN 9 1)
El Dorado: brake computer bug caused recall of all El Dorados (SEN 4 4)
Nuclear reactor design: but Shock II model/program (SEN 4 2)
Reactor overheating, low-oil indicator; two-fault coincidence (SEN 8 5)
Mariner I Atlas booster launch failure DO 100 I = 1.10 (not 1, 10) (SEN 8 5)
SF BART train doors sometimes open on long legs between stations (SEN 8 5)
Various system intrusions; implanted Trojan horses
Cyber command identified users with the same password (SEN 8 3)
VMS tape backup SW trashed directories on disc dumped in image mode (SEN 8 5)
Vancouver Stock Index lost 574 points over 22 months – roundoff! (SEN 9 1)
Gobbling of legitimate automatic teller cards (SEN 9 2)
Chernenko at MOSKVAX: network mail hoax, 1 April 1984 (SEN 9 4)
Hardware / Software
FAA Air Traffic Control: many computer system outages (e.g. SEN 5 3)
F18 missile thrust while clamped, plane lost 20,000 feet (SEN 8 5)
ARPAnet: total collapse [27 Oct 1980] (SEN 9 2 [Rosen])
SF Mini Metro: Ghost Train (SEN 8 3) (Problem STILL occurs now and then.) Harrah's $1.7 Million payoff scam – Trojan horse chip (SEN 8 5)
1984 Rose Bowl scoreboard takeover (Cal Tech versus. MIT) (SEN 9 2)
Computer as catalyst: human frailties
Air New Zealand crash 28 Nov 1979; computer flight data error detected, but pilots not informed; plane flew into Antarctic Mt Erebus (SEN 6 3 & 6 5)
Exocet missile not on expected missile list, detected as friend (SEN 8 3)
Wizards altering software or critical data (various cases)
Embezzlements, e.g. Muhammed Ali swindle [$23.2 Million],
Security Pacific [$10.2 Million], City National, Beverly Hills CA [$1.1 Million]

A couple of years later, HSE also published awareness documents on the safety of programmable electronic systems (PES), including the Out of Control report (HSE 1993), and some earlier studies that looked at the feasibility of providing a validated framework for selecting software engineering techniques. In addition, the sociologist Donald MacKenzie (MacKenzie 1994) compiled and analysed the ACM Risks data, and concluded that "Despite widespread interest in computer system failures, there have been few systematic, empirical studies of computer-related accidents. 'Risks' reports in the Association for Computing Machinery's Software Engineering Notes provide a basis for investigating computer-related accidental deaths. The total number of such deaths, world-wide, up until the end of 1992 is estimated to be 1,100 ± 1,000. Physical causes (chiefly electromagnetic interference) appear to be implicated in up to 4% of the deaths for which data were available, while 3% involved software error, and about 92% failures in human-computer interaction." In the updated (HSE 1995) publication 44% of incidents are attributed to requirements and specification issues.

The nuclear industry has had a major influence on the development of approaches to safety-related computer system development and assurance. From the late 1970s the European Working Group on Industrial Computer Systems (EWICS), a cross-sector pre-standardisation working group, developed a series of guidelines and books that documented best practices. The guidance was subsequently incorporated into the IEC 880 standard on software for nuclear systems (IEC 1986). The experience of EDF and Merlin-Gerin with the first generation of reactor protection systems was fed into the committee. The software engineering approach in the EWICS guidelines (Redmill 1988, 1989) and their book on safety techniques (Bishop 1990) represented the then state of the art.

The UK MoD were, as one might expect, pioneers in the use of critical software and the development of static analysis tools to analyse the code (MAL-PAS) as well as forays into formally proven hardware designs. In the cold war context there were concerns that digital systems should work only when needed: inadvertent operation of military systems could have disastrous consequences. In the light of finding defects in certain operational systems, dramatic changes to the supply chain as well as reductions in MoD scientific personnel, they responded in 1989 with the publication of a new draft interim standard 00-55 (MoD 1989). This used expertise from the nuclear and aerospace industry, MoD and elsewhere to develop a market-leading standard around the requirements for mathematically formally verified software and statistical testing.

It was soon realised – in part because the response by industry to the standard was to classify systems as non-safety critical and outside the remit of 00-55 – that a wider system standard was needed. This led to Def Stan 00-56 (MoD 1991). There was considerable work to take into account strong industry and trade association comments (led by the DTI that developed a detailed trace from all comments to the final issue of the standard).

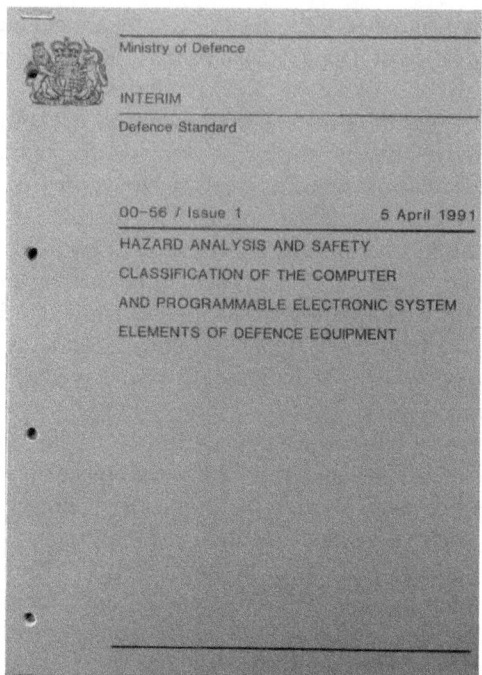

Fig.1: An early Def Stan 00-56

In parallel with the development in the defence sector, the HSE led the production of the IEC generic standards that became known as IEC 61508 (IEC 1998). Draft publications (IEC 1993) emerged in the early 1990s sharing much in common with the defence standards but addressing a wide range of systems and safety criticalities. During their prolonged drafting they developed detail, consistency and international recognition. However, the technical basis of their software aspects remained fixed. The software techniques guidance in IEC 61508 and its software engineering approach were essentially just an extension and internationalisation of EWICS guidance on techniques (Bishop 1990). There are still a number of technical difficulties in IEC 61508 (e.g. how SILs are used) and in the treatment of systematic failure.

Around 1993 the limitations to the claims that could be justified by testing were investigated by NASA (Butler and Finelli 1993), and similar results, involving testing and other evidence, were published by Littlewood and Strigini (1993). The 10-4 limit was one set by pragmatics of testing technology, but did not include the assumption doubt that we might now make explicit (Bloomfield and Littlewood 2007). Earlier empirical work on the limits of reliability of redundant system caused by design-related Common Mode Failure (Bourne et al 1981) had already led to the nuclear industry using claim limits and driving diversity in architectures.

So when the Club was formed in 1992 many of the technical themes that we find today were established:

- The difficulty and complexity of software
- The role of formal methods and static analysis (the debate around the correctness of the Viper microprocessor, the development of the analysis tool MALPAS)
- The role of statistical testing and the difficult interface between software reliability and Probabilistic Risk Assessment (PRA)
- Explicit consideration of the basis for arguing about justification and the 'argument of excellence'
- The importance of good software engineering process and quality
- The limits to what can be realistically claimed
- That not all safety systems are equally critical or have same type of roles so that a generic approach to describing criticality was needed
- The requirements on engineering systems should be graded in some way with criticality
- The importance of standards in defining technical approaches and shaping the market

And, rather unannounced, the World Wide Web had just been born…

2.2 Other developments in the 1990s

In the 1990s the market and range of safety-related computer based systems was growing to reflect the benefit that could be gained from these systems, in response to the technology push of the supply chain and the general global emergence of the digital and connected world.

The size in terms of lines of code of each successive generation of systems was increasing enabled by the continuation of Moore's Law and the development of distributed architectures. While there were still specialised bespoke systems, many assessments and systems now had to cope with a mixture of platforms, languages, architectures and provenance.

As the demand for safety-related systems developed, and as the standards were deployed there was a need for specialist services and advice and as a result, a wide range of different roles developed. Although there was no study at the time, the growth in the safety-related market meant that the range of skill levels and expertise varied enormously within the market place. There was definitely a role for the Club, and the professional institutions, tackling issues of competency and training.

Standards continued to have an important role with the development of the generic IEC 1508 (IEC 1993), the forerunner to IEC 61508 (IEC 1998)[4], and its take up by different sectors. The MoD was also active in the development of standards. In 1997 the 1991 Interim MoD Def Stan 00-55 was revised to become a full standard and became the first standard to explicitly require a software safety case. This was a radical departure from previous standards but offered some flexibility in the justification of the software, important in view of industry comments on the interim standard. The key features of the revised standard were:

- Deterministic reasoning and proof
- Statistical testing
- Importance of a range of attributes (not just correctness)
- Safety cases and reports
- Sound process to provide trustworthy evidence
- Multi-legged arguments and associated metaphors ("belt and braces")
- Systematic approaches and clarity of roles and responsibilities and other recommendations to reduce project risks
- Evidence preferences: *deterministic* evidence is usually to be preferred to statistical; *quantitative* evidence is usually to be preferred to qualitative; *direct* evidence is usually to be preferred to indirect

The nuclear expertise was influential in the original Def Stan 00-55. As with many standards the guidance aspects of 00-55 grappled with how to treat software of lower criticalities: at one extreme everything is required and at the other a minimum set of good practices is expected. Populating the regions in between has been problematic and largely a product of the standards process rather than the scientific one. Indeed the whole issue of empirical validation of standards is rather neglected.

Adelard had an important role in the development of the defence standards and drafted the safety case requirements. The origins of the work go back to the individuals' involvement (Bloomfield, Bishop and Froome) in the days of the Public Inquiry into the Sizewell B Primary Protection System (CEGB 1982). The work is similar to the approach used by Toulmin (Toulmin 1958) although developed somewhat independently. The concepts were first documented (e.g. Claims, Arguments and Evidence (CAE)) in the EU SHIP project (Bishop and Bloomfield 1995) and the work was taken up within a UK nuclear research programme. This led, in 1998, to ASCAD (Bloomfield et al. 1998), a safety case development manual. ASCAD provided the now customary definition of a case as 'a documented body of evidence that provides a convincing and valid argument that a system is adequately safe for a given application in a given environment'. In addition to the Adelard work there was research being done at

---

[4] See the paper by Ron Bell in these proceedings.

York University (McDermid 1994, Kelly 1997) that later led to the Goal Structuring Notation (Kelly 1998 and GSN 2011).

The ASCAD manual incorporated, with permission, considerable work from the UK nuclear research programme:

- On long-term and safety case maintenance
- How to address specific design issues, even the work on reversible computing to provide fault tolerance (Bishop 1997)
- The work on worst-case reliability bounds that provided a generic link between process quality (as measured in residual defects) and reliability in operation. (Bishop 1996)
- Field experience collected from a range of projects and also used in the SOCS report (ACSNI 1997)
- On argument architecture based on analogies and analysis of PWR pressure vessel cases (CEGB 1982).

It also made use of nuclear work on safety culture and work from the EU project REAIMS on organisational learning and human factors (Bloomfield et al 1997). This work continued into the new millennium with work sponsored by HSE on assessing Software of Uncertain Pedigree (SOUP) (Jones et al. 2001, Bishop et al. 2002b), work on multi-legged arguments and confidence (Bloomfield and Littlewood 2003, 2007), software criticality analysis and software integrity analysis (Bishop et al. 2002a). How the field has developed since will have to be a topic for another paper although some details are in (Bloomfield 2010b).

In 1995 the Advisory Committee on the Safety of Nuclear Installations (ACSNI)1 set up the Study Group on the Safety of Operational Computer Systems. The report from this group (ACSNI 1997) addressed the broad principles upon which the evidence and reasoning of an acceptable safety case for a computer-based, safety-critical system should be based and is still a useful and relevant source of information. A contemporary and comprehensive review of the historical development of safety cases in a variety of sectors is provided by (Health Foundation 2012) and a rigorous analysis of assurance case concepts is provided in (Rushby 2015).

3 The new millennium

At the turn of the millennium the web had grown from an estimated 16 million users in 1995 to ~250 million users. It passed the billion mark at the end of

---

[1] Became the Nuclear Safety Advisory Committee (NuSAC) and then disbanded.

2005 and in 2016 has ~3.4 billion users. The possibilities and challenges of this interconnected world were influencing policy makers and the research community but it was world events that provided dramatic drivers to refocus and re-shape safety engineering.

3.1 World events

Changes to safety and how systems are regulated and evaluated are often driven by accidents and disasters (see (Health Foundation 2012) for a summary of some of these and a good perspective on safety cases across sectors). The attacks on 9/11 in 2001 caused a dramatic change in the threat assumptions that we make as well as a graphic illustration of the interdependent and interconnected world we live in. In addition the West review of CPNI (unpublished but see (Weinstein and Wild 2012)), the Buncefield explosion (HMG 2008), the effects of climate change and the related environmental impact from the floods in the UK (Pitt 2008) spurred on interest in interdependencies (Bloomfield 2009a, 2009b) in infrastructure and resilience and this was given further impetus by the Japanese Tsunami of March 2011 and the ensuing disaster at Fukushima Daiichi nuclear plant (IAEA 2015).

Fig.2[1]: Buncefield: one of the largest explosions in peacetime Europe

In addition to these externalities, the evolving technology and systems provide their own challenges.

---

[1] Photo by Rick Martin, taken 10 minutes after the explosion

3.2 Connectivity and the socio-technical

Before considering these challenges, it is worth noting how far the systems had evolved and how complex even the policy world had become. It was not just safety that was of concern but overall dependability. In terms of research, in 2000, in the UK, a number of key Interdisciplinary Research Collaborations (IRC) were set up by EPSRC[1]. Notably, from our perspective, there was the dependability IRC, DIRC[2] led by Cliff Jones with Ian Sommerville, Bev Littlewood, Stuart Anderson, and Alan Burns as the principle investigators. DIRC had five enduring research themes: Responsibility, Structure, Risk, Diversity and Timeliness.

In 2003 we developed for the European Commission, with our AMSD project partners, the 'A Dependability Roadmap for the Information Society in Europe' (Masera and Bloomfield 2003). The main objective was to determine those areas for which research and development would be most beneficial and needed in the coming years. The project did its own analysis but also built on road-mapping reports produced by closely related communities covering security, mobility, privacy and Critical Infrastructure Protection (CIP). The substantive output was the identification of the dependability capabilities that are required and an indication of the gaps between current ability and these future ambitions. What in the early 90's was provided by a 30 page SafeIT report had now become close to 300 pages reflecting the complexity of the subject (and the growth of word processing!).

The road-map identified the key characteristics of the systems envisaged in the scenarios and it is useful to quote these to show how far the world had come from initial concerns in the 1990s. In 2003 these were articulated as:

- scale and complexity – as hardware capabilities improve and costs reduce, there is continuing pressure to attempt to build systems of ever greater scope and functional sophistication, especially for the software components
- boundary-less nature of the systems and interconnectedness – few systems have a clear-cut frontier, and they are drawn in systems within systems, within larger systems; in addition, all nodes, connected through a common underpinning infrastructure that becomes a critical factor, are always reachable and from everywhere, which results in unpredictable emergent behaviours

---

[1] Engineering and Physical Sciences Research Council – a UK funding body for academic research

[2] DIRC http://www.dirc.org.uk/)

- heterogeneity and blurring of human/device boundary e.g. wrist held gadgets, wearable devices, implantable devices
- incremental development and deployment – systems are never finished, evolution is incessant, upgrades, changes in functionality, and new features are being added at a continuous pace
- self-configuration and adaptation – systems are expected to be able to respond to the changing circumstances of the environments where they are embedded
- multiple innovative types of networking architectures and strategies for sharing resources – cloud-like, peer-to-peer, 'on-the-fly' services, etc.
- multiplicity of fault types – in particular the growing danger of malicious faults, both due to individual or organised external attackers, and due to deceitful insiders.

These issues can be summed up by the buzzword laden title of a study for DSTL (Gashi and Bloomfield 2008) on 'Evaluating the resilience and security of boundaryless, evolving socio-technical Systems of Systems'.

The excerpt from the ACM Risks Bulletin in Table 2 provides a contemporary view of the issues.

Table 2. Risks as of October 26th, 2016

Russian Suspected of Hacking U.S. Tech Companies Is Indicted
Radio interference disables cars and cell phones in Evanston
Report on 'Ethics of AI'
As Artificial Intelligence Evolves, So Does Its Criminal Potential
Pittsburgh's new artificially intelligent stoplights could mean no more pointless idling
Re: Self-driving cars shouldn't have to choose who to protect in a crash
Samsung washing machines in Australasia hot issue since 2013
China's Total Information Awareness?
Every LTE call, text, can be intercepted, blacked out, hacker finds
Unneeded Services Foster Botnets and other security problems
Internet becoming unreadable, lighter thinner fonts
Dyn Statement on the 21 Oct 2016 DoS Attack
Hacked Cameras, DVRs Powered Today's Massive Internet Outage
German voting system, for comparison
Re: Undetectable election hacking?
The Right to be Forgotten
Forum on Risks to the Public in Computers and Related Systems
ACM Committee on Computers and Public Policy, Peter G. Neumann, moderator, Volume 29 Issue 88, Tuesday 25 October 2016

The above excerpt highlights autonomy and AI as a current issue as well as the interdependencies that can be exploited because everyday products are connected to the Internet (the so called Internet of Things).

In addition to the challenge of resilience, security and autonomy we must not forget the challenge of 'normal business': the challenge of doing the things we know how to do in a rigorous and competent manner.

So in the following sections we consider these four key challenges in more detail:

1. Security-informed safety
2. Resilience
3. Autonomy
4. Normal business

This is not of course a complete list of challenges. The RAEng report on the connected world (RAEng 2015), the NAS study on 'Sufficient Evidence' (Jackson 2007) and the UK cyber strategy (HMG 2016) provide insights into the range of interlocking issues to be addressed if we were producing a comprehensive roadmap for achieving and evaluating safety and resilience.

4 Four challenges

Challenge 1: Security-informed safety

Traditionally, security and safety have been treated as separate disciplines, with their own regulation, standards, culture and engineering. This approach is increasingly becoming infeasible as there is a growing realisation that security and safety are closely interconnected: it is no longer acceptable to assume that a safety system is immune from malware because it is built using bespoke hardware and software, or that it cannot be attacked because it is separated from the outside world by an 'air gap'. In reality, the existence of the air gap is often a myth (see (DHS 2011), (Baylon 2015)). A safety justification, or safety case, is incomplete and unconvincing without a consideration of the impact of security.

The advent of cyber issues brings enormous challenges and changes to the traditional safety engineering tempo and approach. In Adelard we had a number of projects on safety and security in the 1990s including at the end of the 90s a report on the development of security requirements for CAA safety-related software. It was not until post 9/11 that we examined the issues for a number of agencies. Public output from the NATO study group tasked to look into synergies between dual use technologies is one example of this. We have also recently had permission to publish a summary of the work on rail systems (Bloomfield et al 2016). This ranged from an analysis of potential vulnerabilities in the European Rail Traffic Management System (ERTMS)[1] specifications through to a high-level cyber security risk assessment of a national ERTMS implementation and detailed analysis of particular ERTMS systems on behalf of the GB rail industry.

The impact of cyber issues is exacerbated by the increasing sophistication of attackers, the commoditisation of low-end attacks, the increasing vulnerabilities of digital systems as well as their connectivity – both designed and inadvertent. The December 2015 attack on the Ukrainian power grid demonstrated how vulnerable critical infrastructures can be to malicious actions. A destructive malware wrecked computers of several regional distribution power companies and wiped out sensitive control systems for parts of the power grid, causing power outages and blackouts. The attack was very well-coordinated and comprised of multiple different elements, including a denial-of-service attack to the phone systems, a direct interaction from the adversary, and the malware itself installed on workstations and servers to enable the attack (Assante 2016). This crafted attack can be considered the third public example of targeted intrusions leading to outages or physical damage available to date, along with the well-known Stuxnet and the German Steelworks facility attacks (Lee 2014). Such attacks are just one example of the type of adversaries a holistic safety case needs to address.

---

[1] European Railway Traffic Management System (ERTMS) is a major industrial project that aims to replace the many different national train control, command and signaling systems in Europe with a standardized system.

```
BOOL result; // eax@4

if ( fdwReason && fdwReason == 1 && (DisableThreadLibraryCalls(hinstDLL),
sub_10001186()) )
    result = sub_1000123A(hinstDLL);
  else
    result = 0;
  return result;
}
//----- (10001030) --------------------------------------------------------
int __stdcall StartAddress(LPCWSTR lpString2)
{
  int v1; // eax@2
  UINT v2; // edi@5
  int v4; // [sp-4h] [bp-414h]@4
  int v5; // [sp+0h] [bp-410h]@4
  WCHAR FileName; // [sp+208h] [bp-208h]@4

  if ( lpString2 )
  {
    v1 = lstrlenW(lpString2) + 1;
    if ( v1 > 260 )
      v1 = 260;
    lstrcpynW(&FileName, lpString2, v1);
    if ( sub_100011CE(&FileName, (WCHAR *)&v5) )
    {
      v2 = ((int (__thiscall *)(int, signed int))SetErrorMode)(v4, 32775);
      sub_100010AD((int)&v5, &FileName);
      SetErrorMode(v2);
    }
  }
  return 0;
}

//----- (100010AD) --------------------------------------------------------
signed int __cdecl sub_100010AD(int a1, LPCWSTR lpFileName)
{
  signed int result; // eax@1
  int v3; // ebx@1
  int v4; // edi@2
  const void *v5; // ecx@2
```

Fig. 3. Some of the Stuxnet code - reverse engineered

In our practice, to assess the generic impact of security on safety we use the Claims, Argument, Evidence (CAE) framework (see for example (Bishop 1998, IEC 2011, Netkachova 2015a). This analysis shows that a significant portion of a security-informed safety case needs to address security explicitly (Bloomfield 2013). For example, the following areas are particularly significant from a security perspective and need more scrutiny in a security-informed justification of a safety system.

Table 3. Some security-informed safety issues

- Supply chain integrity.
- Malicious events post deployment,that will also change in nature and scope as the threat environment changes. This will lead to substantial changes to the design and the implementation process.
- Weakening of security controls as the capability of the attacker and technology changes. This may have major impact on proposed lifetime of installed equipment and design for refurbishment and change.

- Security considerations are likely to challenge the effectiveness and independence of safety barriers.
- Design changes to address user interactions, training, configuration, and vulnerabilities. This might lead to additional functional requirements that implement security controls.
- Architectural changes to address end-to-end security and good security design practices.
- Possible exploitation of the device/service to attack itself or others.

There are technical drivers to integrate security into safety analyses – because of the interactions and trade-offs that are necessary to consider. For example, at the requirements stage we might need to consider how the security aspects of the information flow policy when a plant is under attack or there are degraded plant conditions impact the safety. Another type of issue is that we might need to consider at the architecture level whether that highly critical third party component does have sufficient security provenance given its supply chain. There are also softer aspects. Safety assessment involves building trust with the supply chain, visiting their factories and assessing their culture: all aspects highly relevant to security as well as safety.

There are a variety of initiatives to integrate security into hazard analyses. We have been using security- (or cyber-) informed HAZOP to assess architectures of industrial systems (Bloomfield 2016) and in this we adapt this well-known approach with additional security guidewords and an enhanced multidisciplinary team. Another area where there is common ground between security and safety is in static analysis of code. Both security and safety perspectives are needed to assess the likelihood of vulnerabilities being exploited and the effectiveness and consequences of their mitigations.

There are also business drivers to integrating safety and security, as stakeholders do not want to pay twice for assurance, or worse, find they have conflicts between safety and security that significantly impact project timescales and require considerable rework or re-architecting of the system.

Impact on safety regulation

Regulators, whether internal or external to the licensee, require an integrated, principled and defensible cyber strategy that reflects their responsibilities and role. Cyber-related risks are a challenge to regulators for a number of reasons: it is a real and present risk, it touches so many aspects of activity, it is a technically difficult area, and demands a tempo of change and adaptation that is faster than many organisations are used to. On the other hand, it is only one of many sources of risk that regulators have to deal with and address in a holistic manner. Below we describe the methodology we used to define the impact of cyber security on safety regulation for the UK Civil Aviation regulator.

The first part of our approach is to systematically define regulatory objectives, using a Claims Argument Evidence framework, developing from an organisation's strategic vision and legal responsibilities. We then select from these key objectives such as:

1. Processes, procedures and plans
2. Capability
3. Technical approaches for security-informed safety and resilience
4. Confidence in the overall regulatory system
5. Composition and systemic risks
6. Cyber-related events and changes
7. Confidence building in other stakeholders

We then need to consider how the organisation currently meets these objectives and where it would like to be and how it might get there. To develop this regulatory roadmap we first define a regulatory maturity model. Maturity models have two components: they define levels of maturity, or progress, and relate these to properties or 'dimensions' of the capability they are addressing. Not surprisingly, given the relatively specialised nature of cyber and cyber regulation, there are no explicit regulatory models. We therefore developed a model based on the objectives from the CAE analysis and our analysis of appropriate stages of maturity.

The first component of the maturity model identifies a number of levels that reflect intuitional progress or maturity. For example, the original Software Engineering Institute (SEI) Capability Immaturity Model levels focus on process maturity and range from *Initial* (chaotic, ad hoc, individual heroics) - the starting point for use of a new or undocumented repeat process - through to *Optimizing* in which process management includes deliberate process optimization/improvement.

We adapted this approach to safety regulation. Safety regulators should already have an established risk management framework and the impact of cyber will be on the need to adjust and develop it in the light of changing threats, increasing knowledge of vulnerabilities and innovations in implementation and assurance technology. Similarly, decisions should be risk-informed and strategic from an early stage although the detail and confidence in these judgements can improve as the organisation matures.

We therefore propose the levels characterised as follows:

Start-up. At this level the approach to cyber issues is very embryonic in nature. There are initial discussions about cyber-security, but few concrete actions have been taken. Issues have not been identified.

Formative. Responsibility and authority for practices are assigned to personnel. Personnel performing the practice have adequate skills and knowledge but technical approaches are exploratory or prototypical. Systematic programme to address cyber issues is being formulated. Evidence is being gathered to address current state with respect to objectives.

Established. A systematic programme to address cyber issues is being implemented. Stakeholders are identified and involved. Adequate resources are provided to support the process. Technical approach matures enough to the basis of compelling guidance or standards. These are used to guide practice implementation. The programme objectives have been operationalized and are being monitored and managed.

Mature. There is evidence that the programme is implemented and is effective. Significant changes to systems, significant incidents or attacks have been successfully dealt with and the organisation learns from this experience.

The other component of a maturity model is the 'dimensions' of the approach. For those we use objectives derived with our CAE framework.

Having established the framework of the maturity model and the strategic objectives for the regulator we are then able to develop a roadmap based on a detailed analysis of the challenges that the regulator faces.

Challenge 2: Resilience as well as safety

One of the insights from the theory of complex systems and from actual incidents is that complex interconnected systems fail more often than might be expected and that a power-law often governs the relationship between the frequency of incident and the scale of loss. Also, we find in our analysis of security of industrial safety critical systems that availability becomes an issue as many systems (e.g. in rail transport, power plants) are designed to fail-stop. This safety bias makes denial of service attacks relatively easy. If we add to this the increasing threat, motivation and capability of attackers we see that it is reasonable to assume many of our large critical systems may suffer widespread failures. While we might hope for the best, we should plan for incident recovery and adaptation: in other words systems need to be resilient.

A resilient system is an adaptive system, one that responds to change, can survive and prosper when challenged, and can deal with attack and surprises. The term is powerfully suggestive, but we need some clarity if we are to design for it and evaluate it: we need to move from metaphor to usable models.

In trying to make the concept of resilience more operational, we find it useful to distinguish two types:

1.   resilience to design basis threats and events – this could be expressed

in the usual terms of fault-tolerance, availability, robustness, and so forth; and

2. resilience beyond design basis threats events and use – this might be split into known threats that are considered incredible or so infrequent that they're ignored, and unknown threats.

We can often engineer systems successfully to cope with the first type of resilience, but the second type is a more formidable challenge. We might wish to make systems more heterogeneous and connected and with more resources to support the second type, but doing so might make them more expensive and suboptimal in terms of the first type of resilience.

A simple question to ask is 'what is the system?' Infrastructure is often considered to be the basic physical and organizational structures and facilities needed for the operation of a society, but this can be an oversimplification. A significant observation is the importance of 'soft' intangible infrastructures. For instance, trust between individuals, between individuals and organizations, and between these and the state is essential for service delivery. These intangibles are often hidden or ignored but come to the fore in times of crisis and disaster recovery.

As with so many other assets and resources, trust can be built up, destroyed, squandered, and undermined. If we are to understand resilience, we must take into account these essential, yet 'softer' aspects and their relationship to the more tangible ones. We should be cognizant that these soft aspects are just as much the target of security threats as the more obvious physical and cyber systems.

To design and evaluate resilience we need a model-based approach to:

- Address the scale and connectivity of systems
- Deal with uncertainties in structure and to understand and evaluate systemic risks
- Interpret and analyse incidents to guide mitigation and recovery strategies
- Provide rebuttal and commentary on events as appropriate
- For many complex systems, especially the critical ones, it is often impossible to perform live analysis. Instead, a model of a system operating in a simulated environment is constructed.

There are many approaches to modelling infrastructure and interdependencies (see reviews from (Utne 2011), (Bloomfield and Chozos 2009b)). In the Adelard and CSR City, University of London approach (Bloomfield et al. 2010a) – Probabilistic Interdependency Analysis (PIA) – we look at both qualitative and quantitative aspects of assessment. The models are in part probabilistic and part deterministic and include appropriate service models, documenting assumptions about resources, environmental impact, threats and any other factors. The modelled systems are studied with an operational environment where cyber-attacks are introduced with an explicit adversary model.

This type of model-based approach and probabilistic design are fundamental to the evaluation of critical infrastructures and safety-related systems and we need to be sure we can trust these models. The models are complicated and rely on complex software for their calculations. We have experimented with using the CAE assurance framework to support an analysis of their trustworthiness (Netkachova 2015b). The CAE framework supports the justification by helping elicit the claims we are making, identifying the arguments we are using to support or refute the claims and indicating what evidence we have, if any to justify what we are claiming. It provides the basis for challenge and peer review.

Challenge 3: Autonomy - an assurance gap

The third challenge comes from technology and industrial 'push' rather than from externalities and disasters, as well as a desire to improve safety and the quality of life. There is intense interest at the moment in robotic and autonomous systems, both within the technical and engineering communities and more broadly. These systems have always had an appeal as they blend social impact, technology, science fiction, and philosophy with newsworthy speculation and sensationalism in imagined futures. These futures might be arriving much faster than we once thought as there is a convergence of technology (of sensors, actuators, power, sensing and learning) and strong business drivers and social need. The diverse applications discussed include, inter alia, green mobility (just summon a car-pod to take you to see friends or shops), healthcare robots to serve an ageing population, robotic surgery, autonomous freight deliveries, safe and greener motorway driving and working in hazardous and uncertain nuclear environments post-accident (RAS 2014). This is before one considers the defence and security applications.

There is a spectrum of different types of systems that fall under the 'autonomy' umbrella, ranging from incremental improvements to existing automation assistance to fully autonomous future systems. The US National Academy of Sciences study (NAS 2014) on 'Autonomy Research for Civil Aviation: Toward a New Era of Flight' usefully coins the phrase 'increasingly autonomous' (IA) systems and identifies assurance as the one critical, crosscutting challenge. After a detailed examination of the issues it concludes there are four major barriers to ensuring that IA systems will enhance rather than diminish the safety and reliability of the National Airspace system:

- Certification process
- Decision making by adaptive/nondeterministic systems
- Trust in adaptive/nondeterministic IA systems
- Verification and validation

Addressing the assurance of IA and fully autonomous system challenges will need a multidisciplinary approach with social and political engagement as the risks and resilience of these systems, or what we would like from these systems, is by no means clear. As usual there will be those who reap the benefits and those that suffer the risks and it may not be a fair division. We will probably be more vulnerable to widespread chaos and confusion and how we deal with this, and who pays for the systems that recover the situation, needs to be debated (as we know markets can drive out resilience unless incentives or regulations are explicitly designed to prevent this e.g. in the diversity of mobile telecoms backbone).

These autonomous systems will be part of a system of systems, which are ultra large in scale, with complex system behaviour and which will adapt to success, failure, social changes and political interaction. Nevertheless they still need to be engineered to be trustworthy even when they are used in ways not envisaged by their developers. When we see the amount of certification and assessment effort that today goes into a simple device in the nuclear, rail and aerospace industries the task of assuring autonomous systems seems daunting.

However, we know that we need appropriate system architectures and trusted components. We need architectures that limit parts of the system that need to be highly trusted, building on current ideas of protection systems with static safety envelopes. There is also more ecologically oriented work that is relevant: viability domains describe a similar concept to protection envelopes and their mathematical underpinnings may provide some further insights (Deffuant 2011). These concepts could be extended to more dynamic, real time, evaluation of safety invariants based on deeper system properties and would need to address the performative nature of the underpinning models if security was taken into account. These systems can be – rather fancifully – thought of as prototypical explicit ethical systems within an autonomous systems architecture that can be assured[1]. So we would then seek to enforce explicit ethical behaviour, rather than safe and trustworthy behaviours being an emergent property of unanalysable and incomprehensible software. Whether such architectures are feasible is an open question.

Cyber and security issues are an enormous challenge. Recent well-publicised car hacking incidents (Greenberg 2015) show the risks of not addressing these in current systems, let alone in future ones. Research and industrial strategies should address the challenges and benefits of autonomy head on. Both because it is a strategic objective, important in its own right, but also because addressing

---

[1] Autonomy and ethics has generated considerable literature already. An overview of some of the issues is provided in Autonomous Systems: Social, Legal and Ethical Issue, The Royal Academy of Engineering ISBN 1-903496-48-9   August 2009

it will pull through technologies and insights that can be deployed from existing automation to fully autonomous systems.

Part of this research and innovation needed to address the 'assurance gap' should examine the basic principles behind assurance for their applicability to autonomous systems to see if they are still valid and what insights they can bring. We have been involved in a number of initiatives where we try and distil the safety principles for automation systems in the nuclear industry drawing on the work of the International Atomic Energy Agency (IAEA) and the UK Safety Assessment Principles (ONR 2014). This has led to the following principles (Harmonics 2015):

- Effective understanding of the hazards and their control should be demonstrated
- Intended and unintended behaviour of the technology should be understood
- Multiple and complex interactions between technical systems and also human systems to create adverse consequences should be recognised
- Active challenge should be part of decision making throughout the organisation. Needs of all stakeholders to understand and challenge case should be taken into account in its structure and presentation
- Lessons learned from internal and external sources should be incorporated
- Justification should be logical, coherent, traceable, accessible and repeatable with a rigour commensurate with the degree of trust required of the system.

Can these be applied to autonomous systems? It is interesting that their emphasis on understanding, explanation, challenge and learning are quite general and raise the question of who, or what, should have this understanding. So perhaps we should shift our thinking of the dependability of these systems as complex automatic gadgets (e.g. how can we apply DO-178C or IEC 61508 to them) to whether and how their assurance addresses the principles of understanding, explanation, challenge and learning as we reinvent our approach to assurance.

In addition, we might get some more leverage by taking an anthropomorphic viewpoint – whether as a metaphor or quite literally. In the UK, as in many other countries, one just has to pass a single practical driving test to drive a car with some additional theory exam on the rules to be followed. This may take 25-30 hours of driving lessons and to fly a private plane requires 70 hours[1]. This would seem sufficient to take an individual from almost certain failure to some socially acceptable rate of self-harm of 1 in 10,000 hours.

In simple architectural terms we can think of this in the usual specific application vs generic platform distinction. One reason we trust the human 'plat-

---

[1] As claimed in http://www.aopa.org/letsgoflying/faqs.html

form' is because we assume some in-built characteristics of self-preservation and of some learnt ethics. There is also lots of evidence that humans can do 'similar' or more difficult tasks – walk, ride a horse, a bicycle, cross the street in a rush hour and we have evidence, or belief, that the abilities to do this are quite generic. If we have a less trustworthy platform then we require more specific evidence: my blind friend trusts his safety to his guide dog as he commutes into London but the specific training has taken far more than would be required from a human helper. So fully autonomous systems will be easier to assure and licence, once they have a trusted and widely used platform.

So we may see a number of tipping points or transitions. The first is in risk behaviour where the use of an autonomous vehicle is the norm and non-use has to be justified. The second is in assurance when the incremental approaches used to assure 'increasingly autonomous' systems runs out of steam as the automation becomes too complex to assure unless the assurance exploits the self adaptive, explanatory and understanding inherent in these systems. The third is that, as with PC operating systems in the 1990s, whoever owns the trustworthy platform will own the industry. And lastly, when we just give autonomous vehicles a simple driving test, as we do with teenagers, we will know that trust in autonomous transport has really arrived.

Challenge 4: Business as usual

Safety systems encompass a rich and varied landscape from less critical safety-related systems, whose malfunction may potentially compromise safety and might lead to accidents with marginal or negligible severity, to highly critical systems, whose failure or malfunction can result in death and serious injuries to people, damage to property or the environment. Today these systems range from small embedded domestic toys to large multi-organisation command and control systems of systems. Achieving and assuring safety is a specialist activity and rigorous safety analysis is needed to identify and mitigate hazards. In some sectors, notably high hazard ones, achieving and assuring safety is a relatively mature undertaking - although of course we must not be complacent (Dunn 2013, Koopman 2014, FDA) particularly in the context of organizational change, project pressures and resource limitations. The FDA statistics of 710 deaths from infusion pumps in 2005-2009 provide a reminder of what is at stake if we do not deploy appropriate controls, as does the Nimrod accident and the associated report (Haddon-Cave 2009).

These accidents highlight the importance of knowledge management and the need to focus on actual evidence, as well as the overall need to focus on our understanding of the safety of the system not just the information on how hard we have tried. So we must not underestimate the challenge of 'normal business': the challenge of applying what we know in a rigorous and competent manner.

5 Conclusions

As one of us (Bloomfield) was involved with the club from its inception to the present we were asked to provide some reflection on the past 25 years of safety engineering. This we have attempted by taking as our focus work done at Adelard and City, University of London. It is therefore a very personal perspective with all the strengths and limitations that brings. History is important because how we communicate our history —what we include, what we exclude, the events we deem significant – shapes how we think of ourselves as a community of practitioners. It helps define our safety and dependability culture.

This paper provides some background to the establishment of the SCSC in 1992 and the world of safety-related computing and assurance at that time. It describes some of the advances in the 1990s and then considers how since 2000 the dramatic growth in the Internet and the changes to security threats exemplified by the attacks of 9/11 have changed the safety engineering world. The need for change is further illustrated by the impact of climate change, the growing importance of interdependencies and lessons from Fukushima accident and the need for resilience. Changes in technology, in particular the increasing autonomy of systems also provide another driver for change. The paper therefore discusses the challenges of security-informed safety, resilience and autonomy and some of the work done to start addressing them. But there is also an additional challenge we must not forget: although achieving and assuring safety is a relatively mature undertaking we must not be complacent, particularly in the context of organisational change, project pressures and resource limitations. We believe safety engineers need to be broader in their view of the world, deep in their understanding of systems and ambitious in addressing the future challenges we have outlined above. These challenges should not be seen as just potential risks: the changes they represent and the underlying advances in technology and information provide opportunities for innovation so we, as safety engineers, can help build confidence in a computerised and connected world.

Acknowledgments This paper draws heavily on the work of Adelard and City, University of London colleagues especially on previous papers with Nick Chozos, Sofia Guerra, Bev Littlewood, and Robert Stroud.

References

ACARD (1986) Software: a vital key to UK competitiveness. Advisory Council on Applied Research and Development. HMSO

ACSNI (1997) The use of computers in safety-critical applications. Final Report of the Study Group on the Safety of Operational Computer Systems (SOCS) constituted by the Advisory Committee on the Safety of Nuclear Installations. HSE Books, London

Assante (2016), M.J. Assante, "Confirmation of a Co-ordinated Attack on the Ukrainian Power Grid," blog, 9 Jan. 2016; https://ics.sans.org/blog/2016/01/09/confirmation-of-a-coordinated-attack-on-the-ukrainian-power-grid. Accessed 12 November 2016.

Baylon (2015) C. Babylon, R. Brunt, and D. Livingstone, Cyber Security at Civil Nuclear Facilities: Understanding the Risks, Chatham House, Royal Inst. of Int'l A airs, 2015 https://www.chathamhouse.org/sites/files/chathamhouse/field/field_document/20151005

https://www.chathamhouse.org/sites/files/chathamhouse/field/field_document/20151005CyberSecuri tyNuclearBaylonBruntLivingstone.pdf. Accessed 12 November 2016.

Bishop PG (ed) (1990) Dependability of critical computer systems 3. Elsevier Applied Science

Bishop PG, Bloomfield RE (1995) The SHIP safety case. In: Rabe G (ed) Proc SafeComp 95, 14th IFAC Conf on Computer Safety, Reliability and Security, Belgirate, Italy

Bishop PG and Bloomfield RE (1996), "A Conservative Theory for Long-Term Reliability Growth Prediction", IEEE Trans. Reliability, vol. 45, no. 4, Dec. 96, pp 550–560

Bishop PG (1997) Using Reversible Computing to Achieve Fail-safety, ISSRE 97, Nov 1997, Alberquerque, New Mexico, USA., IEEE Computer Society Press, DOI 10.1109/ISSRE.1997.630863

Bishop PG, Bloomfield RE (1998) A methodology for safety case development. In: Redmill F, Anderson T (eds) Industrial perspectives of safety-critical systems. Springer-Verlag

Bishop PG, Bloomfield RE, Clement TP, Guerra ASL (2002a) Software criticality analysis of COTS/SOUP. SAFECOMP 2002, Catania, Italy

Bishop PG, Bloomfield RE, Froome PKD (2002b) Justifying the use of software of uncertain pedigree (SOUP) in safety related applications. 5th Int Symp Programmable Electronic Systems in Safety Related Applications, Cologne see also (Jones 2001)

Bloomfield RE and Froome PKDF (1989), Aspects of the licensing and assessment of highly dependable computer systems, in Measurement for Software Control and Assurance, ISBN 978-1-85166-246-3, Springer 1989.

Bloomfield RE (1990) SafeIT, the safety of programmable electronic systems: a government consultation document on activities to promote the safety of computer-controlled systems. Department of Trade and Industry

Bloomfield RE, Brazendale J (1990) SafeIT2, standards framework. Department of Trade and Industry

Bloomfield RB, Bowers J, Emmet L, Viller S, (1997) PERE: Evaluation and Improvement of Dependable Processes, Safecomp 96: The 15th International Conference on Computer Safety, Reliability and Security, Vienna, Austria October 23--25 1996, Springer London 1997, pp 322—331, ISBN 978-1-4471-0937-2, DOI 10.1007/978-1-4471-0937-2_28

Bloomfield RE, Littlewood B (2003) Multi-legged arguments: the impact of diversity upon confidence in dependability arguments. Proc DSN 2003. IEEE Computer Society

Bloomfield RE, Littlewood B (2007) Confidence: its role in dependability cases for risk assessment. Intl Conf Dependable Systems and Networks, Edinburgh, IEEE Computer Society

Bloomfield RE, Bishop PG, Jones CCM, Froome PKD (1998) ASCAD – Adelard safety case development manual. Adelard LLP.

Bloomfield (2009a), R.E., Chozos, N., Nobles, P: Infrastructure interdependency analysis: Requirements, capabilities and strategy. Adelard document reference: d418/12101/3, issue 1 (2009). http://www.adelard.com/papers/d418v13_public.pdf. Accessed 22 November 2016.

Bloomfield (2009b, R., N. Chozos, and P. Nobles, Infrastructure interdependency analysis: Introductory research review. 2009, Adelard LLP. http://www.adelard.com/papers/d422v10_review.pdf. Accessed 22 November 2016.

Bloomfield et al (2010a) Bloomfield, R. E., Chozos, N., Popov, P. T., Stankovic, V., Wright, D. & Howell-Morris, R. (2010). Preliminary Interdependency Analysis (PIA): Method and tool support (Report No. D/501/12102/2 v2.0). London: Adelard LLP and City University London.

Bloomfield R.E., Bishop P.G. (2010b) Safety and Assurance Cases: Past, Present and Possible Future – an Adelard Perspective, Proceedings of the Eighteenth Safety-Critical Systems Symposium, Bristol, UK, 9-11th February 2010, DOI 10.1007/978-1-84996-086-1_4 ISBN 978-1-84996-085-4, Springer-Verlag London.

Bloomfield (2013), R. E., Netkachova, K. & Stroud, R. Security-Informed Safety: If it's not secure, it's not safe. Software Eng. for Resilient Systems, A. Gorbenko, A. Romanovsky, and V. Kharchenko, eds., LNCS 8166, Springer, 2013, pp. 17–32.

Bloomfield et al (2016) Bloomfield, R. E., Bendele, M., Bishop, P. G., Stroud, R. & Tonks, S. (2016). The risk assessment of ERTMS-based railway systems from a cyber security perspective: Methodology and lessons learned. Paper presented at the First International Conference, RSSRail 2016, 28-30 Jun 2016, Paris, France.

Bourne AJ, Edwards GT, Hunns DM et al (1981) Defences against common-mode failures in redundancy systems. UKAEA Safety and Reliability Directorate

Butler R, Finelli G (1993) The infeasibility of quantifying the reliability of life-critical real-time software. IEEE Trans Software Engineering 19:3-12

CEGB (1982) Sizewell B preconstruction safety report. Central Electricity Generating Board

Deffuant, G Viability and Resilience of Complex Systems: Concepts, Methods and Case Studies from Ecology and Society (Understanding Complex Systems), ISBN-13: 978-3642204227, Springer 2011.

DHS (2011) DHS evidence "Hearing Before The Subcommittee On National Security, Homeland Defense And Foreign Operations Of The Committee On Oversight And Government Reform House Of Representatives One Hundred Twelfth Congress First Session, May 25, 2011, Serial No. 112–55"

Dunn (2013) M. Dunn, "Toyota's Killer Firmware: Bad Design and Its Consequences," EDN, 28 Oct. 2013; www.edn.com

FDA US Food and Drug Administration, Infusion Pump Improvement Initiative, http://www.fda.gov/medicaldevices/productsandmedicalprocedures/generalhospitaldevicesands upplies/infusionpumps/ucm202501.htm. Accessed 12 November 2016.

Gashi, I. & Bloomfield, R. E. (2008). Evaluating the resilience and security of boundaryless, evolving socio-technical Systems of Systems. Centre for Software Reliability, City University London.

Greenberg (2015) A. Greenberg, "Hackers Remotely Kill a Jeep on the Highway— with Me in It," Wired, 21 July 2015; https://www.wired.com/2015/07/hackers-remotely-kill-jeep-highway/. Accessed 12 November 2016.

GSN (2011) GSN Community Standard, Version 1, November 2011, available from www.goalstructuringnotation.info. Accessed 12 November 2016.

Haddon-Cave , C (2009) The Nimrod Review: an independent review into the broader issues surrounding the loss of the RAF Nimrod MR2 aircraft XV230 in Afghanistan in 2006, ISBN 9780102962659, Oct 2009. https://www.gov.uk/government/uploads/system/uploads/attachment_data/file/229037/1025.pd f. Accessed 23rd Nov 2016

Harmonics (2015) Harmonics principles are summarized in http://cordis.europa.eu/result/rcn/168229_en.html. Accessed 12 November 2016.

Health Foundation (2012), Evidence: Using safety cases in industry and healthcare, by the Health Foundation, 90 Long Acre, London WC2E 9RA, ISBN 978-1-906461-43-0 can be downloaded from http://www.health.org.uk/publication/using-safety-cases-industry-and-healthcare. Accessed 12 November 2016.

HMG (2009) The Buncefield Investigation, The Government and Competent Authority's Response, Cm. 749, 2009, ISBN: 978 0 10 174912 1

HMG (2016) National Cyber Security Strategy 2016 -21, retrieved from https://www.gov.uk/government/publications/national-cyber-security-strategy-2016-to-2021. Accessed 12 November 2016.

HSE (1987) Programmable electronic systems in safety related applications. Health and Safety Executive

HSE (1995, 2003), Out of control – a compilation of incidents involving control systems. Health and Safety Executive, First published 1995, Second edition 2003 ISBN 978 0 7176 2192 7

IAEA (2015), The Fukushima Daiichi Accident, Report by the Director General, GC(59)/14 ISBN 978–92–0–107015–9, International Atomic Energy Agency

IEC (1986) IEC 880 Software for computers in the safety systems of nuclear power stations. International Electrotechnical Commission

IEC (1993) Functional safety of electrical/electronic/programmable electronic systems: generic aspects. Part 1: General requirements. Draft standard from IEC Sub-Committee 65A: System Aspects, Working Group 10. International Electrotechnical Commission

IEC (1998) Functional safety of electrical, electronic, and programmable electronic safety related systems. IEC 61508, Parts 1 to 7, 1998 to 2000. International Electrotechnical Commission

IEE (1989) Software in safety related systems. The Institution of Electrical Engineers and the British Computer Society

Jackson D, Thomas M, Millett LI (eds) (2007) Software for dependable systems: sufficient evidence? Committee on Certifiably Dependable Software Systems, National Research Council

Jones C, Bloomfield RE, Froome PKD, Bishop PG (2001) Methods for assessing the safety integrity of safety-related software of uncertain pedigree (SOUP). HSE Contract Research Report CRR 337/2001. Health and Safety Executive, available from http://www.hse.gov.uk/research/crr_pdf/2001/crr01337.pdf http://www.hse.gov.uk/research/crr_pdf/2001/crr01336.pdf

Kelly TP (1998) Arguing safety: a systematic approach to managing safety cases. PhD thesis, University of York

Kelly T and McDermid J (1997), "Safety Case Construction and Reuse using Patterns", Proc 16th Conf on Computer Safety, Reliability and Security (Safecomp '97)

Koopman (2014) P. Koopman, "A Case Study of Toyota Unintended Acceleration and Software Safety," slide presentation, Carnegie Mellon Univ., 18 Sept. 2014; https://users.ece.cmu.edu/~koopman/pubs/koopman14_toyota_ua_slides.pdf

Lee (2014) R.M. Lee, M.J. Assante, and T. Conway, *German Steel Mill Cyber Attack*, ICS Defense Use Case, SANS Industrial Control Systems, 30 Dec. 2014; https://ics.sans.org/media/ICS-CPPE-case-Study-2-German-Steelworks_Facility.pdf

Littlewood B (2000) The use of proofs in diversity arguments. IEEE Trans Softw Eng 26:1022-1023

Littlewood B, Strigini L (1993) Assessment of ultra-high dependability for software-based systems. Comm ACM 36:69-80

Littlewood B, Wright D (2007) The use of multi-legged arguments to increase confidence in safety claims for software-based systems: a study based on a BBN of an idealised example. IEEE Trans Softw Eng 33:347-365

MacKenzie, D. (1994) Computer-related accidental death: an empirical exploration. Science and Public Policy 21(4), August 1994, pp. 233-248. DOI: 10.1093/spp/21.4.233

Masera, M, Bloomfield, R (2003) A Dependability Roadmap for the Information Society in Europe, AMSD Deliverable D1.1, 2003. http://www.adelard.com/resources/reports/amsd.html

McDermid JA (1994) Support for safety cases and safety argument using SAM. Reliab Eng Syst Saf 43:111-127

MoD (1989) Draft Interim Def-Stan 00-55, the procurement of safety critical software in defence equipment. Ministry of Defence

MoD (1991) Interim Def-Stan 00-56, hazard analysis and safety classification of the computer and programmable electronic system elements of defence equipment. Ministry of Defence

MoD (2004) Def Stan 00-56 Safety management requirements for defence systems. Issue 3. Ministry of Defence

NAS (2014) Committee on Autonomy Research for Civil Aviation; Aeronautics and Space Engineering Board; National Research Council, Autonomy Research for Civil Aviation: Toward a New Era of Flight, ISBN 978-0-309-30614-0 , This PDF available from The National Academies Press at http://www.nap.edu/catalog.php?record_id=18815

Netkachova (2015a), K., Netkachov, O., Bloomfield, R. "Tool Support for Assurance Case Building Blocks, Providing a Helping Hand with CAE", Lecture Notes in Computer Science, Computer Safety Reliability and Security, pp 62-71, doi: 10.1007/978-3-319-24249-1_6

Netkachova (2015b), K., Bloomfield, R., Popov, P., Netkachov, O. "Using Structured Assurance Case Approach to Analyse Security and Reliability of Critical Infrastructures", Lecture Notes in Computer Science, Computer Safety Reliability and Security, pp 345-354, doi:10.1007/978-3-319-24249-1_30

ONR (2014) Office for Nuclear Regulation, Safety Assessment Principles for Nuclear Facilities 2014, Edition Revision 0

Pitt   (2008),   "The   Pitt   Review: lessons   learned   from   the   2007   floods" http://webarchive.nationalarchives.gov.uk/20100807034701/http:/archive.cabinetoffic e.gov.uk/pittreview/_/media/assets/www.cabinetoffice.gov.uk/flooding_review/pitt_re view_full%20pdf.pdf

RAEng (2015) Connecting data: driving productivity and innovation, www.raeng.org.uk/connectingdata ISBN: 978-1-909327-22-1

RAS (2014) See for example RAS 2020, Robotics and Autonomous Systems Strategy, RAS Sig 2014

Redmill F (ed) (1988) Dependability of critical computer systems 1. Elsevier Applied Science

Redmill F (ed) (1989) Dependability of critical computer systems 2. Elsevier Applied Science

Rushby (2015) The Interpretation and Evaluation of Assurance Cases, Technical Report SRI-CSL-15-01, July 2015

Toulmin SE (1958) The uses of argument. Cambridge University Press

Utne (2011), Utne, I.B., P. Hokstad, and J. Vatn, A method for risk modeling of interdependencies in critical infrastructures. Reliability Engineering and System Safety, 2011. 96(6): p. 671-678

Weinstein S., Wild C. (2012) The United Kingdom's Centre for the Protection of National Infrastructure: An Evaluation of the UK Government's Response Mechanism to Cyber Attacks on Critical Infrastructures in Law, Policy, and Technology by Eduardo Gelbstein, Pauline C. Reich Publisher: IGI Global Release Date: June 2012, ISBN: 9781615208319

Software Handling of Hardware Errors

Chris Hobbs

QNX Software Systems

Abstract *Developing embedded systems for safety-critical markets is not easy. Over the past decade, detecting and handling the errors arising from increasingly unreliable hardware and increasingly complex, multi-threaded software has made this even more difficult. This paper describes a software architecture that separates various aspects of the system design, providing increased and tuneable immunity to random software and hardware errors*

Introduction

This paper describes a software architecture that has been found effective in addressing two problems:

1. The first problem stems from the reduction in die sizes of integrated circuits, making them increasingly susceptible to EMI (Armstrong 2012), the secondary effects of cosmic rays (Schroeder et al. 2009) and internal crosstalk (Kim et al. 2014). When hardware errors were infrequent, it was acceptable to move a system to its Design Safe State whenever an error was detected. However, many embedded systems providing functional safety experience several hardware errors daily, and moving to the safe state as frequently as that is unacceptable, because many have availability goals in addition to reliability ones.
2. The second problem concerns the increasing level of sophistication required in the computational algorithms. Sophisticated algorithms have been used in railway systems (e.g. brake control) for many years, but such sophistication is now starting to appear in advanced driver assistance systems (ADAS) in cars and in more complex medical devices. This increase in sophistication means that two distinct skills are now required to produce a software product: the mathematical skill to implement stable and convergent algorithms, and the statistical skill to determine how much replication and diversity is needed to satisfy the system's robustness requirements.

Ideally it should be possible to handle these two problems independently: the mathematical algorithm should not have to address dependability and the dependability strategy should be independent of the complexity of the algorithm.

The architecture outlined in this paper addresses both of these problems.

2 New Questions for new Problems

This section considers the problems described above, and suggests that the questions we have traditionally asked must be changed.

2.1 Refocusing the Hardware Question

Because of the high frequency of occurrences, it is no longer acceptable to move directly to the system's Design Safe State when a hardware error (e.g. a bit flip in memory) is detected. (Schroeder et al. 2009) indicates that such history is important for predicting future errors, so data about such errors should be recorded for statistical purposes, but it is perhaps better that the application not be informed of such errors. Most bit flips have no effect on the safe operation of the system.

This change of focus means that the important question is no longer "has a hardware error occurred?" Rather it is "has a hardware or software error affected the integrity of the safety-critical computation?"

One useful simplification resulting from this change of emphasis is that hardware errors and software heisenbugs[1] can be treated identically: "have they affected system safety?"

2.2 Refocusing the Dependability Question

Diversity and replication have been used to increase dependability since at least 1834. Readers of this paper are probably (over-) familiar with the quotation:

> The most certain and effectual check upon errors which arise in the process of computation, is to cause the same computations to be made by separate and

---

[1]   A "heisenbug" is a non-reproducible bug often caused by subtle timing races between threads. Such bugs change in behaviour when debug code is added and differ in their effects each time they occur. They can be extremely difficult to locate and fix because it is necessary to trace back from the failure, through the error, to the fault through two layers of N-to-M mappings.

independent computers; and this check is rendered still more decisive if they make their computations by different methods. (Lardner 1834 – the quotation appears on page 278).

They may be less familiar with Lardner's following sentence:

> It is, nevertheless, a remarkable fact, that several computers, working separately and independently, do frequently commit precisely the same error.

This last observation is one of the criticisms levelled against applying hardware lock-step computation to a safety-critical application. When copies of an application are run in lock-step with each other, any heisenbug encountered by one replica will also simultaneously hit other replicas; there is no resilience against heisenbugs. Other disadvantages of hardware lock-step include:

- the level of replication being defined statically by the hardware.
- the inability to meet Lardner's challenge by having diverse implementations, only replication being supported. As described in section 5 below, diversification is particularly powerful when using ASIL decomposition in accordance with ISO 26262 (ISO 2011).

However, it is accepted that true lock-step processing is convenient in terms of development time: one copy of the program is produced and several copies can be run in lock-step without further development.

This paper addresses the question: "can the convenience of lock-step processing be retained, while avoiding the constraints it imposes?"

2.3 Refocusing the Design Question

Because the algorithmic complexity of the application and the statistical complexity of the system dependability require different skills, it is increasing important that the question: "How can we build the required dependability into this application?" be replaced with "How can we separate the design of application from the design of the system dependability?"

This separation is particularly useful in a system where the level of dependability of a particular subsystem must be dynamically changed during operation, depending on the system state: between, for example, cruising on a motorway, driving through a city centre in heavy traffic and parking at a supermarket.

The separation also makes it possible to reüse a legacy application that was written without consideration of dependability.

3 The Reference Embedded Environment

Embedded systems normally have the characteristics illustrated in Figure 1:

1. a computation engine reads values from sensors: speed, acceleration, drug flow, distance, position, camera image, lidar, keyboard, etc.;
2. the engine performs some form of computation;
3. the engine, directly or through other engines, commands actuators which cause something to happen in the external world: brakes to be released, drug flow to stop, a corner to be turned, etc.

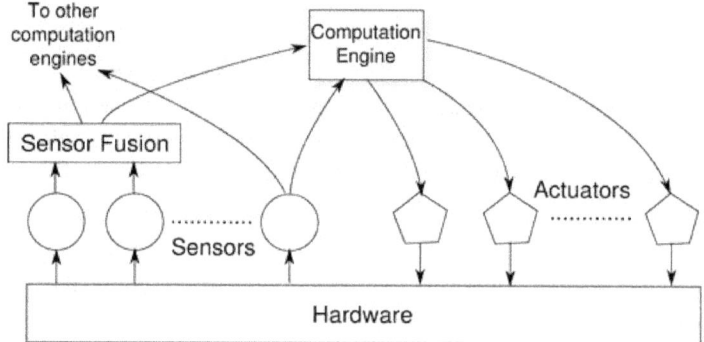

Fig 1. Logical View of a Typical Embedded System

Sensors may be shared between computation engines when the same input is required by two subsystems and sensor fusion is increasingly common where the values of several sensors are combined before being presented to the computation engine. Figure 1 is the reference architecture that is used in the remainder of this paper.

When the designer wishes to provide a higher level of availability or reliability[1] than is shown in Figure 1, the computation engine may be replicated or diversified and a check made that both copies make the same decision. Hardware lock-step can implement the replication pattern, but not the diversification pattern, and has the effect of increasing reliability (as both servers have to agree) at the cost of decreasing availability (if either server fails, the system fails).

---

[1]   Availability is a measure of how often the system responds, reliability a measure of how often the response is correct. The term "dependability" is used in this paper to describe whichever is important for the particular system under consideration.

4 Loosely-Coupled, Lock-Step Architecture

Replicating a computation engine, other than by using hardware lock-step, normally leads to the problem of synchronising the various replicas. New protocols have to be introduced to allow the replicas to remain in step and this complicates the implementation, particularly if the number of replicas is dynamic, changing with the current operating state. One advantage of the architecture proposed in this paper is that the middleware handles this synchronisation without the applications needing to be changed.

4.1 Virtual Synchrony

Given the advantage of true lock-step processing, the question arises as to whether its primary characteristics can be retained while abstracting the concept to remove its constraints. This question was addressed in (Birman and Joseph 1987) by introducing the concept of *virtual synchrony*:

> In a virtually synchronous environment, routines can be programmed and will behave as if distributed actions were performed instantaneously and in lock-step. .... It will appear to any observer -- any process using the system -- that all processes observed the same events in the same order.

Virtual synchrony has been applied to many mission-critical applications, but rarely to embedded systems, an exception being described in (Ferrari et al. 2013), and rarely to systems that must meet the requirements of safety standards such as IEC 61508 (IEC 2010) and ISO 26262. This paper describes an experiment to see to what extent the virtual synchrony protocols can be applied to an embedded system and what level of protection they provide.

4.2 The Challenges of the Embedded Environment

There are numerous open-source and commercial implementations of virtual synchrony middleware, but these are mainly aimed at providing extremely high throughput in a server farm environment, where very large amounts of data pass over extended and unreliable networks, and where relatively simply operations are performed on the data. They are also focussed on supporting hot software upgrades by allowing a group to remain operational while instances within the group are changed. This is generally not needed in an embedded system with a particular duty and maintenance cycle.

Within the embedded world of a safety-critical device in a car, medical device or train, the volume of data from the sensors is comparatively small, perhaps a few hundreds of bytes per millisecond from speed or location sensors, travelling over relatively reliable links formed by limited networks (e.g. within a car), backplanes, tracking on a board, or even between cores on a multi-core processor chip. Some sensors, for example cameras, produce data at much higher rates than this, but in these cases it may not be necessary for the actual data to flow through the virtual synchrony middleware. Instead they may be stored and only control messages be transferred.

However, the operations performed on the data are complex: determining whether or not a target in an image from a front-facing camera is a person, deciding when to apply the brakes and with what pressure, etc.

One simplification valid for many embedded systems is that, unlike the server farm, state does not need to be persistent across executions. When a system such as that in Figure 1 starts up, old sensor data is unlikely to be needed; the system starts afresh each time the ignition is turned on.

This paper describes a implementation by QNX Software Systems of middleware providing virtually synchronous mechanisms in an embedded environment. The purpose of the investigation was to see whether the virtual synchrony architecture could be mapped efficiently to that environment. QNX has termed this middleware "loosely-coupled, lock-step" code, emphasising that synchronisation points exist (lock-step), but these are not hard as they would be with hardware lock-step (loosely-coupled).

4.3 The Virtually Synchronous Environment

Figure 2 illustrates the same system as Figure 1, but using a virtual synchrony architecture as viewed by an external observer. Viewed internally, the different elements operate as in true synchronism.

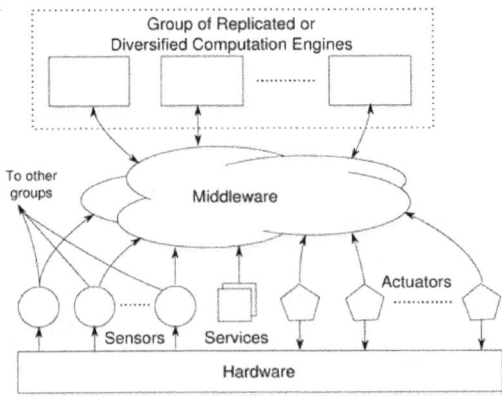

Fig. 2. Virtual Synchrony Extension

Within this environment, instances of the computation engines join process groups, and the management of group membership is a primary responsibility of the loosely-coupled, lock-step middleware. A group is a set of members that together mimic a single, more dependable abstract process. These members are typically spread across different processors or different cores within one processor to ensure that they do not run in genuine lock-step and to ensure that there is no reliance on a single piece of hardware. The middleware shown in Figure 2 is also completely distributed as shown below in Figure 5 and therefore is also resilient against and single failure. Group membership overcomes some particular categories of failures, including heisenbugs affecting a minority of the elements.

A particular engine may be a member of several groups and a group may contain any number of engines. The members of a group may be physically distinct and implemented on different processors. For this reason, and to avoid single points of failure, it is essential that the system operate without access to a common (global) clock.

Members of a group may also be diverse. The term "diversity" has become almost synonymous with "N-version programming" in some quarters, but the diversity supported by this architecture is more likely to be an active monitor as described in section 5 below or the use of different compilers, or the same compiler with different optimisation levels, to compile identical source code and thereby form diverse implementations. This does not provide the same level of diversity as an active monitor or N-version programming, in part because the same libraries may be used, but this multiple-compiler technique is particularly advantageous when the source code is in a language such as C++ where demonstrating the correctness of the compiler, as required by ISO 26262 or IEC 61508, may be very difficult.

When a new instance joins a group, its state is synchronised with the other members of the group, after which point it receives notification of events (group membership changes, information from sensors, etc.) according to a chosen ordering scheme.

In most cases, the appropriate ordering is the so-called "total" ordering: every group member sees every event in exactly the same order. The absence of a globally-visible clock means that some other technique must be applied to allow the group members to agree whether event A happened before or after event B. With this guarantee, viewed externally, each group member starts in the same state and processes the same events in exactly the same order. All members must, therefore, reach the same synchronisation points, albeit at different times. This behaviour is not hard synchronism as would result from hardware lock-step, rather this is loosely-coupled, lock-step where each step is the completion of the processing of a particular event by the slowest, non-failed, member of a group.[1]

Figure 2 contains an element labelled "services". In a virtually synchronous environment, the members of the group remain in step by processing the same events in the same order. It is therefore essential that a particular group member not make a unilateral decision based on some data unavailable to other group members. This could lead to divergence of state and therefore inconsistent later responses. Timer ticks, for example, must been seen as events that are handled through the virtual synchrony middleware. It would be inappropriate for one group member to see a timer tick *before* a particular sensor value, while another group member saw it *after*. The middleware provides this sequencing as long as such inputs are treated as services. Other services can include persistent (NVRAM) storage and clock synchronisation.

As in Figure 1, the sensors provide data to what they believe is a single computation engine. These data are, however, intercepted by the middleware and presented to each engine instance with the predefined order guarantee. In the simplest implementation, each engine is unaware of the other group members and performs the complex calculations. When the actuators query the server group, depending on the required dependability level, they can request one, some or all the responses from group members. Unless all responses are requested, the middleware discards the unwanted ones, thereby improving system performance.

---

[1]   Other, looser, event orderings are appropriate when it is acceptable that different members of the group receive messages in different orders. In this case, an additional level of entropy is introduced into the system, increasing resilience against heisenbugs arising from the precise sequencing of messages.

The choice of the number of responses requested allows the designer of the actuator to select a point on the spectrum between high availability[1] at the cost of reliability (1ooN: one response) and high reliability at the cost of availability (NooN: all responses). This point on the spectrum can be changed dynamically depending on the current system environment: when a car is travelling at high speed on a motorway, perhaps a move is made towards availability rather than reliability, whereas when it is travelling slowly in a city, reliability might take precedence.

The actuator can then make the necessary choice of what to do given the responses it has received.

4.4 Hierarchical Systems

The system illustrated in Figure 2 is particularly simple because it has a single layer of clients (sensors and actuators) and servers (computation engines). In practice it is expected that the servers themselves would act as clients to other groups of servers. This complicates the dependability calculation for the system, but, because of the isolation provided, does not alter the implementation of the server's components. Computation engines can also be members of multiple groups.

5 Exploiting Decomposition

A common design pattern for safety-critical systems is that of the "safety bag" or "active monitor" (see, for example, IEC 61508-3, 7.4.2.11). In this pattern, a complex algorithm is used to calculate the optimal response of the system and this calculation is overseen by a much simpler monitor. The monitor can preserve the safety of the overall system by placing limits on the output of the more sophisticated algorithm. As the paragraph from IEC 61508-3 says, 'The combination of the two processors gives higher confidence that the end-to-end safety function ... will be achieved.'

The proposed architecture supports two configurations of the safety-bag pattern; see Figures 3 and 4.

---

[1]   The term "availability" is used as a measure of how often the system responds to an input; the term "reliability" as a measure how often the response is correct. These two system properties are antagonistic (improving one reduces the other) and in some situations availability may be more important, in other situations reliability.

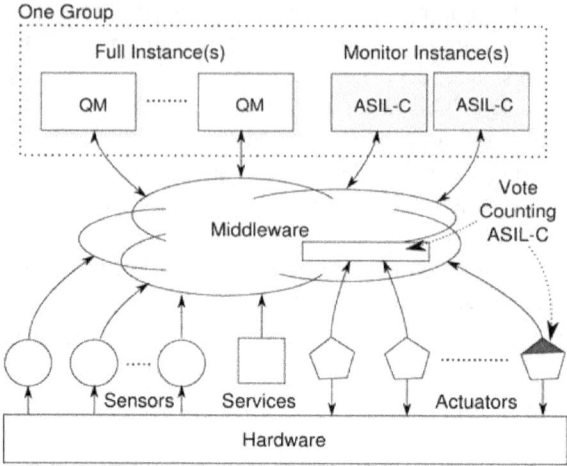

Fig. 3. Active Monitors: Configuration 1

Figure 3 illustrates how the architecture given in Figure 2 can be used to support active monitoring using diverse components within one group. The terms "QM" and "ASIL-C" are levels of requirement defined in ISO 26262 where QM imposes the fewest requirements and ASIL-D the most. In a system like this with components of different criticalities, ISO 26262 and IEC 61508 have strict rules about isolation. Some of this isolation is provided by the underlying operating system, but this architecture supports that by requiring no global clock, Lamport clocks being used instead (Lamport 1978), and by allowing components to run on different processors or different cores on the same processor.

In Figure 3, diverse instances of the computation engines have joined the same group. It is common to find that, because of their complexity, the optimal implementations cannot be certified to a high ASIL. If much simpler monitors, developed to a higher ASIL, join the group as well, they will receive the same events in the same order and will be able to respond with possibly non-optimal, but "safe", values. Given the sensor inputs, the monitor might, for example, perform a simple calculation and find that any brake pressure value between 23 and 47 is acceptably safe; being simple, it cannot calculate a precise value.

The corresponding actuator might then receive messages from the monitors with the range 23 to 47 and values from the main calculation engines of 37.896. This value would be accepted. If the monitors disagreed or the precise values were out of range, then the actuator would move to its safe state. As shown in Figure 3, the actuator itself may compare the various responses from the group members or, in the case of static policies unmodified by system state, a statically-configured component of the loosely-coupled, lock-step middleware may perform this function.

Figure 4 illustrates a more classical safety-bag pattern. For simplicity in drawing, the middleware is not shown in Figure 4, rather just the directions of flow. In this case, the optimal implementations form a group and receive the input from the sensors. Having calculated the corresponding outputs, these are passed, not to the actuators, but to a group of monitors. The monitors have seen the same sensor values as the optimal implementations and so can determine whether the output values are sane (in which case they are passed to the actuators) or not (in which case an imprecise, but safe, value can be substituted, or the system can be moved to its safe state).

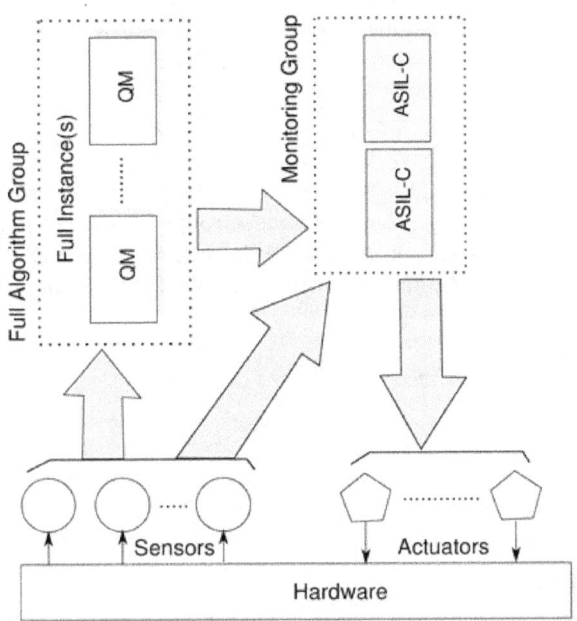

Fig. 4. Active Monitors: Configuration 2

If the middleware itself were developed to a high ASIL or SIL, then either of these patterns would substantially de-risk the independent safety review of the complete system.

6 Implementation

The descriptions given so far in this paper have been somewhat abstract. This section gives more attention to the actual implementation details and describes the exploratory development being carried out.

6.1 Client and Server Libraries

Figure 5 is a version of figure 2, redrawn to illustrate the actual implementation. The abstract cloud has been replaced by two libraries, linked respectively with the servers (computation engines) and the clients (sensors and actuators). These libraries provide an application programming interface (API) to hide all of the details of the loosely-coupled, lock-step operation from the application code.

The server API permits a server to join one or more groups, identified by strings, and then receive and respond to events. When a server joins a group that already has members, it is provided with the state by the middleware transferring it from an existing group member. Groups do not have to be defined *a priori*: the first server to join a group effectively creates that group, it being assumed that there is no initial state needed by the first member.

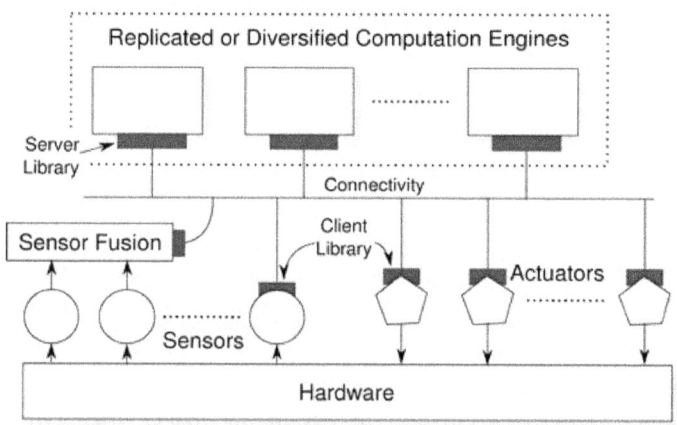

Fig. 5. Implementation

Adding a call to the function to join the group and providing a interface that allows state to be extracted may be the only modifications needed to an existing application. While the application can, for load-sharing purposes, request details of the number of members and its sequence number within a group, there is generally no need for it to be aware of its groups.

The client API is even simpler, as the client application does not belong to a group. The API permits the client to send requests to a group with a particular ordering, and to receive back one, some, or all of the responses from the group members.

Because the complexity of the virtual synchrony algorithms is hidden by the libraries, themselves developed in accordance with a high ASIL or SIL, modifying an existing program to operate within this environment is particularly simple.

6.2 Communications Channel

Figure 5 shows that some form of connectivity is required between the various instances. In a safety-critical device this would possibly be implemented as a "black channel" or, if it were between processes on the same processor, as trusted, shared memory. The loosely-coupled, lock-step architecture makes few demands on the guarantees offered by the communications path and a Data Distribution Service (DDS) could form a useful intermediate layer. At the physical layer, in a car, this connectivity might be a CAN bus or Ethernet. The first implementations have used QNX's synchronous messaging system as the communications channel.

One problem that the loosely-coupled, lock-step architecture must address is that of breaks in connectivity. As (Tushar et al. 1996) demonstrated, it is impossible for a group membership system to distinguish between a communications failure that bifurcates the group and the near simultaneous failure of several members of the group. This logical impossibility has to be addressed in a system like that of Figure 5 by an imperfect, but acceptable, pragmatic work-around: when several members of the group apparently fail simultaneously, the members of the smaller segment close down. This prevents two subgroups forming and diverging. If this is not an acceptable work-around, the protocols could, in principle, be augmented in line with (Amir and Tutu, 2002), but this has not been explored in the code currently under development.

6.3 First Implementations

The first prototype prepared made use of the SPREAD 4.4 toolkit --- see section III of (Amir et al. 2005). SPREAD was extremely easy to port to the embedded environment and the support from the SPREAD team was excellent. However, it is clear that whatever middleware is used, it needs to be certified against standards such as IEC 61508, ISO 26262 and EN 50128, and this could be difficult with preëxisting software such as SPREAD.

For this reason, QNX built a second implementation based directly on the original virtual synchrony algorithms contained in (Birman 96). Because a global clock source cannot be assumed, particularly if the components are distributed across different processors (e.g. different ECUs in a car), Lamport clocks as described in (Lamport 1978) are used to allow agreement to be reached on the order in which two events have happened. The second implementation is presently being subject to various tests comparing its behaviour with non-protected systems in the face of random errors.

To exercise the middleware code, this system has a set of simulated sensors producing values for the speed of a car and the distances to an obstacle ahead and the car behind. The simulated actuators light the brake lights (to warn the car behind) and apply brake pressure. Experiments are taking place with different numbers of replicas of the computation engine and with a diverse system as illustrated in Figure 3.

Hardware errors, both random and targeted at particular data structures, are being generated by software and, in early 2017, these experiments will be repeated in the Los Alamo nuclear facility with an ARM-processor board under bombardment from a neutron stream. This will simulate speeded-up exposure to the secondary effects of cosmic rays. During these experiments, low-level traces from the operating system's instrumented kernel will be taken while various levels of protection are evaluated. The traces will then be analysed using software generated at the University of Waterloo, Ontario.

Software heisenbugs are much easier to produce. Various computation engines have been programmed to "fail" and produce the wrong answer with preset probabilities. When this occurs, the offending engine is forced off-line, is restarted and resynchronised with the group.

More results, including those from the nuclear facility, will be available at the symposium presentation.

6.4 Moving Beyond the Initial Implementation

Based on early customer feedback, QNX is considering a production system, either as an extension of the initial implementation, or by making use of the Derecho software from Cornell University (Behrens et al. 2016). As with SPREAD, the question of certification will have to be addressed if the latter path is chosen and discussions are taking place with Cornell on the possible hardening of Derecho against random hardware and software bugs by adding data and time diversity to the trigger conditions on which Derecho relies. Building on earlier work at Cornell, this would potentially lead to a proven-correct middleware layer.

7 Summary

The loosely-coupled, lock-step architecture described in this paper applies the proven technique of virtual synchrony to a number of problems encountered when developing a safety-critical, embedded device. In summary:

1. it provides resilience against random hardware errors, an increasing problem with modern integrated hardware.
2. it provides resilience against random software errors, heisenbugs, an increasing problem with multi-threaded code running on multi-core processors.
3. when an instance fails, for example because of a random error, the system continues to operate, possibly in degraded mode, and the restart of an instance can be handled automatically by the middleware. This resilience provides a level of "fail-operational" behaviour.
4. it separates two aspects of the design, the mathematical algorithms and the system resiliency, allowing these to be undertaken (and verified) independently of each other by teams with the appropriate skills.
5. the underlying virtual synchrony algorithms can be proven correct, as recommended for ASIL C and D in table 6 of ISO 26262-8 and for SIL 2 to 4 in table A.4 of IEC 61508-3. The availability of such proof reduces the certification work as described in paragraph 7.4.7.2 of IEC 61508: 'Where the development uses formal methods, formal proofs or assertions ... tests may be reduced in scope'.
6. it allows the level of resiliency required for a particular subsystem to be set during development and to be tuned dynamically for the current environment: motorway cruising, in-city driving, parking, etc. There are numerous parameters that can be adjusted for setting the resiliency level, including: the number of replicas of the server, the number of identical responses required before a response is accepted and the level of diversity in the implementations.
7. it supports diversity among computation engines, in particular supporting the "active monitor" or "safety bag" concept that allows a safety-critical design to be cleanly partitioned ("ASIL decomposition" in the terms of ISO 26262). Another application of diversity is to increase confidence in the compiler, particularly if it is written in a language such as C++, by passing the same source code through different compilers.

At the top of this paper, three refocused questions were posed. It is useful to note how the proposed architecture addresses these:

Question 1: *Can we determine whether a hardware or software error affected the integrity of the safety-critical computation rather than just whether such an error occurred?*

The use of replicated or diversified implementations, as supported by the proposed architecture, provides significant protection against random errors, whether caused by hardware or software. Any error that occurs which does not result in different answers from the different servers, is an error that should be logged, but otherwise can be ignored.

Question 2: *Can the convenience of true, lock-step processing be exploited, while avoiding the constraints it imposes?*

The convenience of true, hardware-derived, lock-step processing arises from the simplicity of taking an implementation and replicating this without (much) further thought. This is a proper subset of the architecture described in this paper.

Question 3: *Can the design work associated with the mathematical complexity of the computation engine, be isolated from the statistical complexity of providing adequate dependability? In particular, can the required level of dependability be modified dynamically without changing the application code?*

The architecture described in this paper goes a long way towards this goal. A server implementation does not need to be aware of its group membership, although in some instances performance improvement can result if it does. Again, it may be advantageous for a client to anticipate and compare multiple answers from the group members rather than simply taking the first returned result, but this is not necessary. In particular, retrofitting existing implementations into the proposed architecture is particularly straightforward.

This means that the algorithm design and implementation can largely ignore the dependability design, the algorithm being implemented as though it will be a single module. If a safety-bag technique is to be used as shown in Figure 3, then the monitors are also implemented without regard to replication.

The dependability design then takes these components with estimated failure rates and modes, and models the overall system dependability in the various modes of operation using conventional (Bayesian) fault-tree analysis. Adding necessary resilience found to be necessary can be carried out without affecting the algorithmic code.

Acknowledgments   I would like to acknowledge Kerry Johnson's contribution to the ideas described in this paper and QNX's support in implementing the proposed architecture. All of the work builds on Ken Birman's development of the virtual synchrony protocols and I have had very helpful discussions with Ken on adapting his team's ideas for the embedded world. My colleagues Adam Mallory and Patrick Lee have provided very useful insights into the structuring of this paper.

References

Amir Y, Nita-Rotaru C, Stanton, J, Tsudik, G (2005) Secure Spread: An Integrated Architecture for Secure Group Communication. IEEE Transactions on Dependable and Secure Computing (TDSC), vol. 2, no. 3, pages 248-261, September 2005.

Amir, Y. and Tutu, C (2002) From total Order to Database Replication. Proceedings. 22nd International Conference on Distributed Computing Systems, 2002.

Armstrong K (2012) Including Electromagnetic Interference (EMI) in Functional Safety Risk Asessments. Proceedings of the 20th Safety-Critical Systems Symposium. ISBN 978-1-4471-2493-1.

Behrens J, Birman K, Jha S, Milano M, Tremel E, Bagdasaryan E, Gkountouvas T, Song W, van Renesse R (2016) Derecho: Group Communication at the Speed of Light. Paper currently under submission. September 2016.

Birman K (1996) Building Secure and Reliable Network Applications. ISBN 1-884777-29-5.

Birman K, Joseph T (1987) Exploiting Virtual Synchrony in Distributed Systems. ACM SIGOPS Operating Systems Review: Volume 21 Issue 5, November 1987.

Ferrari F, Zimmerling M, Mottola L, Thiele L (2013) Virtual Synchrony Guarantees for Cyber-Physical Systems. IEEE 32nd International Symposium on Reliable Distributed Systems (SRDS), 2013.

IEC (2010) Functional Safety of Electrical/Electronic/Programmable Electronic Safety-Related

Systems. IEC 61508, Second Edition.

ISO (2011) Road Vehicles: Functional Safety, ISO 26262, First Edition

Kim Y, Daly R, Kim J, Fallin C, Lee J, Lee D, Wilkerson C, Lai K, Mutlu O (2014) Flipping bits in memory without accessing them: An Experimental Study of DRAM Disturbance Errors. 41st International Symposium on Computer Architecture, 2014, Minneapolis, MN, USA, June 14-18, 2014

Lamport L (1978) Time, Clocks, and the Ordering of Events in a Distributed System. Communications of the ACM, volume 21, number 7, 1978

Lardner D (1834) Babbage's Calculating Engine. The Edinburgh Review, July 1834, pages 263 to 327.

Schroeder B, Pinheiro E, Weber, W (2009) DRAM Errors in the Wild: a Large-Scale Field Study. Proceedings of the Eleventh International Joint Conference on Measurement and Modeling of Computer Systems.

Tushar D C, Vassos H, Toueg S, Charron-Bost B (1996) On the Impossibility of Group Membership. Proceedings of the Fifteenth Annual ACM Symposium on Principles of Distributed Computing.

Closing the Gap – The Formally Verified Optimizing Compiler CompCert

Daniel Kästner[1], Xavier Leroy[2], Sandrine Blazy[3], Bernhard Schommer[4], Michael Schmidt[1], Christian Ferdinand[1]

1: AbsInt GmbH, Saarbrücken, Germany
2: Inria Paris-Rocquencourt, Le Chesnay, France
3: University of Rennes 1 - IRISA, Rennes, France
4: Saarland University, Saarbrücken, Germany

Abstract *CompCert is the first commercially available optimizing compiler that is formally verified, using machine-assisted mathematical proofs, to be free from miscompilation. The executable code it produces is proved to behave exactly as specified by the semantics of the source C program. CompCert's intended use is the compilation of safety-critical and mission-critical software meeting high levels of assurance. This article gives an overview of the design of CompCert and its proof concept, summarizes the resulting confidence argument, and gives an overview of relevant tool qualification strategies. We briefly summarize practical experience and give an overview of recent CompCert developments.*

1 Introduction

Modern compilers are highly complex software systems which contain many highly tuned and sophisticated algorithms; however these can contain bugs. Studies like (NULLSTONE Corporation 2007, Eide and Regehr 2008) and (Yang et al.2011) have found numerous bugs in all investigated open source and commercial compilers, including compiler crashes and miscompilation [1] issues. Although such *wrong-code* errors can be detected in the normal software testing stage it does not typically include systematic checks for them. When they occur in the field, they can be hard to isolate and to fix.

Whereas in non-critical software functional software bugs tend to have bigger impact than miscompilation errors, the importance of the latter dramatically increases in safety-critical systems. Contemporary safety standards such as DO-178B/C, ISO-26262, or IEC-61508 require identification of potential hazards and to demonstrate that the software does not violate the relevant safety goals. Many verification activities are performed at the architecture, model, or source code level, but all properties demonstrated there may not be satisfied at the executable code level when miscompilation happens. This is true, not only for source code review but also for formal, tool-assisted verification methods such as static analysers, deductive verifiers, and model checkers. Moreover, properties asserted by the operating system may be violated when its binary code contains wrong-code errors induced when compiling the OS. In consequence, miscompilation is a non-negligible risk that must be addressed by additional, difficult and costly verification activities such as more testing and more code reviews at the generated assembly code level.

The first attempts to formally prove the correctness of a compiler date back to the 1960's (McCarthy and Painter 1967). Since 2015, with the CompCert compiler, the first formally-verified optimizing C compiler is commercially available. What sets CompCert apart from any other production compiler, is that it is formally verified, using machine-assisted mathematical proofs, to be exempt from miscompilation issues. In other words, the executable code it produces is proved to behave exactly as specified by the semantics of the source C program. This level of confidence in the correctness of the compilation process is unprecedented. In particular, using the CompCert C compiler is a natural complement to applying formal verification techniques (static analysis, program proof, model checking) at the source code level: the correctness proof of CompCert C guarantees that all safety properties verified on the source code automatically hold as well for the generated executable.

Usage of CompCert offers multiple benefits. First, the cost of finding and fixing compiler bugs and shipping the patch to customers can be avoided. The testing effort required to ascertain software properties at the binary executable level can be reduced. Whereas in the past for highly critical applications compiler optimizations were often completely switched off, using optimized code now becomes feasible.

The paper is structured as follows: in section 2 we give a top-level overview of the CompCert compiler and its tool flow. Section 3 summarizes the main code generation and optimization stages of CompCert and its annotation concept. The formal CompCert proof is outlined in section 4. Section 5 presents the

---

[1] Miscompilation means that the compiler silently generates incorrect machine code from a correct source program.

Valex tool for a posteriori validation of assembly and linking phases. Section 6 describes the reference interpreter provided by CompCert for testing and semantic validation purposes. Section 7 summarizes the confidence argument for CompCert and adequate tool qualification strategies. Section 8 summarizes experimental results and practical experience obtained with the CompCert compiler.

2 CompCert Overview

An overview of the CompCert-based workflow is given in Fig. 1. The input to the compilation process is a set of C source and header files. CompCert itself focuses on the task of compilation and includes neither preprocessor, assembler, nor linker. Therefore it has to be used in combination with a legacy compiler tool chain. Since preprocessing, assembling and linking are well-established stages there are no particular tool chain requirements; as an example CompCert has successfully been used with the GCC and Diab compilers.

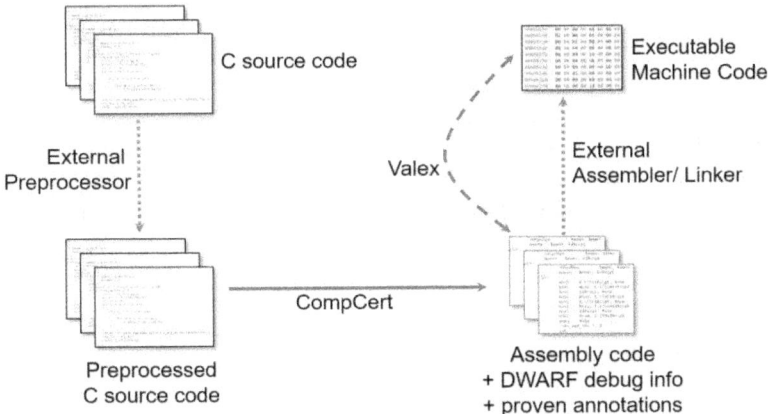

Fig. 1. CompCert Workflow

While early versions of CompCert were limited to single-file inputs, Comp-Cert now also supports separate compilation (cf. Sec. 4.3). It reads the set of preprocessed C files produced by the legacy preprocessor, performs a series of code generation and optimization steps (cf. Sec. 3.1) and produces a set of assembly files enhanced by debug information.

CompCert generates DWARF2[2] debugging information for functions and variables, including information about their type, size, alignment and location.

---

[2] cf. DWARF Debugging Standard Website (http://dwarfstd.org).

This also includes local variables so that the values of all variables can be inspected during program execution in a debugger. To this end CompCert introduces a dedicated pass which computes the live ranges of local variables and their locations throughout the live range.

The generated assembly code can contain formal CompCert annotations which can be inserted at the C code level and are carried throughout the code generation process. This way, traceability information, or semantic information to be passed to other tools can be transported to the machine code level. Since they are fully covered by the CompCert proof the information is reliable and provides proven links between the machine code and the source code level (cf. Sec. 3.2).

After assembling and linking by the legacy tool chain the final executable code is produced. To increase confidence in the assembling and linking stages CompCert provides a tool for translation validation, called Valex, which performs equivalence checks between assembly and executable code (cf. Sec. 5).

2.1 Availability

The CompCert sources can be downloaded from Inria[3] free of charge; the current state of the development can be viewed on Github[4]. In addition, a released distribution with long-term support is available from AbsInt, either as a source code download or as pre-compiled binary for Windows and Linux. The package also contains pre-configured setup files for the compiler driver to control the cooperation between the CompCert executable and the external cross compiler required for preprocessing, assembling and linking.

3 CompCert Design

Like other compilers, CompCert is structured as a pipeline of compilation passes, depicted in Fig. 2 along with the intermediate languages involved. The 20 passes bridge the gap between C source files and object code, going through 11 intermediate languages. The passes can be grouped in four successive phases, described in the following sections.

---

[3] http://compcert.inria.fr/download.html
[4] https://github.com/AbsInt/CompCert

3.1 CompCert Phases

Parsing Phase 1 performs preprocessing (using an off-the-shelf preprocessor such as that of GCC), tokenization and parsing into an ambiguous abstract syntax tree (AST), and type-checking and scope resolution, obtaining a precise, unambiguous AST and producing error and warning messages as appropriate. The parser is automatically generated from the grammar of the C language by the Menhir parser generator, along with a Coq[5] proof of correctness of the parser (Jourdan et al.2012). Optionally, some features of C that are not handled by the verified front-end are implemented by source-to-source rewriting over the AST. For example, bit-fields in structures are transformed into regular fields plus bit shifting and masking. The subset of the C language handled here is very large, including all of MISRA-C 2004 (Motor Industry Software Reliability Association 2004) and almost all of ISO C99 (ISO 1999), with the exception of variable-length arrays and unstructured, non-MISRA `switch` statements (e.g. Duff's device[6]).

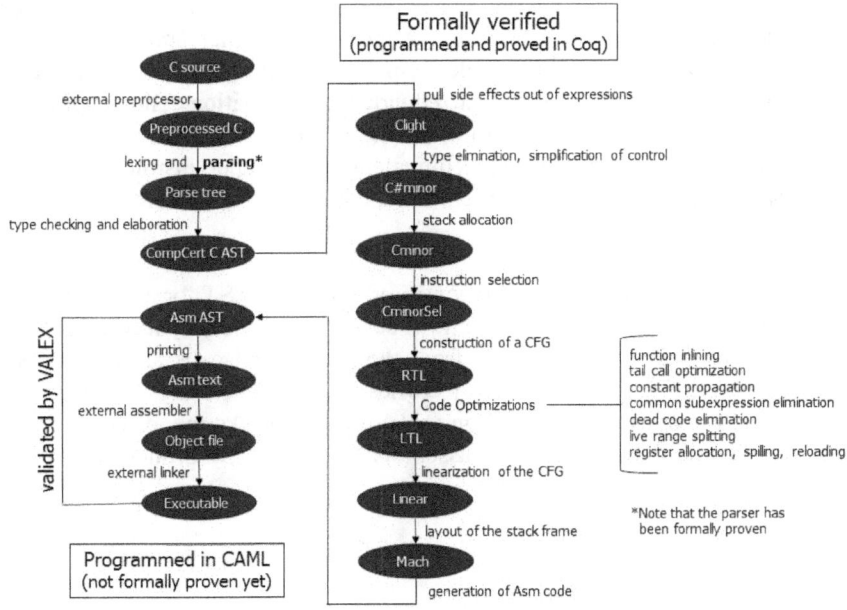

Fig. 2. CompCert Phases

---

[5] Coq is a formal proof management system. It provides a formal language to write mathematical definitions, executable algorithms and theorems together with an environment for semi-interactive development of machine-checked proofs (http://coq.inria.fr).

[6] http://en.wikipedia.org/wiki/Duff's_device

C front-end compiler The second phase first re-checks the types inferred for expressions, then determines an evaluation order among the several permitted by the C standard. This is achieved by pulling side effects (assignments, function calls) outside of expressions, turning them into independent statements. Then, local variables of scalar types whose addresses are never taken (using the & operator) are identified and turned into temporary variables; all other local variables are allocated in the stack frame. Finally, all type-dependent behaviours of C (overloaded arithmetic operators, implicit conversions, layout of data structures) are made explicit through the insertion of explicit conversions and address computations. The front-end phase outputs Cminor[7] code.

Back-end compiler The third phase comprises 12 of the passes of CompCert, including all optimizations and most dependencies on the target architecture. It bridges the gap between the output of the front-end and the assembly code by progressively refining control (from structured control to control-flow graphs to labels and jumps) and function-local data (from temporary variables to hardware registers and stack-frame slots). The most important optimization performed is register allocation, which uses the sophisticated Iterated Register Coalescing algorithm (George and Appel 1996). Other optimizations include function inlining, instruction selection, constant propagation, common subexpression elimination (CSE), and redundancy elimination. These optimizations implement several strategies to eliminate computations that are useless or redundant, or to turn them into equivalent but cheaper instruction sequences. Loop optimizations and instruction scheduling optimizations are not implemented yet.

Assembling The final phase of CompCert takes the AST for assembly language produced by the back-end, prints it in concrete assembly syntax, adds DWARF debugging information coming from the parser, and calls an off-the-shelf assembler and linker to produce object files and executable files. To improve confidence, the translation validation tool Valex re-checks the executable file produced by the linker against the assembly language AST produced by the back-end.

3.2 CompCert Annotations

CompCert provides a general mechanism to attach free-form annotations with formal semantics (plain text possibly mentioning the values of variables) to C

---

[7] Cminor is a simple, untyped intermediate language featuring both structured (if/else, loops) and unstructured control (goto).

program points. The annotations are transported throughout compilation, all the way to the generated assembly code, where variable names are expressed in terms of machine code addresses and machine registers. A simple example is the annotation:

```
__builtin_annot("x is %1 and y is %2", x, y);
```

The formal semantics of such an annotation is that of a *pro forma* "print" statement: when executed, an observable event is added to the trace of I/O operations which records the text of the annotation and the values of the argument variables x and y. In the generated machine code, annotations produce no instructions, just an assembler comment or debugging information consisting of the text of the annotations where the escapes (%1 and %2) are replaced by the actual locations (in registers or in memory) where the argument variables x and y were placed by the compiler. Hence we obtain:

```
# annotation: x is r7 and y is mem(word,r1+16)
```

if x was allocated to register r7 and y was allocated to a stack location at offset 16 from the stack pointer r1.

A first advantage of this mechanism is that it provides proven traceability: the link between machine-level storage cells and source-level variables is covered by the proof. Another typical use of annotations is to track pieces of code such as library function symbols. We can put annotations at the beginning and the end of every library function symbol, recording the values of its arguments and result variables. The semantic preservation proof therefore guarantees that symbols are entered and finished in the same order and with the same arguments and results, both in the source and generated code. This ensures in particular that the compiler did not reorder or otherwise alter the sequence of symbol invocations present in the source program -- a guarantee that cannot be obtained by observing system calls and volatile memory accesses only.

A third application of the annotation mechanism is to enable WCET tools to compute more precise worst-case execution time (WCET) bounds. Indeed, WCET tools like aiT (Ferdinand and Heckmann 2004) operate directly on the executable code, but they sometimes require programmers to provide additional information (e.g. the bound of a while loop) that cannot easily be reconstructed from the machine code alone. When using CompCert, such information can be safely extracted from annotations inserted at the source code level. A tool automating this task was developed by Airbus: it generates a machine-level annotation file usable by the aiT WCET Analyser. Compiling a whole flight control software from Airbus (about 4 MB of assembly code) with CompCert resulted in significantly improved performance in terms of WCET bounds and code size (Bedin Franca et al.2012).

4 The CompCert Proof

The CompCert front-end and back-end compilation passes are all formally proved to be free of miscompilation errors; as a consequence, so is their composition. The property that is formally verified is *semantic preservation* between the input code and output code of every pass.

4.1 Operational Semantics

To state the semantic preservation property with mathematical precision, we give formal semantics for every source, intermediate and target language, from C to assembly. These semantics associate to each program the set of all its possible behaviours. These behaviours indicate whether the program terminates (normally by exiting or abnormally by causing a run-time error such as dereferencing the null pointer) or runs forever. Behaviours also contain a trace of all observable input/output actions performed by the program, such as system calls, annotations as described in Sec. 3.2., and accesses to "volatile" memory areas that could correspond to a memory-mapped I/O device.

Technically, the semantics of the various languages are specified in small-step operational style as labelled transition systems (LTS). A LTS is a mathematical relation $currentstate \xrightarrow{trace} nextstat$ that describes one step of execution of the program and its effect on the program state. For assembly languages, program states are just the contents of processor registers and memory locations. For higher-level languages such as C, program states have a richer structure, including memory contents, an abstract program point designating the statement or expression to execute next, environments mapping variables to memory locations, as well as an abstraction of the stack of function calls.

A generic construction defines the observable behaviours from these transition systems, by iterating transitions from an initial state (the initial call to the main function): $S_0 \xrightarrow{t1} S_1 \xrightarrow{t2} \ldots$ Such sequences of transitions can go on infinitely, denoting a program that runs forever, or stop on a state S_n from which no transition is possible, denoting a terminating execution. The concatenation of the traces $t_1.t_2\ldots$ describes the I/O actions performed. Several behaviours are possible for the same program if non-determinism is involved. This can be internal non-determinism (e.g. multiple possible evaluation orders in C) or external non-determinism (e.g. reading from a memory-mapped device can produce multiple results depending on I/O behaviours).

4.2 Semantic Preservation

To a first approximation, a compiler preserves semantics if the generated code has exactly the same set of observable behaviours as the source code (same termination properties, same I/O actions). This first approximation fails to account for two important degrees of freedom left to the compiler. First, the source program can have several possible behaviours: this is the case for C, which permits several evaluation orders for expressions. A compiler is allowed to reduce this non-determinism by picking one specific evaluation order. Second, a C compiler can "optimize away" run-time errors present in the source code, replacing them by any behaviour of its choice. (This is the essence of the notion of "undefined behaviour" in the ISO C standards.) As an example consider an out-of-bounds array access:

```
int main(void)
{ int t[2];   // feasible indices are 0 and 1
  t[2] = 1;   // out of bounds
  return 0;
}
```

This is undefined behaviour according to ISO C, and a run-time error according to the formal semantics of CompCert C. The generated assembly code does not check array bounds and therefore writes 1 in a stack location. This location can be padding, in which case the compiled program terminates normally, or can contain the return address for "main", smashing the stack and causing execution to continue at address 1, with unpredictable effects. Finally, an optimizing compiler like CompCert can notice that the assignment to t[2] is useless (the t array is not used afterwards) and remove it from the generated code, causing the compiled program to terminate normally.

To address the two degrees of flexibility mentioned above, CompCert's formal verification uses the following definition of semantic preservation, viewed as a refinement over observable behaviours:

Definition 1 (Semantic Preservation): *If the compiler produces compiled code C from source code S, without reporting compile-time errors, then every observable behaviour of C is either identical to an allowed behaviour of S, or improves over such an allowed behaviour of S by replacing undefined behaviours with more defined behaviours.*

The semantic preservation property is a corollary of a stronger property, called a simulation diagram that relates the transitions that C can make with those that S can make. First, 15 such simulation diagrams are proved independently, one for each pass of the front-end and back-end compilers. Then, the diagrams are composed together, establishing semantic preservation for the whole compiler.

The proofs are very large, owing to the many passes and the many cases to be considered – too large to be carried using pencil and paper. We therefore use machine assistance in the form of the Coq proof assistant. Coq gives us means to write precise, unambiguous specifications; conduct proofs in interaction with the tool; and automatically re-check the proofs for soundness and completeness. We therefore achieve very high levels of confidence in the proof. At 100,000 lines of Coq and 6 person-years of effort, CompCert's proof is among the largest ever performed with a proof assistant.

4.3 Separate Compilation and Linking

In Definition 1, semantic preservation is stated in terms of *whole programs*: the source program S is compiled in one run of the compiler to an executable program C, whose semantics is then related to that of S. This is not how compilers are used in practice: the source program is composed of several *compilation units* residing in different files; each unit is *separately compiled* to an object file; finally, the executable is obtained by *linking* together the object files.

The implementation of CompCert supports this familiar separate compilation scenario (the -c command-line option). However, until release 2.7, the proof of semantic preservation did not cover this scenario, leaving open the possibility that CompCert could miscompile when used for separate compilation. Kang et al. (Kang et al.2016) found an example of this problem in CompCert 2.5. Consider the declaration:

```
int * const p;
```

If this is the only declaration of p in the program, it gets initialized to the default value for a pointer, namely the null pointer. Since p is const, it keeps this value through the execution of the program. The alias analysis of CompCert 2.5 was building on those observations to conclude that p has an empty points-to set. Memory accesses were then optimized under this assumption. All this is correct in a whole-program scenario, where the compiler sees that the declaration above is the only declaration of p. However, after separate compilation of the unit containing the declaration above, the unit can be linked with another unit that declares p with a non-null initialization:

```
int x; int * const p = &x;
```

In the resulting executable program, p is not the null pointer, and the optimizations performed by CompCert 2.5 can be wrong.

This particular issue was fixed in CompCert 2.6 by making the alias analysis more conservative. However, we still missed formal evidence that all CompCert optimizations are correct in the presence of separate compilation. To this end, and following the approach invented by Kang et al. (Kang et al.2016), CompCert 2.7 strengthens the statement and proof of semantic preservation to take separate compilation and linking into account:

Definition 1 (Semantic preservation with separate compilation): *Consider n source compilation units $S_1, ... S_n$ that compile separately to compiled units $C_1, ..., C_n$ without reporting compile-time errors. Assume that the source units link together without error to a whole source program $S = S_1 \oplus ... \oplus S_n$. Then, the compiled units link without errors to a whole compiled program $C = C_1 \oplus ... \oplus C_n$. Moreover, every observable behaviour of C is either identical to or improved upon an allowed behaviour of S, as in Definition 1.*

This approach of Kang et al. relies on a notion of syntactic linking between two or more compilation units, written $\oplus$ in the definition above, that extends the operations traditionally performed over object files by linkers to all the source, intermediate, and target languages of CompCert. For instance, in the case of the source CompCert C language, syntactic linking is defined by considering all declarations of identically-named global variables and functions. If the declarations are compatible, as in extern int x and int x = 1, the most precise declaration is retained (int x = 1). If two declarations are incompatible, such as int x = 1 and int x = 2, syntactic linking fails.

A limitation of this approach is that it describes only linking between compilation units written in the same language, but not linking between, say, a C source file and a hand-written assembly file. Formalizing and reasoning upon such cross-language linking and interoperability is a difficult, active research problem (Ahmed 2015, Neis et al.2015, Stewart et al.2015).

5 Translation Validation

Currently the verified part of the compilation tool chain ends at the generated assembly code. In order to bridge this gap we have developed a tool for automatic translation validation, called *Valex*, which validates the assembling and linking stages a posteriori.

Fig. 3. Translation Validation with Valex

Valex checks the correctness of the assembling and linking of a statically and fully linked executable file P_E against the internal abstract assembly representation P_A produced by CompCert from the source C program P_S. The internal abstract assembly as well as the linked executable are passed as arguments to the Valex tool. The main goal is to verify that every function defined in a C source file compiled by CompCert and not optimized away by it can be found in the linked executable and that its disassembled machine instructions match the abstract assembly code. To that end, after parsing the abstract assembly code Valex extracts the symbol table and all sections from the linked executable. Then the functions contained in the abstract assembly code are disassembled. Extraction and disassembling is done by two invocations of exec2crl, the executable reader of aiT and StackAnalyzer (Abs 2016). Apart from matching the instructions in the abstract assembly code against the instructions contained in the linked executable Valex also checks whether symbols are used consistently, whether variable size and initialization data correspond and whether variables are placed in the right sections in the executable.

Currently Valex can check linked PowerPC executables that have been produced from C source code by the CompCert C compiler using the Diab assembler and linker from Wind River Systems, or the GCC tool chain (version 4.8, together with GNU binutils 2.24).

6 The Reference Interpreter

The CompCert compiler also provides an interpreter that can execute simple C programs without compilation. More precisely, preprocessing, parsing and initial elaboration are performed; the resulting CompCert C abstract syntax is then executed by interpretation.

This is a *reference* interpreter, meaning that it implements exactly the formal semantics of CompCert C against which CompCert is proved correct. In particular, all behaviours that are undefined in the formal semantics are reported as such by the interpreter. In contrast, compiling a program that invokes undefined behaviour often causes this behaviour to become defined or be optimized away, making it impossible to observe by running the compiled executable. Likewise, the reference interpreter can explore all evaluation orders allowed by the CompCert C formal semantics, while CompCert, as a compiler, implements only one of the possible evaluation orders. This makes the reference interpreter very useful to explore the CompCert C semantics and test C code fragments for undefined behaviours.

Here is an example of such an exploration. Consider the program:

```
#include <stdio.h>
int x[2] = { 12, 34 };
int main(void)
{
    int i = 65536 * 65536 + 2; // will overflow
    printf("i = %d\n", i);
    printf("x[i] = %d\n", x[i]);
}
```

Running it with the -interp -quiet options through CompCert, we obtain:

```
i = 2
Stuck state: in function main, expression
        printf(<ptr __stringlit_2>, <loc x+8>)
Stuck subexpression: <loc x+8>
ERROR: Undefined behaviour
```

The first line (i = 2) is the output of the printf statement. It shows that the arithmetic overflow in the computation of i is not undefined behaviour in CompCert C but is defined modulo 2^{32}. The following lines diagnose an undefined behaviour, namely accessing the array x outside of its bounds. (Here, <loc x + 8> means dereferencing the memory location 8 bytes past the beginning of x.) A -trace option is available which provides a full trace of interpretation, showing every execution step taken and every intermediate state.

Technically, the reference interpreter is obtained and proved correct as follows. In Coq, a computable function step from execution states to sets of (observable events, execution states) pairs is defined, then proved sound and complete with respect to the transition relation of the CompCert C operational semantics:

$$S \overset{t}{\longrightarrow} S' \Leftrightarrow (t, S') \in step(S)$$

The step function is then extracted to OCaml[8] code and linked with hand-written code that iterates step to form transition sequences. By default, only one successor state in step(S) is deterministically chosen, but the -random

---

[8] https://ocaml.org

option causes this choice to be made randomly between all possible successors, and the -all option triggers an exhaustive breadth-first exploration instead.

There are some limitations with using CompCert in reference interpreter mode. First, the only standard C library functions supported are printf, malloc and free. Hence, the only programs that can currently be interpreted are self-contained tests with fixed inputs. Second, interpretation is 10^5 to 10^6 times slower than execution of compiled code, unless exhaustive exploration is requested, in which case interpretation is exponentially slower.

Despite these limitations, we found the reference interpreter of CompCert useful: first, to animate the formal semantics of CompCert C, helping build confidence in it; second, to test code fragments for undefined behaviours.

7 The Confidence Argument

As described in Sec. 4 all of CompCert's front-end and back-end compilation passes are formally proved to be free of miscompilation errors. These formal proofs bring strong confidence in the correctness of the front-end and back-end parts of CompCert. These parts include all optimizations – which are particularly difficult to qualify by traditional methods – and most code generation algorithms. As described in Sec. 2.1 the source code and the corresponding proofs are freely available, as is the proof assistant Coq. So the source code is amenable to manual review, the proof is reproducible for everybody and can be manually reviewed as well.

The formal proofs do not cover the following aspects:

1. The preprocessing phase
2. The correctness of the specifications used for the formal proof, i.e. the formal semantics of C and assembly,
3. Elements of the parsing phase, mostly lexing, type checking and elaboration
4. The assembly and linking phase.

Those aspects can be handled well by traditional qualification methods, i.e. via a validation suite, to complement the formal proofs. A validation suite for CompCert is currently in development and will be available from AbsInt.

Especially the parsing phase (cf. item 3) can be seen as a straightforward code generation pass which does not include any optimizations and only performs local transformations. Since the internal complexity of this stage is low, systematic testing provides good confidence. CompCert can print the result of parsing in concreteC syntax, facilitating comparison with the C source.

However, it is possible to provide additional confidence beyond the significance of the validation suite, in particular for items 1 and 4. The CompCert reference interpreter described in Sec. 6 can be used to systematically test the C semantics on which the compiler operates. Likewise, the Valex validator described in Sec. 5 provides confidence in the correctness of the assembling and linking phase. It performs translation validation of the generated code which is a widely accepted validation method (Pnueli et al.1998).

At the highest assurance levels, qualification arguments may have to be provided for the tools that produce the executable CompCert compiler from its verified sources, namely the "extraction" mechanism of Coq, which produces OCaml code from the Coq development, combined with the OCaml compiler. We are currently experimenting with an alternate execution path for CompCert that relies on Coq's built-in program execution facilities, bypassing extraction and OCaml compilation. This alternate path runs CompCert much more slowly than the normal path, but fast enough so that it can be used as a validator for selected runs of normal CompCert executions.

In summary, CompCert provides unprecedented confidence in the correctness of the compilation phase: the 'normal' level of confidence is reached by providing a validation suite, which is currently accepted best practice; the formal proofs provide much higher levels of confidence concerning the correctness of optimizations and code generation strategies; finally, the Valex translation validator provides additional confidence in the correctness of the assembling and linking stages.

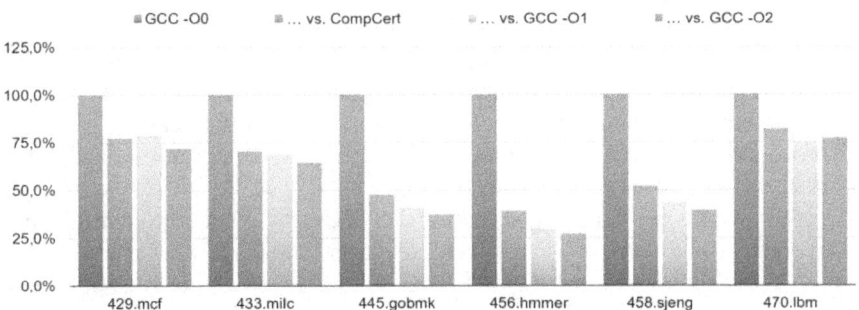

Fig. 4. Execution Time Comparison for SPEC Benchmarks on PowerPC

8 Practical Experience

CompCert targets the following three architectures: 32-bit PowerPC, ARMv6 and above, and IA32 (i.e. Intel/AMD x86 in 32-bit mode with SSE2 extension).

The result of the SPEC CPU2006[9] benchmarks measured on a PowerPC G5 are illustrated in Fig. 4 and Fig. 5, where Fig. 4 shows the execution time of the generated code and Fig. 5 its size. The experiments show that the code generated by CompCert runs about 40% faster than the code generated by GCC without optimizations, approximately 12% slower than GCC 4 at optimization level 1, and 20% slower than GCC 4 at optimization level 2.

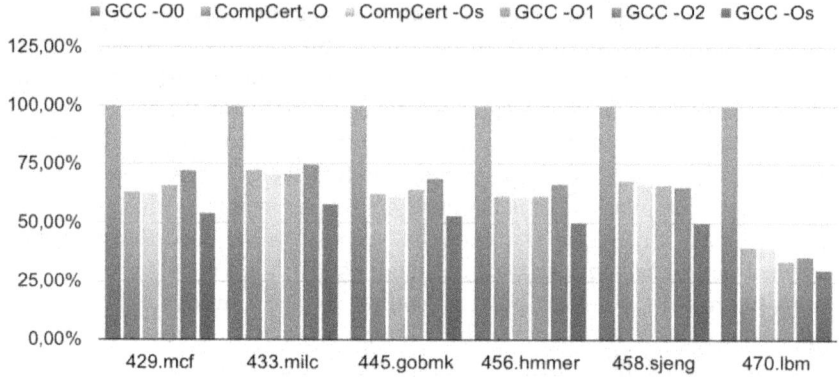

Fig. 5. Code Size Comparison for SPEC Benchmarks on PowerPC.

Regarding code size, the code generated by CompCert in modes -Os is about 1% smaller than in mode -O; it is about 40% smaller than the code generated by GCC -O0, similar to the code size of GCC -O1 (less than 1% difference), 5% smaller than GCC -O2, and about 20% larger than the code of GCC -Os.

Since SPEC is a general-purpose compiler benchmark we also considered another benchmark which is more oriented towards embedded computing. This suite comprises computational kernels from various application areas: signal processing, physical simulation, 3d graphics, text compression, and cryptography.

The results are similar than with the SPEC benchmarks: executing the code generated by CompCert -O reduces execution time to 48% compared to GCC -O0, GCC -O1 achieves 45%, and GCC -O2 42%. Hence the code generated by CompCert runs about 52% faster than the code generated by GCC without optimizations, approximately 11% slower than GCC 4 at optimization level 1, and 23% slower than GCC 4 at optimization level 2.

Regarding code size, the code generated by CompCert in modes -Os here is less than 1% smaller than in mode -O; it is about 17% smaller than the code generated by GCC -O0, 4% larger than the code size of GCC -O1, similar to GCC-O2 (difference smaller than 1%), and about 5% larger than the code of GCC -Os.

---

[9] http://www.spec.org/cpu2006

In general, due to lack of aggressive loop optimizations, performance is lower on HPC[10] codes involving dense matrix computations. This is also the main reason for the difference in execution time between CompCert and GCC with high optimization levels.

The performance of CompCert on ARM is similar to the PowerPC architecture. On IA32, due to its paucity of registers and its specific calling conventions, CompCert is approximately 20% slower than GCC 4 at optimization level 1 on the benchmark suite.

9 Conclusion

CompCert is a formally verified optimizing C compiler: the executable code it produces is proved to behave exactly as specified by the semantics of the source C program. Experimental studies and practical experience demonstrate that it generates efficient and compact code. Further requirements for industrial application, notably the availability of debug information, and support for Linux and Windows platforms have been established. Explicit traceability mechanisms enable a seamless mapping from source code properties to properties of the executable object code. We have summarized the confidence argument for CompCert, which makes it uniquely well-suited for highly critical applications.

References

[Abs 2016] AbsInt GmbH, Saarbrücken, Germany. *AbsInt Advanced Analyzer for PowerPC*, April 2016. User Documentation.

[Ahmed 2015] Amal Ahmed. Verified compilers for a multi-language world. In *SNAPL 2015: 1st Summit on Advances in Programming Languages*, volume 32 of *LIPIcs*, pages 15–31. Schloss Dagstuhl - Leibniz-Zentrum fuer Informatik, 2015.

[Bedin Franca *et al.*2012] Ricardo Bedin Franca, Sandrine Blazy, Denis Favre-Felix, Xavier Leroy, Marc Pantel, and Jean Souyris. Formally verified optimizing compilation in ACG-based flight control software. In *ERTS 2012: Embedded Real Time Software and Systems*, 2012.

[Eide and Regehr 2008] Eric Eide and John Regehr. Volatiles are miscompiled, and what to do about it. In *EMSOFT '08*, pages 255–264. ACM, 2008.

[Ferdinand and Heckmann 2004] Christian Ferdinand and Reinhold Heckmann. aiT: Worst-Case Execution Time Prediction by Static Programm Analysis. In René Jacquart, editor, *Building the Information Society. IFIP 18th World Computer Congress*, pages 377–384. Kluwer, 2004.

[George and Appel 1996] Lal George and Andrew W. Appel. Iterated register coalescing. *ACM Trans. Prog. Lang. Syst.*, 18(3):300–324, 1996.

[ISO 1999] ISO. International standard ISO/IEC 9899:1999, Programming languages – C, 1999.

---

[10] High-Performance Computing

[Jourdan *et al.*2012] Jacques-Henri Jourdan, FranÃ§ois Pottier, and Xavier Leroy. Validating LR(1) parsers. In *ESOP 2012: 21st European Symposium on Programming*, volume 7211 of *LNCS*, pages 397–416. Springer, 2012.

[Kang *et al.*2016] Jeehoon Kang, Yoonseung Kim, Chung-Kil Hur, Derek Dreyer, and Viktor Vafeiadis. Lightweight verification of separate compilation. In *POPL 2016: 43rd symposium on Principles of Programming Languages*, pages 178–190. ACM, 2016.

[McCarthy and Painter 1967] John McCarthy and James Painter. Correctness of a compiler for arithmetic expressions. In *Mathematical Aspects of Computer Science*, volume 19, pages 33–41, 1967.

[Motor Industry Software Reliability Association 2004] Motor Industry Software Reliability Association. MISRA-C: 2004 – Guidelines for the use of the C language in critical systems, 2004.

[Neis *et al.*2015] Georg Neis, Chung-Kil Hur, Jan-Oliver Kaiser, Craig McLaughlin, Derek Dreyer, and Viktor Vafeiadis. Pilsner: a compositionally verified compiler for a higher-order imperative language. In *ICFP 2015: 20th International Conference on Functional Programming*, pages 166–178. ACM, 2015.

[NULLSTONE Corporation 2007] NULLSTONE Corporation. NULLSTONE for C. http://www.nullstone.com/htmls/ns-c.htm, 2007.

[Pnueli *et al.*1998] Amir Pnueli, Michael Siegel, and Eli Singerman. Translation validation. In *TACAS'98: Tools and Algorithms for Construction and Analysis of Systems*, volume 1384 of *LNCS*, pages 151–166. Springer, 1998.

[Stewart *et al.*2015] Gordon Stewart, Lennart Beringer, Santiago Cuellar, and Andrew W. Appel. Compositional CompCert. In *POPL 2015: 42rd symposium on Principles of Programming Languages*, pages 275–287. ACM, 2015.

[Yang *et al.*2011] Xuejun Yang, Yang Chen, Eric Eide, and John Regehr. Finding and understanding bugs in C compilers. In *PLDI '11*, pages 283–294. ACM, 2011.

Using Formal Proof to meet Executable Object Code and Coverage Objectives in DO-333

N J Tudor, C M O'Halloran

D-RisQ Ltd

Malvern, UK

Abstract   *This paper describes a technology proof of concept for the automated verification of Executable Object Code (EOC) using Formal Methods. The project called FEVER was carried out by D-RisQ Ltd in conjunction with Lemma1, both small companies with expertise on various facets of formal verification. The target use of FEVER is within embedded systems that will be require safety certification, specifically targeting unmanned systems. The rationale being that if a route to certification using formal development from requirements to EOC could be shown, then the perceived untenable amount of testing for such systems could be drastically reduced. We chose to use the aerospace software guidance DO-178C, the Formal Methods Supplement DO-333 and Tool Qualification DO-330 as these set out the relevant Objectives for a formal development. The work required the use of a formalised version of the ARM 7 Instruction Set Architecture. This was captured in a language called HOL and was based upon work carried out by Cambridge University. The project used source code for a simple decision making system written in C in order to develop the technology.*

1. Introduction

1.1.   Use of DO-333

During 2005, SC205/WG71 was set up with the aim of refreshing the de facto software standard for civil aerospace in order to produce RTCA DO-

178C/EUROCAE ED-12C (DO-178C, 2011). As part of the activities, a specific examination of the need for technology specific supplements was undertaken. This included 'Model Based Design and Verification' and 'Object Oriented Technologies' as well as 'Formal Methods'; this last supplement is referred to as DO-333 (DO-333, 2011). DO-333 establishes the use of formal proof to replace testing based Objectives. The supplement highlights those Objectives throughout the software development life cycle from requirements to Executable Object Code (EOC) and coverage that formal methods may be used to meet instead of, or in addition to, review and test. This paper is focussed on a technology that could show how it might be possible to meet EOC Objectives as outlined in Annex A to DO-333, Table FM.A-6 and hence, in context of a formal development, also meet coverage Objectives as outlined in Table FM.A-7.

1.2.   The Project

The project undertaken by D-RisQ was called Formal Executable Object Code Verification (FEVER) and was focussed on a proof of concept study to prove C code developed from a Simulink diagram and compiled to ARM 7 Instruction Set Architecture (ISA). The example chosen for the project was a simple autonomous system. This example was specifically selected in order to try to tackle one of the major problems in autonomous systems; i.e. that the current approaches to software certification are test oriented. With a test oriented approach to verification, there is perceived to be an extremely large test task, potentially too large to be practically viable. Hence if it can be automatically proven that the EOC satisfies its requirements with a minimised use of test, then there might be a credible route to the certification of such systems.

2.   Approach

2.1.   At Least as Good as DO-178B

When SC-205/WG-71 was formed the principle of showing that the new documentation was "at least as good as DO-178B" was adopted. This meant that there was no need to "raise the bar", but that it also must not be lowered. Any approach to meeting existing objectives had to be shown to be at least equivalent. This is nothing new as DO-178B is defined as an 'Acceptable Means of Compliance' and an Applicant has to show how the process they adopt will produce software that will meet the Objectives. Within Section 12.3 of DO-178C, there is a section on Alternative Methods in which it is stated "An alternative method should be shown to satisfy the objectives of this document or the applicable supplement". This gives the opportunity to propose FEVER, noting that we should not 'lower the bar'.

2.2.   *Example System*

The original aim had been to use an open Stateflow (The Mathworks, 2016) model of an autonomous system, but for the purposes of development of the EOC verification technology, the Source code produced was too large and complex. Consequently, a much simpler set of C code was selected from which automation could be developed. This enabled a more focussed approach to covering source code and EOC constructs and to understand how to cope with optimisations. The aim for D-RisQ is to develop a complete formal process from requirements to EOC which will enable claims to be made for certification.

3.   Proof vs. Test

3.1.   *Do we need to test?*

We are proposing the use of formal verification to replace test using DO-333. Section 6.7 sets out what needs to be shown for EOC verification. Note that the opening line of this Section states "Verification of the Executable Object Code is primarily performed by testing". This line was inserted to ensure that politically in the WG71/SC205 group the approach to using FM would have a fall back to test. There have been cases of applicants trying to avoid undertaking the expensive testing required to meet coverage analysis Objectives. It was therefore highly desirable to ensure that coverage in some form had to be undertaken. Indeed, there were significant voices that would not have accepted the possibility of avoiding test to meet these Objectives. However, the fundamentals of the mathematics behind computing, point to the possibility that the essentially binary nature of software could be proven to implement requirements instead of undertaking tests. There may be other reasons for testing the EOC within the context of DO-178C, and these will be examined later in the paper.

3.2.   *Verification Objectives Overview*

3.2.1.   *DO-178C Objectives*

The basis for any certification using a Supplement will always have to use DO-178C as the Supplements, including DO-333, do not cover every topic and cross refers to the core text in DO-178C. There are various areas such as configuration management, planning and certification liaison that may be varied because of the use of FM but nothing that fundamentally changes the Objectives. Verification is a different matter as this is what formal methods are primarily focussed upon. The diagram in Fig outlines the software certification artefacts and where verification is needed to meet Objectives within software development. Figure 1 does not show the Additional FM Objectives (see Figure 1) necessary in Tables FM.A3, 4 and 5.

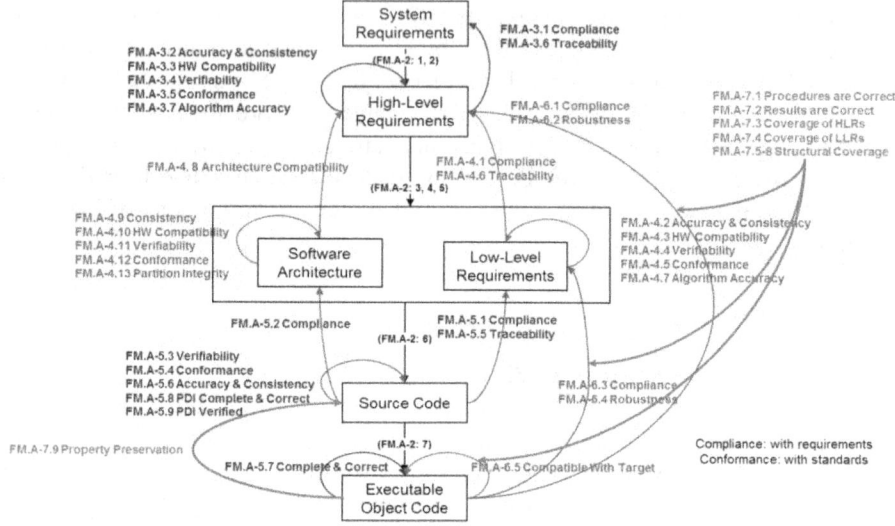

Fig. 1. Verification Objectives Tables A-2 to A-7 to DO-333

3.2.2. DO-333 vs DO-178C Objectives

DO-333 has equivalent Objectives to DO-178C. These are used to override DO-178C when certification credit is being sought through the use of FM. Annex A to DO-333 is a set of tables which are a short summary of where evidence of certification credit has to be made against a particular software Level A-D (white dot ◯) and identifies where independence is required (black dot ●). While developing DO-333, it became apparent that there was a need for some extra Objectives in certain parts of the life cycle that would ensure that there would be a correct use of FM. These extra Objectives are necessary to ensure that the verification Objectives that are claimed to be met through the use of a particular technique can be relied upon. As shown in Table, is an extract from Table FM.A3 which shows the four extra Objectives (FM8-11) and how they are met by DO-333 (paragraph number references preceded by FM) and the required level of independence.

Table 1. Extract from DO-333, Table FM.A3

Objective			Activity Ref	Applicability by Software Level			
	Description	Ref	Ref	A	B	C	D
FM 8	Formal analysis cases and procedures are correct.	FM 6.3.6.a FM 6.3.6.b	FM6.3.6	●	○	○	
FM 9	Formal analysis results are correct and discrepancies explained.	FM 6.3.6.c	FM 6.3.6	●	○	○	
FM 10	Requirements formalization is correct.	FM 6.3.i	FM6.3.i	●	●	○	
FM 11	Formal method is correctly defined, justified, and appropriate.	FM.6.2.1	FM.6.2.1a FM.6.2.1b FM.6.2.1c	●	○	○	○

These four extra Objectives are the same for other parts of the life cycle. The table at Table shows the extra formal methods Objectives and where in the life cycle they are required. It can be observed that there are no specific Objectives for EOC. This is because at the time of writing DO-333, it was not thought possible to gain acceptance of proof instead of test to meet any of the Table FM.A-6 Objectives. However, any approach to the use of FM to meet the EOC Objectives would be expected to have provided similar evidence. Section FM.6.7 outlines the Objectives that have to be met for EOC verification and the criteria for doing so.

Table 2. The Extra FM Objectives in the Life Cycle

Objective		Reference Section	
		Objective	Activity
FMA-3.8, HLR FMA-4.14, LLR FMA-5.10, Source Code	Formal analysis cases and procedures are correct	FM.6.3.6.a FM.6.3.6.b	FM.6.3.6
FMA-3.9, HLR FMA-4.15, LLR FMA-5.11, Source Code	Formal analysis results are correct	FM.6.3.6.c	FM.6.3.6
FMA-3.10, HLR FMA-4.16, LLR FMA-5.12, Source Code	Requirements formalisation is correct	FM.6.3.i	FM.6.3.i

FMA-3.11, HLR FMA-4.17, LLR FMA-5.13, Source Code FMA-7-10, Coverage	Formal Method is correctly defined, justified and appropriate	FM.6.2.1	FM.6.2.1.a FM.6.2.1.b FM.6.2.1.c

HLR – High level Requirements, LLR – Low Level Requirements

4. Approach to FEVER

4.1.   Foundation Research

The FEVER project is based upon research carried out by Cambridge University (PFVOC, 2013) that in turn was based upon a formalisation in Cambridge's Higher Order Logic (HOL) (HOL, 2016) of the ARM 7 Instruction Set Architecture (ISA) (Fox, 2012) (Myreen, 2008) (Myreen, 2012). D-RisQ also had a set of semantics for a subset of the C90 programming language expressed in Z (O'Halloran, 2007).   D-RisQ has also previously exploited the ProofPower Z theorem prover (Arthan, 2005), (Lemma1, 2016) in conjunction with Lemma 1 to prove the correctness of source code automatically generated from Simulink (O'Halloran, 2013); this tool is called CLawZ.   Lemma 1 also had the ProofPower HOL version of the tool which meant that there were good links between the companies and the technologies. Therefore the foundation technologies and formal methods appear to be a good starting point for EOC verification.

4.2.   Extra Objectives

As discussed in 3.2.2 there are no extra Objectives to be met for EOC, specifically Table A-6. However, should a formal technique be adopted for EOC verification then it stands to reason that the same extra Objectives should be met. It would certainly help with arguments that need to be made concerning robustness of the code, claiming equivalence to structural coverage and, in particular, absence of dead code (Cavalcanti, 2013). If proof of EOC is proposed, then applying the four objectives is logical to show some "equivalence".

4.3.   Definition the Formal Method

A formal language has to be shown to have precise, unambiguous, mathematically defined syntax and semantics that can be reduced to simple mechanistic steps, i.e. in this context formal means that it can be done by a machine. It has to be shown that the foundation of the language is based on a sound and, if possible, complete mathematical theory. If sound then nothing that is untrue can arise from its use. If it is complete then everything that is true can be demonstrated within the system. Propositional logic is an example of a system that is sound and complete but at the cost of limited expressiveness. Other mathematical systems trade expressiveness for completeness, but seldom for soundness. Programming language subsets used for mathematical verification will have a defined syntax along with the formal semantics to give a sound language for demonstrating correctness. There should be an accompanying justification as to why the use of the language is sound and hence formal models can be constructed of the software artefact that can be relied upon to be sound. This means that analysis never asserts that a property is true when it is not true.

One of the formal languages that we use is Higher Order Logic with a machine implementation called the HOL4 theorem prover (HOL, 2016). The HOL4 interactive theorem prover is a proof assistant for higher-order logic: a programming environment in which theorems can be proved and proof tools implemented. Built-in decision procedures and theorem provers can automatically establish many simple theorems. HOL is particularly suitable as a platform for implementing combinations of deduction, execution and property checking. With the support of DSTO Australia, Dr Andrew Pitts was commissioned to validate HOL's definitional principles, i.e. the basis of building theories starting from HOL's handful of axioms and inference rules. He produced mathematical proofs that they could not introduce inconsistency.

4.4.   *Justification and Appropriateness of the Formal Method*

Justifying the appropriateness of the selected method parallels the approach for the adoption of a programming language. The limitations of the technique should also be made known and any assumptions relating to its use should be described and justified. This can be the manner in which aspects of the analysis are constrained (e.g., the treatment of real numbers), or could be about the environment in which the software will eventually run such as assumptions relating to the target computer or about data range limits. HOL is characterised by a number of key features: a simple core logic, support for a growing diversity of proof styles and a large body of user-supplied theories and proof tools that are, or could be, proven with respect to the core logic. For a proof assistant, HOL has a large user community. There have been nine HOL Users Workshops, which have gradually evolved from informal get-togethers to elaborate international meetings with refereed papers. In short the theory and implementations are mature and stable with a significant user community that have validated its use.

4.5.   *Use of the Formal Method*

Having justified the formal methods, checking that its use is correct is the next step. We have to justify why the [requirements] formalisation is correct from

Section 6.3.i of DO-333. i.e. justifying why the translation to the formal description is correct. There are a number of facets to this stage of claim and detail is expected to be described in the Tool Qualification material that will be needed to meet DO-330 (DO-330, 2011).

4.5.1. *Instruction Set Architecture (ISA)*

The approach by Cambridge was to check the behaviour of a specific instruction from the ISA against the behaviour of the formal version in HOL which builds the reference for each instruction. The instruction runs on the processor (in this case using the ARM9 processor running the ARMv7 ISA) and has to be correct, otherwise we have a real problem because either the ISA is incorrect or the build of the processor is not in accordance with specification. Therefore the reference can be relied upon. If the HOL expression exhibits the same behaviour it must be a correct interpretation. Bearing in mind that this is at the level of instructions such as 'fetch', 'put', and the like, it is easy to ensure that the HOL behaviour is exactly the same as the ISA defines it. All we then effectively need to do is to pick up the formalisation for each instruction from the actual EOC.

4.5.2. *Extracting Verification Conditions*

An existing D-RisQ tool, called SPG, has been applied to a series of simple C functions in order to validate it. See Figure 2, SPG generates a formal specification. This is used as a conjecture that it was the original formal specification from which the code had been refined. The correctness (or soundness of the specification) is established by generating a refinement conjecture as though it had been incrementally refined top-down from the formal specification to the code and then proving this conjecture.

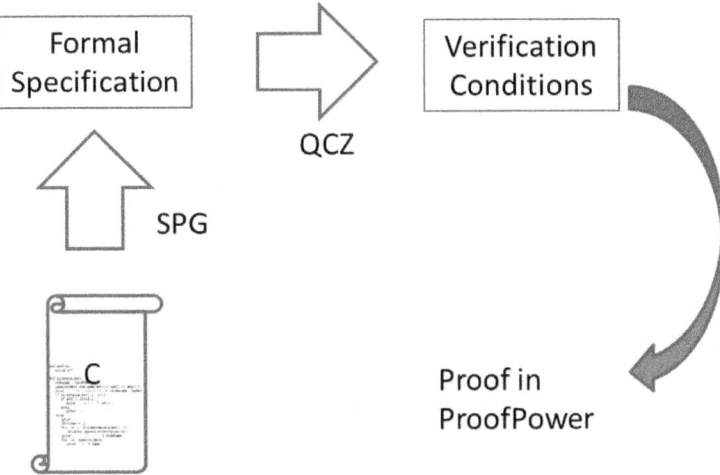

Fig. 2. Overview of Verification Condition generation and Proof

In order to do this, firstly, another tool called QCZ, processes the generated refinement to in turn generate verification conditions. If the verification conditions can be proven by a theorem prover (in this case ProofPower) then the formal specification is correct. In the case of a specification generated from C code by SPG this means that for every input, the C code will terminate in a state specified by a predicate generated by SPG. This does not mean that the C code reflects the intent of the originator of the code, it only means that the specification generated by SPG is accurate. For the task of verifying binary code against a formal specification of the C code, the accuracy of the formal specification with respect to that C code is crucial. This process exploits Carroll Morgan's Theory of Refinement (Carroll, 1990) and has previously been extensively used to verify the Typhoon flight control system code.

4.5.3.   Use of Z

In parallel with the development of HOL90, International Computers Limited (ICL) created their own commercial version of HOL, now called ProofPower and now supported by Lemma1 Ltd. This was targeted at in-house and commercial use, especially for security applications. ProofPower supports exactly the same logic as the other HOL systems, but has different proof infrastructure that evolved to meet the needs of the targeted applications (e.g. customised theorem-proving support for the Z notation and a verification condition generator for two common languages called QCZ and DAZ that support C and Ada respectively).

The Z notation (Davies, 1996) is a mature formal specification language used for describing and modelling computing systems. It is targeted at the clear specification of computer programs and computer-based systems in general and is used worldwide for this purpose. Z is based on the standard mathematical notation used in axiomatic set theory, lambda calculus, and first-order predicate logic. All expressions in Z notation are typed, thereby avoiding some of the paradoxes of naive set theory. Z contains a standardized library (called the mathematical toolkit) of commonly used mathematical functions and predicates.

The specification of the C source code generated by SPG is represented in Z and the relation between the Z specification and the C code is expressed in the wide spectrum language QCZ (mentioned earlier). The QCZ tool processes this representation as a refinement relation and generated the verification conditions discussed in 4.5.2.

4.6.   Independence

There is a requirement to have independence for the Objectives, especially at the higher software Levels. This is achieved though the independent semantics brought to the interpretation of the ISA to HOL and likewise for the Source code. The justification for the correctness of the interpretation and the appropriateness of the technique is described in 4.5.

4.7.   Tackling the Objectives

4.7.1. Table A-6 Objectives

As discussed in 3.1, there was not the political will to allow an automatic replacement of testing as the primary verification method for Executable Object Code. This paper outlines a possible technique for replacement of some test through the use of proof. It is anticipated that tests in the target hardware will be required in order to cover some of the effects of hardware on the software. There are a number of options available for EOC verification potentially using a combination of techniques.

4.7.2. Options for Verification of EOC

The diagram in figure 3 describes the possible verification paths for EOC as described within DO-330, Section 6.7. It shows that for test related activities there are direct compliance and robustness Objectives against both Low Level Requirements (LLR) and High Level Requirements (HLR). For a formal approach it shows that there is an alternative approach which sets an Objective for property preservation between Source and EOC and then Objectives for Source code against LLR and HLR.

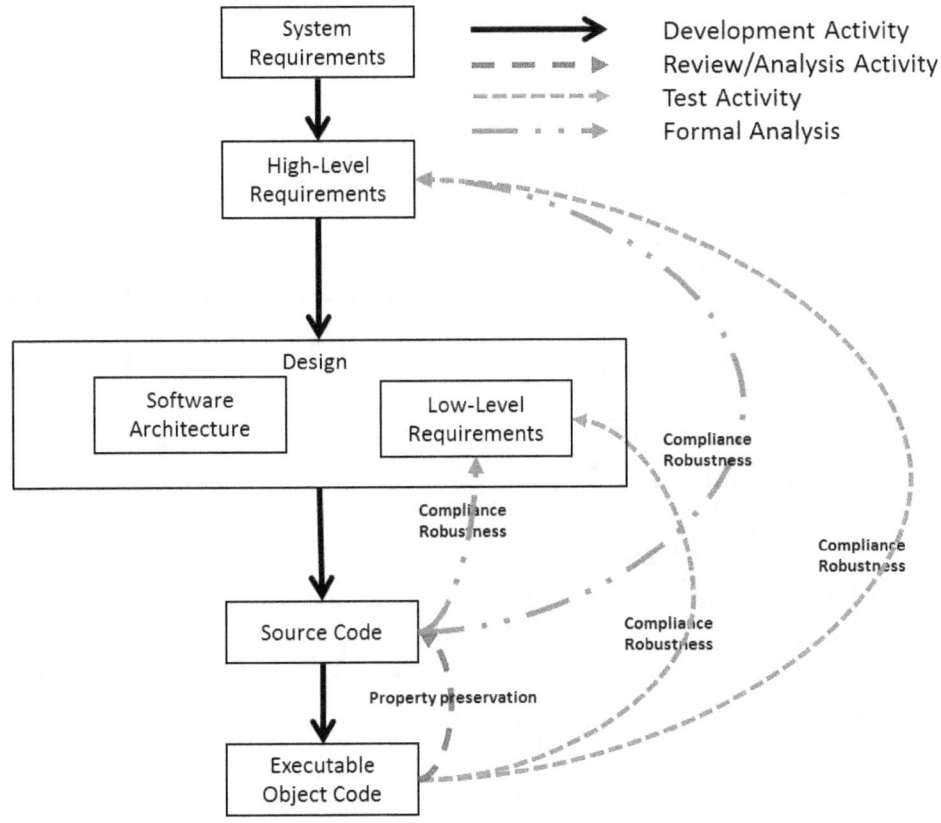

Fig. 3. Possible Verification Paths for EOC with respect to
Source Code and Requirements

4.7.3. *Objective Compliance – EOC*

Table below is an extract from Table FM.A-6 which shows the paragraph references for the specific Objectives. Note that for Objective FM.A-6-5 there are cross references to DO-178C paragraph references. Each Objective has to be met and claims made based upon the verification paths outlined in figure 3.

Table 3. Extract From Table FM.A-6, DO-333

	Objective		Activity	Applicability			
	Description	Ref	Ref	A	B	C	D
1	Executable Object Code complies with high-level requirements	FM.6.7.a FM.6.7.c	FM.6.7 FM.6.5	○	○	○	○
2	Executable Object Code is robust with high-level requirements	FM.6.7.b FM.6.7.c	FM.6.7 FM.6.5	○	○	○	○
3	Executable Object Code complies with low-level requirements	FM.6.7.d FM.6.7.c	FM.6.7 FM.6.5	●	●	○	○
4	Executable Object Code is robust with low-level requirements	FM.6.7.1b	FM.6.7 FM.6.5	●	○	○	
5	Executable Object Code is compatible with target computer	6.4.e FM.6.7.e	6.4.1.a 6.4.3.a FM.6.7	○	○	○	○

4.7.4. Alternative Approach - FEVER

The semantic gap between a formal representation of Source Code to a formal representation of HLR is somewhat hard to bridge; the formal semantic gap between Source Code and LLR is much easier to bridge. However, if it could be shown that the LLR are proven to be correct with respect to the HLR, then we can use the commutative function of basic mathematics to argue an alternative approach (see figure 4) and achieve the same level of confidence. This approach should also meet the independence requirements for each activity.

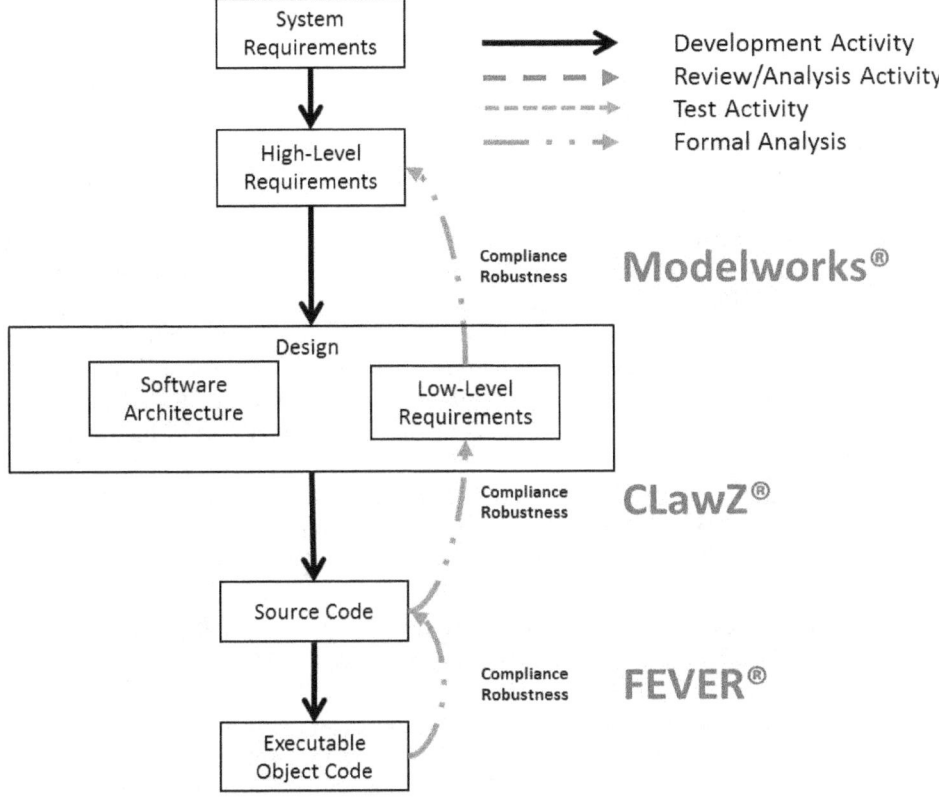

Fig. 4. Alternative Verification Paths for EOC using Commutation of Proof

We have replaced the 'property preservation' approach that was based upon a review/analysis activity with an explicit proof activity (which is actually an analysis activity). Although it is not explicitly required in DO-333, there is a need to meet the 4 Extra Objectives first (as outlined in 3.2.2) and then we can claim that the EOC satisfies its specification and hence the requirements.

4.7.5. *Meeting the Aim of EOC Verification through a Formal Approach*

The aim of EOC verification activities is to verify the correct:

- operation of the EOC in the target computer environment;
- inter-relationships between software requirements and components;
- implementation of the software requirements and software components within the software architecture, and

- implementation of software Low Level Requirements.

The above list is in order of precedence and if we establish correct operation in the target environment, then we have established all of these points. Our approach, using the ISA, therefore gives the basis for the establishment of correct operation in the target. It is also necessary to show that the formal analysis cases are based primarily on the software requirements and that they are developed to not only verify correct functionality, but also establish conditions that reveal potential errors. This is where the selection of the formal technique has to be justified as a focus on the sources of potential error is needed. For object code, the error source is the introduction of functionality by the compiler which would not be a correct implementation of the source code and hence the requirements. The FEVER toolset underpinned by the use of ProofPower provides this guarantee. This allows claims for credit against all of the Table A-6 Objectives, but will leave some small amount of testing related to hardware interactions with the software. While not all aspects of the verification were carried out within FEVER, the context and the limits of claims that could be made against each EOC Objective for Table FM.A-6 is below.

4.7.5.1.   *Compliance & Robustness with the High Level Requirements (HLR) Objectives FM.A-6-1 & 2*

These two Objectives require demonstration of the ability of the software to respond to both normal and abnormal inputs and conditions. The Objectives focus on the external interface to the software, so examination of input ranges, system initialisation, time related functions, state transition and logic equations. From FM.6.7.f, we can show both of these Objectives can be met if the HLR have been formally expressed and we have a formal model of the executable. For the FEVER project we did not develop formal high level requirements. However, we have developed a technology to enable this and it can be used to formally verify that a design implements requirements expressed in structured English.

4.7.5.2.   *Compliance & Robustness with the Low Level Requirements (LLR) Objectives FM.A-6-3 & 4*

Again, the two objectives require demonstration of the ability of the software to respond to both normal and abnormal inputs and conditions. The focus for LLR is very similar, but now checking that the understanding of the HLR has been correctly transitioned to the LLR. It will also check derived requirements. We did not develop a meaningful set of LLRs for the FEVER project, but as outlined in 4.7.5.1, we do have a technology to express designs in a model and to prove source code against it. Hence we should be able to make claims to meet these Objectives using FM.6.7.f and consequently have no need to undertake other forms of verification to meet this Objective.

4.7.5.3. *Compatibility with the Target Computer Objective FM.A-6-5*

Verification to meet this objective concentrates on error sources associated with the software operating within the target computer environment, and on the high-level functionality. The types of errors that can occur, and hence should be verified to be absent, include incorrect interrupt handling, execution time, hardware response to transients, resource exhaustion, stack overflow and violations of software partitioning. Achieving at least some of this Objective is straightforward as we are using a formal specification of the actual ISA for the processor; this is specifically required within DO-333 at Section 6.7.e (1) and (2). However, there will remain some testing in order to verify aspects of the software operating in the hardware, but all functional behaviour, such as stack overflow, can be verified using FEVER. This is because the formal model of the ISA will not cover, for example, how the hardware responds to transients and hence what the effect will be on the software.

Table 4. Summary Claims

	Objective		Activity	Applicability by Software Level				
	Description	Ref	Ref	A	B	C	D	
1	Executable Object Code complies with high-level requirements.	FM.6.7.a FM.6.7.c	FM.6.7 FM.6.5	O	O	O	O	F
2	Executable Object Code is robust with high-level requirements.	FM.6.7.b FM.6.7.c	FM.6.7 FM.6.5	O	O	O	O	F
3	Executable Object Code complies with low-level requirements	FM.6.7.d FM.6.7.c	FM.6.7 FM.6.5	●	●	O		F
4	Executable Object Code is robust with low-level requirements	FM.6.7.b FM.6.7.c	FM.6.7 FM.6.5	●	O	O		F
5	Executable Object Code is compatible with target computer	6.4 FM.6.7.e	6.4.1.a 6.4.3.a FM.6.7	O	O	O	O	P

4.8.   *Claims Objectives FM.A-7*

The final verification Objectives are contained in Table FM.A-7. For the use of DO-333, the 9 DO-178C Objectives are completely replaced with 7 specifically applicable to the use of formal methods (DO333, 2011). The aim of this set of Objectives is to ensure that every aspect of the software and requirements has been appropriately verified. This is commonly referred to as 'coverage' and is typically an analysis activity, meaning that results of verification are measured in some way in order to give an acceptable level of confidence that the EOC will function as expected.

Coverage analysis is a 2 step process, coverage of requirements (both High Level Requirements and Low Level Requirements) and code structure. There firstly has to be a claim that the verification of all the requirements is complete. Structural coverage is a notion tied to testing and therefore some equivalence to test is required rather than an explicitly testing related activity. Structural coverage is designed to show that there is no unintended functionality and to establish a level of quality for the test data. They are intended with increasing rigour to detect:

- Right requirements, incorrect wrong code;
- Wrong requirements vs code;
- Wrong test cases for the requirements;
- Dead code

The curious Objective FM.5-8 in (DO333, 2011) means that the 3 Objectives in DO-178C Table A-7 statement, decision and MC/DC[1] coverage, as well as the data and control coupling Objective can be met through the appropriate use of formal methods. Use of a fully justified formal proof can show the first 2 of the list above as formal methods can be complete and exhaustive if the requirements are also formally expressed (see 4.7.1). Of course this assumes that the requirement set is complete. Formal Methods can also force the avoidance of the wrong proof for the requirements used as a basis for the project. This is because the proofs are independent of the requirements through to code and the processes, outlined in 4.7.3, describe a context for the use of FEVER. Dead code normally has to be detected by a review or the use of some comparison technology. However, within FEVER, it is also possible to prove the absence of dead code. Other forms of 'extraneous' code, e.g. de-activated code.

5. Tool Qualification

Naturally, for any non-trivial application, some form of tool support will be required and it is highly likely that these will need to be qualified. Broadly speaking, formal methods provide verification so qualification is likely to be as a Verification Tool as it was known under DO-178B, or under DO-178C and the new Tool Qualification document (DO-330), a Category 3 Tool. This means achieving Tool Qualification Level 5 (TQL 5).

---

[1] Modified Condition/Decision Coverage – an analysis of testing coverage of code structure used as a 'stop criteria' for testing in DO-178C for Level A code.

6. Conclusion

The FEVER project is aimed solely at the verification of EOC against the C code from which it was compiled. It is only currently aimed at the ARM 9 ISA, but the principles, having been established could be extended to other ISA. Proof failures would mean that the compilation was not correct with respect to the source code. Recognising that there is not a similar set of four additional Objectives for EOC verification, as there are in other verification areas within DO-333, for FEVER it was proposed that the existing four should be applied to EOC. In this manner, the proposed certification approach is an alternate means of compliance and will have to be judged on its merits in due course. The project did not start from a formal expression of requirements. As a result, it would be difficult to argue that FEVER in isolation helped meet all the EOC verification and coverage Objectives in Tables FM-6 and Table FM-7. However, a discussion on how they can be met using other D-RisQ technologies in conjunction with FEVER has been described. There remains certain aspects of EOC verification that requires test as there are hardware influences on the software which cannot be verified through other techniques. Ultimately, verification of EOC is likely to be a mix of techniques, formal and otherwise, using the most appropriate technique in order to provide the requisite evidence.

Acknowledgements The project was based upon a small amount of work undertaken for DSTL and further funding obtained from Innovate UK. Magnus Myreen and Thomas Sewell also provided vital assistance with HOL based de-compilation and the semantics of the graph language into which it is de-compiled.

References

DO-178C, 2011. RTCA DO-178C/EUROCAE ED-12C Software Considerations in Airborne Systems and Equipment Certification.

DO-333, 2011. RTCA DO-333/EUROCAE ED-216 Formal Methods Supplement to DO-178C/ED-12C and DO-278A/ED-109A.

The Mathworks, 2016. www.themathworks.com

PFVOC, 2013. "Assessing Practical Formal Verification of Object Code" contract DSTLX1000079634.

HOL, 2016. https://hol-theorem-prover.org/

Fox, 2012. Anthony Fox. Directions in ISA specification. In Lennart Beringer, Amy P. Felty, editors, Interactive Theorem Proving (ITP), Springer, 2012.

Myreen, 2008. Magnus O. Myreen. Formal verification of machine-code programs, PhD thesis Cambridge University December 2008.

Myreen, 2012. Magnus O. Myreen, Michael J. C. Gordon, Konrad Slind. Decompilation into Logic — Improved, (Formal Methods in Computer-Aided Design 2012).

O'Halloran, 2007. C. M. O'Halloran, and C. H. Pygott, Formalising C and C++ for Use in High Integrity Systems. The Safety of Systems: Proceedings of the Fifteenth Safety-critical Systems Symposium, Bristol, UK, 13--15 February 2007.

Arthan, 2005. R. D. Arthan and R. B. Jones (2005) Z in HOL in ProofPower, BCS FACS FACTS.

Lemma1, 2016. www.lemma1.com

O'Halloran, 2013. Automated Verification of auto-code from Simulink. Automated Software Engineering: Volume 20, Issue 2 (2013), Page 237-264 (DOI) 10.1007/s10515-012-0116-5.

Cavalcanti, 2013. Ana Cavalcanti, Steve King, Colin O'Halloran and Jim Woodcock: Test-data generation by proof, Journal of Formal Aspects of Computing. 2013.

DO-330, 2011. RTCA DO-330/EUROCAE ED-215 Software Tool Qualification Considerations.

Davies, 1996. Jim Davies and Jim Woodcock. Using Z: Specification, Refinement and Proof. International Series in Computer Science. Prentice Hall. ISBN 0-13-948472-8. 1996.

My 36 Years in System Safety Engineering: Looking Backward, Looking Forward

Nancy G. Leveson

Massachusetts Institute of Technology[1]

Abstract *A personal view of where we have come from in system safety, where we are now, and what is needed going forward.*

1 Introduction

I certainly never planned to go into safety engineering. In fact, I'd never heard of it until September 1980 when I got a call from an engineer at a large aerospace company. They were developing a torpedo with about 15 microprocessors on it, which was a lot of computing power for weapons (or most other engineered systems) in those days. They had what he called a "software safety" problem and asked if I would help them with it. I thought I would spend about six months solving the problem and then go back to traditional computer science research. That was 36 years ago, but I think I finally have a solution. That solution requires changing the way we do system safety engineering to account for the differences in both the complexity and the intensive use of software in today's engineered systems.

Figure 1 shows a timeline for when the traditional safety analysis techniques were created. Note that all of the techniques used today were created at least 50 years ago, before computing became a major part of engineering. The world of engineering has had a technical revolution in the meantime, but the same techniques are still used, with engineers attempting to "shove" software into a para-

---

[1] Aeronautics and Astronautics Dept., Massachusetts Institute of Technology. leveson@mit.edu

digm for which it does not fit. In this paper, I try to explain why these attempts are like trying to fit a square peg into a round hole, and I suggest an alternative. Then I suggest some important problems that need to be solved going forward.

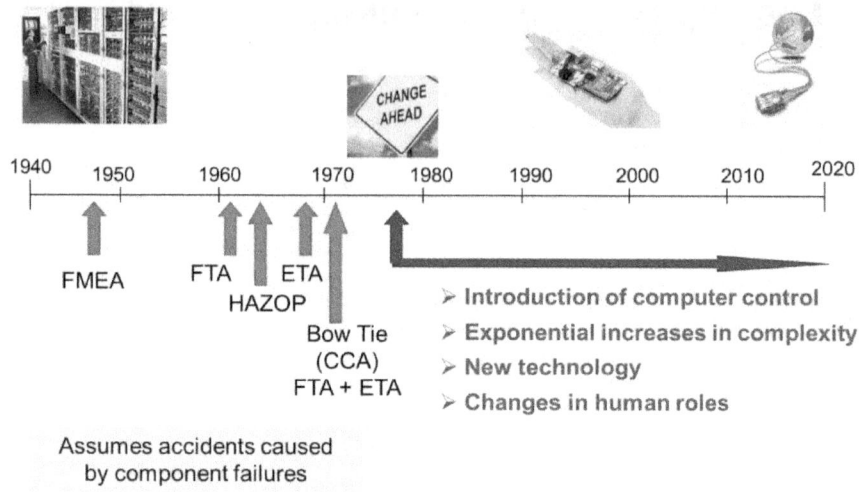

Fig. 1. A timeline for the creation of safety analysis techniques.

1.1 Why Our Efforts are not as Effective as they Should Be

There are several factors that are making our efforts less effective than they need to be in order to prevent the many preventable accidents we are having today.

Inappropriate analysis and design techniques are being used for the systems built today: We are still primarily using analysis approaches that were developed for electromechanical systems and not the digital and other types of technology that are increasing percentages of the systems we are building. Focusing all or most effort in safety analysis and assessment on the electromechanical components and then doing nothing about the other types of technology other than saying "do a good job engineering these systems" or providing a specific level of rigor in their development is essentially, like the ostrich, sticking his head in the sand. We need new, more powerful safety engineering approaches to deal with complexity, coupling, and new types of system components and causes of accidents.

Efforts focused only on the technical components of systems: Beyond ignoring software and digital components, safety engineering typically ignores or only superficially handles operator error, management decision making, and safety culture. While clearly human factors experts and management experts do focus on these factors, they are often handled separate from the system safety effort. As a result, our safety engineering results tackle only a small part of the safety problem. It makes no sense that we blame most accidents on the human operators and then ignore them or treat them only superficially (usually using probabilities) in our system safety assessments. That only guarantees that the assessments are going to be wildly incorrect and dangerous if used for decision making. On the other hand, social scientists focus on humans or management without consideration of the huge impact that the design of the system has on the behavior of the operators and the impact of the technology used on management decision making. Problems stemming from the design of the software, such as mode confusion, situation awareness, distraction, etc. cannot be solved effectively by focusing only on the design of the interface between the operators and the automation. We are designing software and automation today where an operator error is inevitable and then blaming the results on the operator.

Inadequate attention to operations versus development: Safety (and other) engineers often assume an ideal use environment and ignore the fact that systems will sometimes be used in less than ideal conditions and that the conditions under which they operate will change over time. In manufacturing, we sometimes build systems that are difficult to assemble and then blame the workers when they get hurt assembling them. While engineering in general is starting to recognize the importance of manufacturability and usability in engineering design, I see safety engineering paying little attention to these problems. Workplace safety is usually ignored by system safety engineers (although obviously not by those people concerned with workplace safety).

Safety efforts start too late: Eighty to ninety percent of safety-critical decisions are made in early system concept formation (Frola and Miller 1985), while most safety engineering analysis techniques require a relatively complete design as input. For software and for overall system design, if the safety requirements are not available at the beginning of the effort, it is almost always not possible to undo earlier decisions when problems are found. The only practical alternatives for overcoming safety deficiencies late in development is (1) to add costly and often ineffective redundancy or (2) to transfer the problems to the operators in terms of adding operational procedures and requirements that cannot be efficiently or effectively implemented without unacceptably decreasing productivity. Safety cannot be "added" to an unsafe design. Focusing on after-the-fact safety cases or safety assurance is a way to keep a lot of people employed doing relatively useless activities while fooling ourselves that we are doing something about safety.

Safety efforts are superficial, isolated, or misdirected: Safety is often isolated from the engineering design effort. For example, safety efforts may be located in the quality assurance organization, where they inherit the relatively low status of quality assurance engineering versus design engineering. As mentioned, we spend too much time and effort on safety assurance in general, and not in building safety in from the beginning. The focus tends to be on making arguments that systems *are* safe rather than on *making* them safe. By the time the assurance or safety cases are made, the pressures on getting the systems out the door or certified lead to *confirmation bias*, where evidence is sought to show the system *is* safe. Safety must be built into a system from the beginning, it cannot be argued in. Safety engineers should be focusing ninety percent of their efforts on identifying how to build safety into the system starting in the early concept formation stages.

Inadequate risk assessment: Most risk assessment techniques are based on the assumption that accidents are caused by random failures. Today's accidents involving system design errors, unsafe software behavior, and even human error cannot be assessed in this way. For practicality reasons, probabilistic risk assessments usually also assume that the causal failure events are independent, which is rarely true. I have read hundreds of accident reports in the past 36 years and have been involved in investigating accidents myself. In almost all of these accidents, there was a probabilistic risk analysis prior to the accident that showed an extremely small probability of it happening, often in exactly the way it actually did. In these cases, the causal scenario was identified and could have been prevented but was judged to be not worth doing anything about because the assumed likelihood of the events was too low. Often cases are omitted from consideration before determining whether the events could be eliminated or mitigated at relatively low or negligible cost. In some cases, the complacency generated by the very low likelihood assumption led people to ignore precursors when the supposedly impossible events actually occurred but did not, in those cases, lead to a loss. Our risk assessment procedures today ignore the fact that psychologists have shown that humans are terrible at estimating likelihood. At the same time, there have been almost no scientific evaluations of the accuracy of probabilistic risk assessment methods (Rae et.al. 2012, Goerlandt 2016, Leveson 1995)

Clearly, failure rates for standard parts with extensive historical use can be identified. The problem is that this information provides almost no help in evaluating risk in most systems today:

- Accidents are not necessarily caused by component failures,
- Operator error is determined by the context in which operators are working and cannot be represented by a simple probabilistic calculation,
- Software is almost always newly created for a system and thus does not have any useful history (and history would not be appropriate anyway because software used in a different context has no relation to its safety in the new context), and

- None of this accounts for the fact that coupling between components and system design errors (not random failure) is plays a major role in accidents today.

Assessing and communicating about risk is an important and legitimate need and should be a priority in safety engineering research. Pretending that risk assessments based on probability and likelihood estimates actually have legitimacy, however, diverts attention away from a badly needed search for better approaches that do not rely on unknowable probabilities and likelihood estimates.

Limited learning from events: We are not learning enough from accidents and losses. The only thing worse than having a major loss is to not learn as much as possible from it in order to avoid a repetition. For a variety of reasons, some of which involve politics or self-interest, accident causes are usually oversimplified. Unless there is an obvious hardware failure, accidents are usually blamed on the human operators or sometimes on someone low in relative status. In fact, most accidents involve a large number of factors from technical to sociological. Most accident reports, however, seem to fall in the trap of "root cause seduction." There is no such thing as a "root cause." Believing in a root cause or maybe two root causes appeals to our desire for control. If accidents can be reduced to a root cause, then we simply need to remove that cause (which often is declared to be the operator), and the problem is solved. As a result we often fix the symptoms of the real problem but not the process that led to those symptoms. In this way, we end up playing a sophisticated game of "whack-a-mole," get caught up in continual firefighting mode, and have what is essentially the same accident over and over.

The problems are exacerbated by focusing on finding someone or something to blame. The investigation focuses on "who" and not "why." Blame is the enemy of safety.

Summary: We are doing things that require great effort and resources but has never been shown to be cost effective or worthwhile. Scientific evaluations of the traditional analysis techniques and approaches to system safety are almost nonexistent even though engineers have depended on them for close to a century.

What should we be doing instead? Clearly there can be many different opinions about the path forward to more effective system safety engineering. The starting point should be to understand the problem before proposing solutions, particularly solutions that are simply minor changes to what we are doing today.

1.2 Understanding the Problem

Traditional approaches to safety analysis assume that accidents are caused by component failures and, therefore, they focus on reliability analysis using techniques such as fault tree analysis (FTA), event tree analysis (ETA), hazard and operability analysis (HAZOP), and failure modes and effects criticality analysis (FMECA). The goal of these analysis methods is to identify chains of directly related failure events that together will lead to an accident or loss event. The failures identified may be single or multiple. After the component failure scenarios are identified, engineers use fault tolerance or fail-safe techniques to protect against hazards caused by the identified failures. Because failures are targeted in the analysis, redundancy is commonly introduced to reduce the probability of component or functional failure.

This approach to safety made sense for the pure electro-mechanical systems of the past where system components were effectively decoupled, allowing relatively simple interactions among components. System design errors could, for the most part, be identified by testing and corrected in the design during system development. What remained after development were random hardware failures. Operational procedures could be completely specified and human error mostly involved skipping a step or performing a step incorrectly. Component reliability and safety were, therefore, closely related in these systems.

This situation is now changing. Software is an integral part of most systems, allowing enormously more complexity to be introduced. Humans increasingly are assuming supervisory roles over automation, which requires cognitively complex human decision making and not just following predefined procedures.[1] Serious accidents today often result from interactions among components and not just individual or even multiple component failures. It is not just a matter of adding new technology; rather changes in engineering have created new causes of accident.

The extensive use of software is contributing to the problems. Computers have very different characteristics than the electro-mechanical devices they are replacing. Beyond the fact that the usual concept of "failure" does not make much sense when applied to software (which is design abstracted from the physical realization of that design), certainly the "failure modes" of software (if that term has any meaning in this context) are very different than those of hardware. In addition, while software design errors may exist that result in the software not implementing the stated requirements, the role of software in past accidents and safety-related incidents almost always has involved inadequate software requirements. The software can be perfectly "reliable" (will do the

---

[1] While totally autonomous vehicles are in the news, vehicles without any type of human supervision (including "pilots" separate from the vehicles or humans programming the autonomy before a mission) are going to be relatively few in number for quite some time.

same thing continually given the same inputs) or correct with respect to the specified requirements, but may still be unsafe.

What failed in the following accident? Some Navy aircraft were ferrying missiles from one location to another. One pilot executed a planned test by aiming at the aircraft in front and firing a dummy missile. Nobody involved knew that the software was "smart" and designed to substitute a different missile if the one that was commanded to be fired was not in a good position. In this case, there was an antenna between the dummy missile and the target so the software decided to fire a live missile located in a better position instead. Safety is not equivalent to reliability in complex systems.

The problems are similar for human operators. Assumptions about the role of operators in safety have always been oversimplified. Most human factors experts now accept the fact that behavior is affected by the context in which it occurs and humans do not "fail" in a random fashion (see, for example, Dekker 2006, Flach et. al. 1995, Norman 2002, Rasmussen 1997], as usually assumed by traditional hazard analysis and human reliability techniques.

The basic problem is complexity. Complexity has increased in current advanced engineering systems to the point where all the potential interactions among system components cannot be anticipated, identified, and guarded against in design and operations. *Component interaction accidents* (as opposed to *component failure accidents*) are occurring where no components have failed, but a system design error results in accidents caused by previously unidentified and unsafe component interactions and component requirements specification flaws. Hazard analysis techniques based on reliability theory and assumptions that accidents are caused by component failure cannot identify component interaction accidents. Figure 2 shows the problem.

In addition, the increasing complexity in designs and operations requires that hazard analysis be usable early in the design process when fundamental design decisions are being made. Aeronautical engineers learned 50 years ago (Miller 1985, Leveson 1995) that safety could be designed into the aircraft in the same way as structural integrity, stability, and other system properties, and system safety was born. For complex systems, this design process needs to start in the early design stages. Seventy to ninety percent of the safety-critical decisions are made in the concept development stage of a system (Frola and Miller 1985, Leveson 1995). Traditional hazard analysis approaches need to start from a design as input and thus cannot be used in these early stages of a project. The same is true for designing security into a system. Something new is needed. That "something" will involve more powerful and inclusive models of accident causality that consider more than simply component failures. Those models will need to be based on a foundation other than classic reliability theory. System theory is a potential foundation for the future.

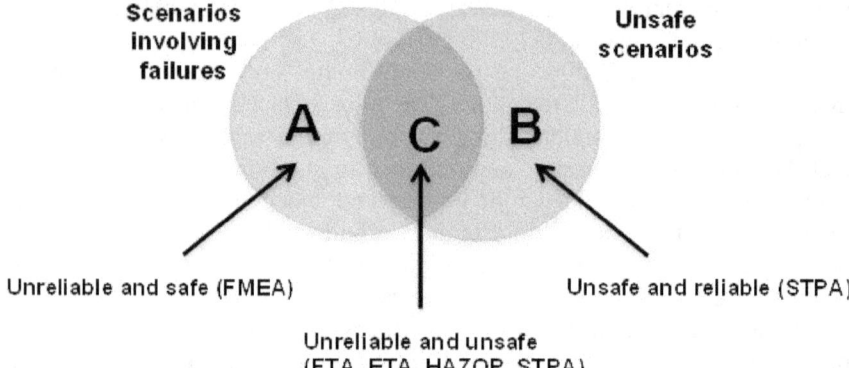

Fig. 2. Reliability and safety are often confused. The circle on the left (green) represents scenarios (events) involving failures while the one on the right (pink) represents unsafe scenarios (that involve an accident). The intersection of A and B is set C, which represents scenarios where unsafe conditions are the result of failures. FMEA and FMECA can identify the scenarios labeled A and C. FTA, ETA, HAZOP, and FMECA can identify the scenarios labeled C. STPA can identify the scenarios labeled B and C. Preventing component and functional failures is not enough to ensure system safety; all unsafe scenarios must be considered including those not involving component or functional failures.

2 System Theory and its Application to System Safety

After becoming increasingly frustrated with the lack of progress in safety engineering and the accidents that are being missed by traditional analysis techniques, about 12 years ago I began developing a new, more comprehensive model of accident causation based on system theory. The model is called STAMP (System-Theoretic Accident Model and Processes) (Leveson 2012). Information about systems theory is provided first and then STAMP is described.

2.1 An Introduction to System Theory

System theory dates from the forties and fifties and was a response to the limitations of classic analysis techniques in coping with the increasingly complex systems that started to be built after World War II (Ashby 1956, Checkland 1981), particularly defense systems. Norbert Wiener (1948) applied the ideas to control and communications engineering and called it *Cybernetics* while Ludwig von Bertalanffy (1950) developed similar ideas in biology. Von Bertalanffy suggested that the emerging ideas in various fields could be combined into a general theory of systems, and that name (rather than Cybernetics) is more commonly used today.

In the traditional approach to dealing with complexity, sometimes referred to as *divide and conquer*, systems are divided up into distinct parts so that the parts can be examined separately and later the results of analyzing each separate component are combined to represent an analysis of the whole: Physical aspects of systems are decomposed into separate physical components while behavior is decomposed into discrete events over time.

Physical aspects → separate physical (or sometimes functional) components

Behavior → discrete events over time

This decomposition, formally called *analytic reduction*, assumes that such separation is feasible, that is:

1. Each component or subsystem operates independently
2. Analysis results are not distorted when the components are considered separately
    a. Components or events are not subject to feedback loops and other nonlinear interactions
    b. The behavior of the components is the same when examined singly as when they are playing their part in the whole.
3. The principles governing the assembling of the components into the whole are straightforward, that is, the interactions among the subsystems are simple enough that they can be considered separate from the behavior of the subsystems themselves.

These are reasonable assumptions for many of the physical parts of the universe and for most pure electromechanical systems. Some system theorists have described these systems as displaying *organized simplicity* (Figure 3). Such systems can be separated into non-interacting subsystems for analysis purposes: the precise nature of the component interactions is known and interactions can be examined pairwise.

Other types of systems display what system theorists have labeled *unorganized complexity*. They lack the underlying structure that allows reductionism to be effective. They can, however, often be treated as aggregates: They are complex, but regular and random enough in their behavior that they can be studied statistically. This study is simplified by treating them as a structureless mass with interchangeable parts and then describing them in terms of averages. The basis of this approach is the law of large numbers: The larger the population, the more likely that observed values are close to the predicted average values.

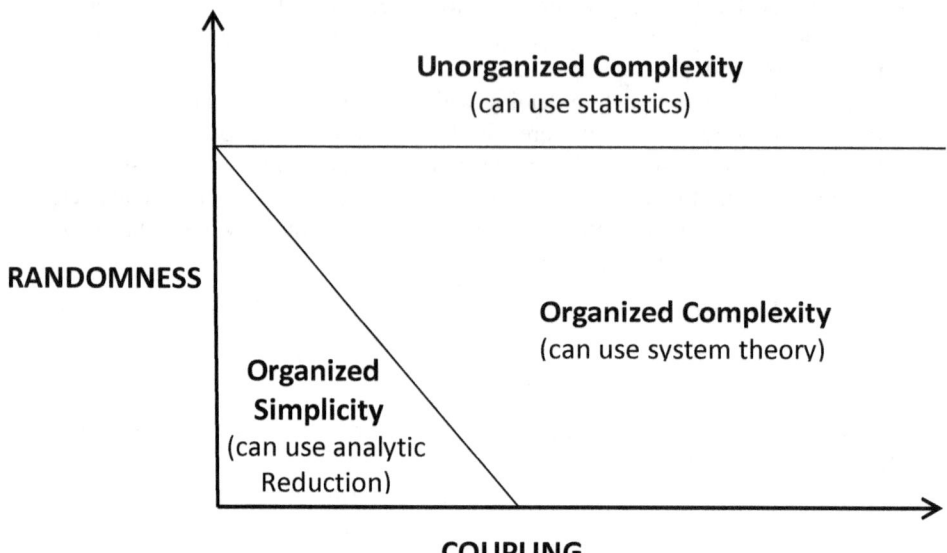

Fig. 3. Three categories of systems (adapted from Gerald Weinberg,
An Introduction to General Systems Theory, John Wiley, 1975)

A third type of system exhibits what system theorists call *organized complexity*. These systems are too complex and the system components are too coupled for complete analysis and too organized for statistics; the averages are deranged by the underlying structure (Weinberg 1975). Many of the complex engineered systems of the post-World War II era, as well as biological systems and social systems, fit into this category. Organized complexity also represents particularly well the problems that are faced by those attempting to build complex software, and it explains the difficulty computer scientists have had in attempting to apply formal analysis and statistics to software.

System theory was developed for this third type of system. The system approach focuses on systems taken as a whole, not on the parts considered separately. It assumes that some properties of systems can only be treated adequately in their entirety, taking into account all facets and relating the social to the technical aspects (Ramo 1973). These system properties derive from the relationships among the parts of systems: how the parts interact and fit together (von Bertalanffy 1950). Thus the system approach concentrates on the analysis and design of the whole as distinct from the components or parts. This holistic approach provides a means for studying systems exhibiting organized complexity.

Emergence is a basic concept in system theory. Some properties in complex systems are *emergent*, that is, they are not just a "sum" of individual component behavior but arise in the interactions among the components. If the interactions are simple enough, that is, the behavior of one component has no or limited impact on the behavior of others (i.e., they are sufficiently decoupled), then the components can be analyzed separately and the analysis results combined to represent a sufficient approximation of the behavior of the whole. But as the interactions become more complex and coupled, emergent properties arise.

Figure 4 depicts the principle of emergence. A system or process is made up of components (the shaded boxes). The components interact in both direct and indirect ways. Emergent system or process properties arise from these interactions. The concept of emergence gives rise to the often quoted basic system theory principle that in complex systems "the whole is greater than the sum of the parts."

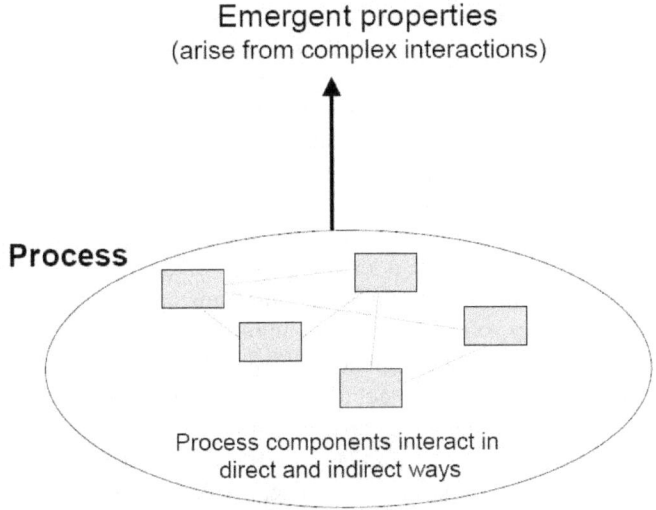

Fig. 4. Emergent properties arise from interactions among system components

Safety is an example of an emergent property, as is security and many other important system properties. Looking only at a "valve" and asking whether a system will be safe that uses that valve is an unanswerable question: System safety will depend on how the valve is used in a system and how it interacts with other components and in the operation of the whole. Note that the reliability of the valve can be determined. But the "safety" of the valve, without consideration of its role in the whole, is limited to the hazards that can be considered by looking only at the valve, such as whether it has sharp edges by which a person could be cut. System safety is determined by the relationship between the valve and the other system components as well as the system environment.

A gun, when discharged out on a desert with no other humans or animals for hundreds of miles may be safe and reliable. When discharged in a crowded mall, the reliability will not have changed, but the safety most assuredly has.

The same is true for software functions, despite the current attempt to assign a "safety integrity level" to specific software. Software is a pure abstraction; it has no sharp edges and cannot catch on fire or explode. In fact, there are no hazards associated with software functions or modules in isolation. The Ariane 4 spacecraft's inertial reference software was safe when used in that spacecraft design, for example, but contributed to the loss of the Ariane 5 when it was reused on the newer spacecraft, which has a higher takeoff trajectory (Lions 1996).

A similar argument can be made for system operators. Pilot procedures to execute a landing might be safe in one airport or in one set of circumstances but not in another. Note that the pilot may reliably execute the landing procedures for a plane at an airport where those procedures are unsafe.

Emergent properties associated with a set of components are, in system theory, related to constraints upon the degree of freedom of those components' behavior. In other words, to control emergent properties, the interactions among the components must be controlled in some way. Figure 5 depicts this principle. In an air traffic control system, allowing each aircraft to optimize its path will not optimize emergent system properties such as collision avoidance and throughput. By enforcing constraints on the behavior of individual aircraft in the controlled airspace, collisions can be avoided and overall system throughput optimized. As another example, lift in an aircraft is a system property, that is, it depends on the behavior and interaction of multiple components of the aircraft.

Using system theory, system safety can be reformulated as a system _control_ problem rather than simply a component _failure_ problem: Mishaps or losses occur when component failures, external disturbances, and/or potentially unsafe interactions among system components are not handled or controlled adequately to enforce the required safety constraints. Examples of safety constraints include: two aircraft must not violate minimum separation, power must never be on when the access door is open, pressure in a deep water well must be controlled, aircraft must maintain sufficient lift to remain airborne unless landing, the public health system must prevent exposure of the public to contaminated water and food products, and runway incursions and operations on wrong runways or taxiways must be prevented. Another way of saying this is that accidents occur when the system safety constraints required to avoid hazards are not satisfied during system operation.

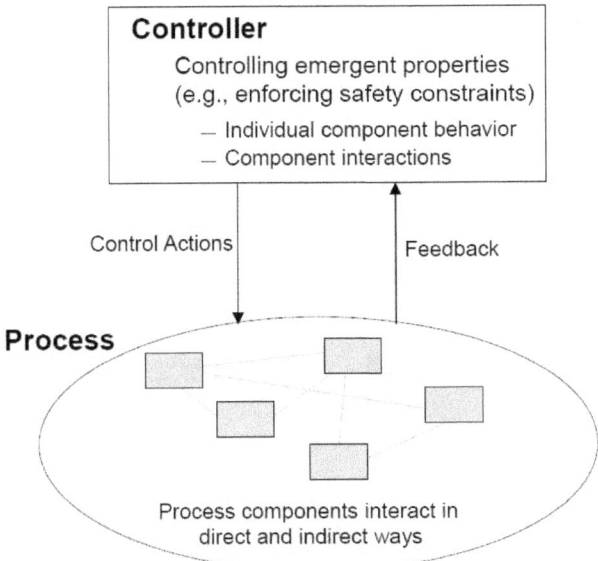

Fig. 5. To prevent hazards and accidents, safety constraints must be enforced on individual component behavior and on component interactions. This concept of control incorporates the basic control theory concept of a feedback control loop as shown above. Controls, however, can also be enforced through design features or other types of controls.

Controls can take a wide variety of forms and may be managerial, organizational, physical, operational, or manufacturing. Note that component failures are included here as a cause of mishaps, but additional causes not involving component failure must also be considered. Standard forms of design features used to prevent or mitigate (control) component failures, such as redundancy, may be used to prevent component failure-based hazards but probably will not prevent component interaction accidents. Indeed, the added complexity caused by the redundancy may even contribute to accidents. In the systems approach to safety, the old model of accident causality is simply being extended by using system theory to include also the causes of accidents that arise in the interactions among components.

To summarize, in a system-theoretic view of safety, the emergent safety properties are controlled or enforced by a set of safety constraints related to the behavior of the system components. Safety constraints specify those relationships among system variables or components that constitute the non-hazardous or safe system states. If the hazard is two aircraft on a collision course, then the safety constraint is that aircraft must never be on a collision course or the paths must be corrected before a collision occurs. If the hazard is inadvertent firing or detonation of weapons, then the safety constraint is that weapons must never be fired or detonated inadvertently. The safety constraints are just the inverse of the hazards, that is, the safety constraints are the requirements that the hazardous conditions must be prevented from occurring or mitigated if they do occur. Accidents (or losses) result from interactions among system components that violate those constraints—in other words, from a lack of enforcement of constraints on component and system behavior.

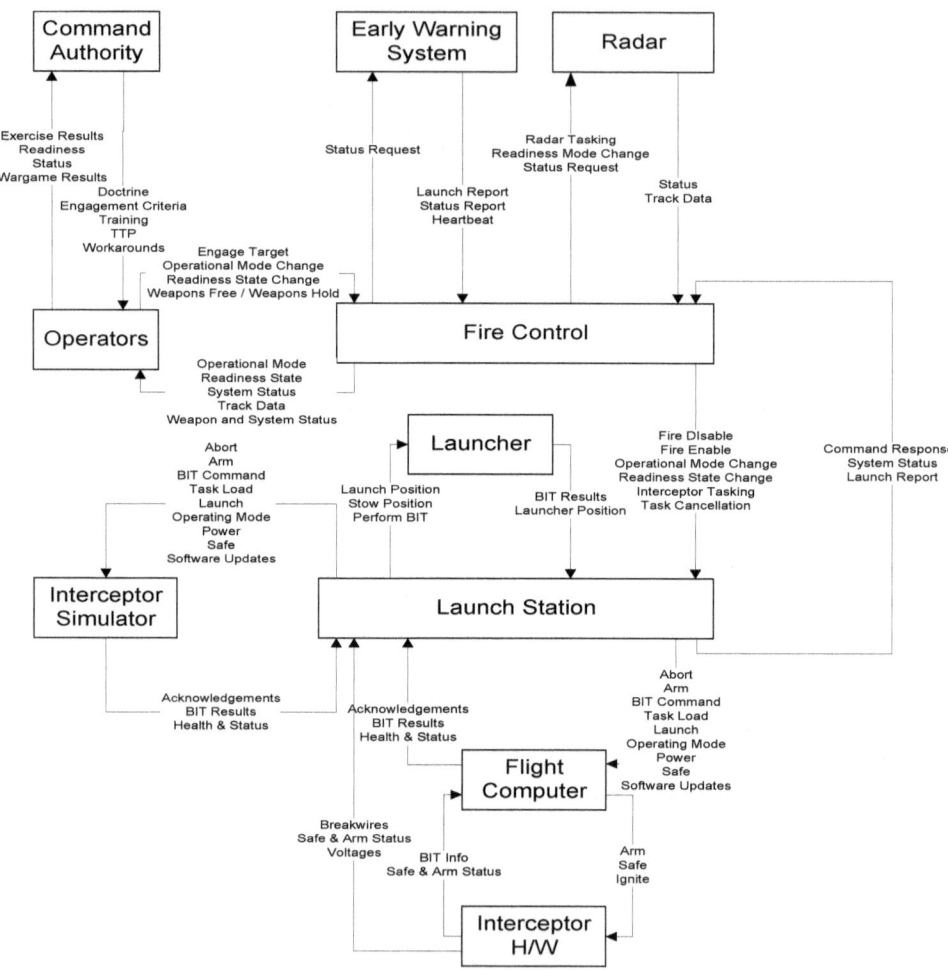

Fig. 6. Safety Control Structure for a fictional Ballistic Missile Defense System (from (Pereira, et. al. 2006))

The controls that enforce the system safety constraints are embodied in the *hierarchical safety control structure.* Hierarchies are another basic concept in systems theory. At any given level of a hierarchical model of complex systems, it is often possible to describe and understand mathematically the behavior of individual components when the behavior is completely independent of other components at the same or other levels. But emergent system properties (such as safety) do not satisfy this assumption and require a description of the acceptable interactions among components. These interactions are controlled through the imposition of constraints upon the behavior of the components.

Figure 6 shows an example of a high-level hierarchical safety control structure for a fictional ballistic missile defense system (Pereira et.al. 2006). Notice that it includes operators and the command authority and not just the physical components in the system. The control structure can include management, government regulators, insurance companies, the courts, stakeholders such as user associations, etc. Because the hazard analysis techniques operate directly on the control structure model, all of these components can be included in the hazard analysis, that is, in the identification of how accidents can occur and who or what contributes to them.

In the control structure shown in Figure 6, the command authority controls the behavior of the operators by providing doctrine, engagement criteria, training, etc. The operators control the behavior of the Fire Control computer, which gets inputs from radars and the early warning system. The Fire Control computer provides instructions to and controls the Launch Station, which controls both the Launcher and the Flight Computer, and so on. Note that the safety control structure is not an architectural design model but a classic functional control structure

In system theory, every controller contains a model of the controlled process, called a *process model* in Figure 7. For human controllers, this model is usually called a *mental model.* This process model or mental model includes assumptions about how the controlled process operates and the current state of the controlled process. It is used by the controller to determine what control actions are necessary to keep the system operating effectively and safely. A simple example is a thermostat that uses a model of the controlled space, including current temperature, desired temperature (set point), etc. to determine what actions to take to keep the temperature at the desired set point.

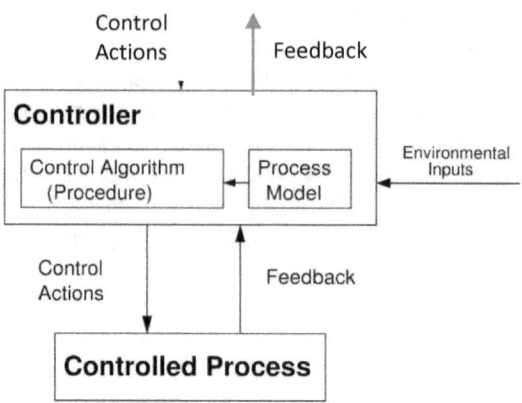

Fig. 7. A typical control loop showing a process model

Accidents in complex systems often result from inconsistencies between the process model used by the controller and the actual process state, which results in the controller providing unsafe control actions. For example, the autopilot software thinks the aircraft is climbing when it really is descending and applies the wrong control law; a military pilot thinks a friendly aircraft is hostile and shoots a missile at it; the software thinks the spacecraft has landed and turns off the descent engines prematurely; radar data is misinterpreted as an incoming threat and an interceptor is launched.

Part of the challenge in designing an effective safety control structure is providing the feedback and inputs necessary to keep the controller's model consistent with the actual state of the controlled process. As stated earlier, an important component in understanding accidents involves determining how and why the controls are ineffective in enforcing the safety constraints on system behavior. Often a controller's unsafe behavior results from the process model used by the controller being incorrect. In hazard analysis based on system theory, the required process models and feedback to maintain their consistency with the state of the controlled process are identified as well as what information must be in the process model for safe control to occur and the requirements for updating it.

Inaccurate process models can explain a large number of accidents involving software. The process model in software may simply be a few variables representing the condition of the controlled process, but it serves a critical function in how the software behaves. Ensuring accurate information in the software process model is an important part of the safe design for a system containing software.

The same is true for accidents related to human errors. The system-theoretic approach being described here provides a potentially much more effective way of identifying safety-critical operator errors and their causes (so they can be eliminated or mitigated) than does treating human error like random machine failure, which is common in the traditional reliability approach to safety. For example, the control model allows more sophisticated analysis of human factors in accidents including things like situation awareness flaws (i.e., a flawed process model), mode confusion, and distraction.

2.2 STAMP (System-Theoretic Accident Model and Processes)

STAMP (System-Theoretic Accident Model and Processes) is simply the application of system theory to safety, treating safety as an emergent property (Leveson 2004a, 2005, 2012). As stated above, traditional hazard and risk analysis methods are based on the assumption that mishaps are caused by chains of component failures, each failure related in a direct way to the ones preceding and following it, for example, the moisture gets into the tank, which causes corrosion, which in turn causes the tank to burst when pressure reaches a critical level.

STAMP extends the types of accidents and causes that can be considered by including non-linear, indirect, and feedback relationships among events. In this way, the traditional causality model is extended to consider new types of accident causes brought about by component interactions (rather than just component failures), both direct and indirect; cognitively complex human mistakes; management and organizational flaws; software errors (particularly requirements errors); etc. For example, firing missiles in a particular order might upset the weight and balance of the aircraft, which could lead to instability. Firing missiles in an unsafe order, in turn, could be related to the intensity of the engagement and potential pilot distraction or flawed situation awareness. In each of these potential causes of aircraft instability, no failure of the individual aircraft components may have occurred but the problems are in the interactions among control actions and environmental events.

Formally, STAMP treats mishaps as dynamic processes that result from an adaptive feedback function that fails to maintain safety as performance changes over time to meet a complex and changing set of goals and values. The mishap or loss itself results not simply from component failure or human error (which are symptoms rather than root causes) but from the inadequate control (i.e., lack of enforcement) of safety-related constraints on the development, design, construction, and operation of the entire sociotechnical system. It includes changes over time as most systems are not static but are continually adapting or being adapted to respond to new situations. Thus a mishap is a complex process involving interactions among system components, including people, societal and organizational structures, engineering activities and physical system components.

Instead of focusing only on the *events* that occur prior to a loss in order to determine why it occurred and how to prevent future occurrences, the entire dynamic mishap or loss *process* is investigated. In STAMP, violation of constraints may result from environmental disturbances or conditions, system component failures, or unsafe interactions among the system components. Inadequate control actions can be traced to:

1. A lack of designed controls, at the physical, social, and organizational levels
2. Inadequate operation of the existing controls, perhaps due to:
    a) Social and political factors (context)
    b) Controller process models that do not match the state of the process being controlled and lead to unsafe control actions on the part of the controller.
    c) Unsafe design or changes to the controls
    d) Component failure
3. Degradation and changes in the safety-control structure over time, both planned and unplanned
4. Unsafe coordination of control actions among multiple controllers.

2.3 Using STAMP in System Safety Engineering

STPA (System-Theoretic Process Analysis) is a hazard analysis method based on STAMP. It is performed using the system control structure and based on the following concepts. There are, in general, four ways that a controller can behave unsafely (see Figure 7):

1. The control action provided leads to a hazard.
2. The lack of a control action leads to a hazard.
3. A potentially safe control action is provided too early, too late, or in the wrong sequence
4. A continuous control action is provided for too long or too short a duration

In addition, systems can get into hazardous states when a required control action is provided but not executed for some reason (e.g., a physical failure, a delay, etc.). The latter is the typical type of causal factor included in the traditional failure-based hazard analysis techniques. Therefore, STPA produces a superset of causal scenarios in that it includes the causal scenarios generated by the older hazard analysis techniques but includes many more that are not.

Briefly, the STPA process involves:

1. Constructing the safety control structure (including the potential control actions, feedback, and process models),
2. Identifying potential unsafe control actions,
3. Constructing the scenarios that could lead to the unsafe control actions.

Standard control theory concepts are used to construct the scenarios. Figure 8 shows some of the potential causes of unsafe control that may be involved in the scenarios.

To better understand the difference between the classic analytic reduction approach to safety and the approach implied by system theory, an example may be helpful. Figure 9 shows a functional decomposition (analytic reduction) of an aircraft taken from SAE ARP 4761. The hazards are stated as the failure of the required functions, such as failure of wheel braking after landing.

The process described in ARP 4761 then involves calculating the probability of failure of each of the leaf nodes in the tree (using Fault Tree Analysis or other similar techniques) and combining them to get the probability of failure of the functions at the level above and so on until a probability of failure of the aircraft is derived.

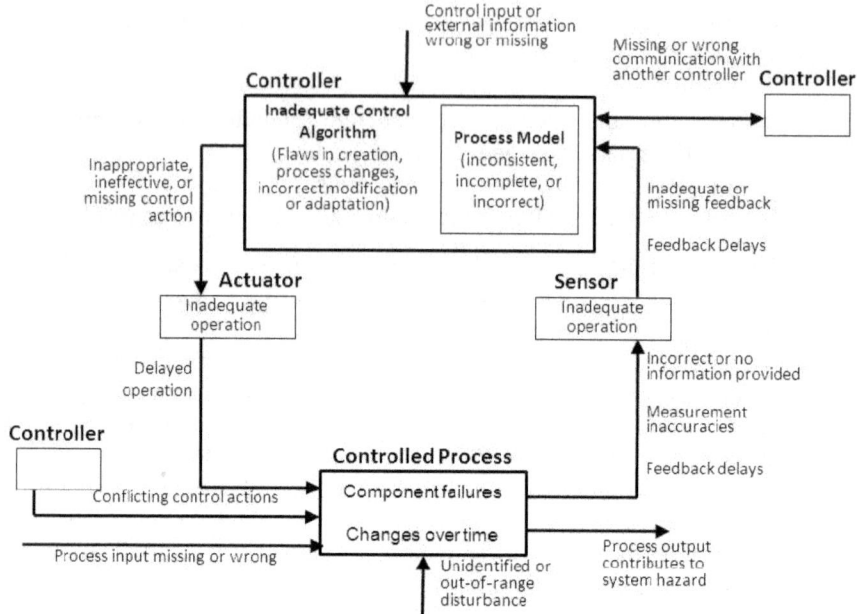

Fig. 8. Some control flaws that can lead to unsafe control. Note that component failures are included but more types of control flaws than component failures are considered.

Air Function Tree

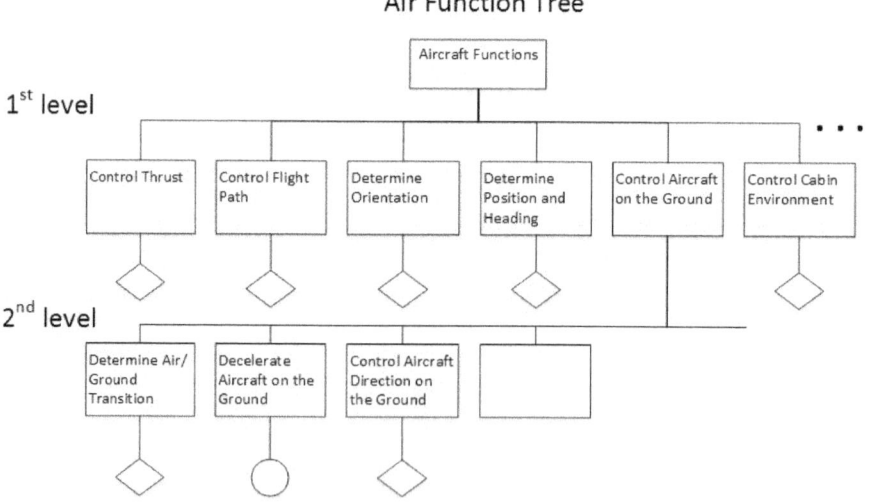

Fig. 9. Two levels of a classic functional decomposition for an aircraft (from SAE ARP 4761)

In contrast, STPA starts with a control structure model. The accidents and high-level hazards are first identified and then the control structure to prevent those hazards is modelled. For example:

A-1. Loss of life or serious injury to aircraft passengers or people in the area of the aircraft

A-2. Unacceptable damage to the aircraft or objects outside the aircraft

System-level hazards related to these losses include:

H1: Insufficient thrust to maintain controlled flight

H2: Breach of airframe integrity

H3: Controlled flight into terrain

H4: An aircraft on the ground comes too close to moving or stationary objects or inadvertently leaves the taxiway

H5: etc.

The high-level control structure used by STPA for the aircraft is shown in Figure 10. The responsibilities assigned to each component (controller) in the safety control structure are modelled as well as the control actions available to the controller, the feedback the controller receives, and models of the controlled process used by the controller to select appropriate control actions. Note the difference between Figure 9 (representing classic analytic reduction) and Figure 10 (representing system theory concepts). Figure 9 simply decomposes the required functions into separate boxes to be analyzed individually, with the individual results later combined. Figure 10 assigns the functions to feedback control loops to be analyzed as a whole.

As with the decomposition shown in Figure 9, the high-level control structure will be refined to be more detailed. An example for ground movement is shown in Figure 11. At an early stage of development, design details may not have been determined, but the safety constraints and causal scenarios for violating them can be identified for that level of refinement. These constraints (requirements) can then be refined when further design detail is created. If there is a need to back up to a higher level of design (e.g, the designers change their minds), it is easy to return to the higher-level of requirements generated by STPA and proceed from there.

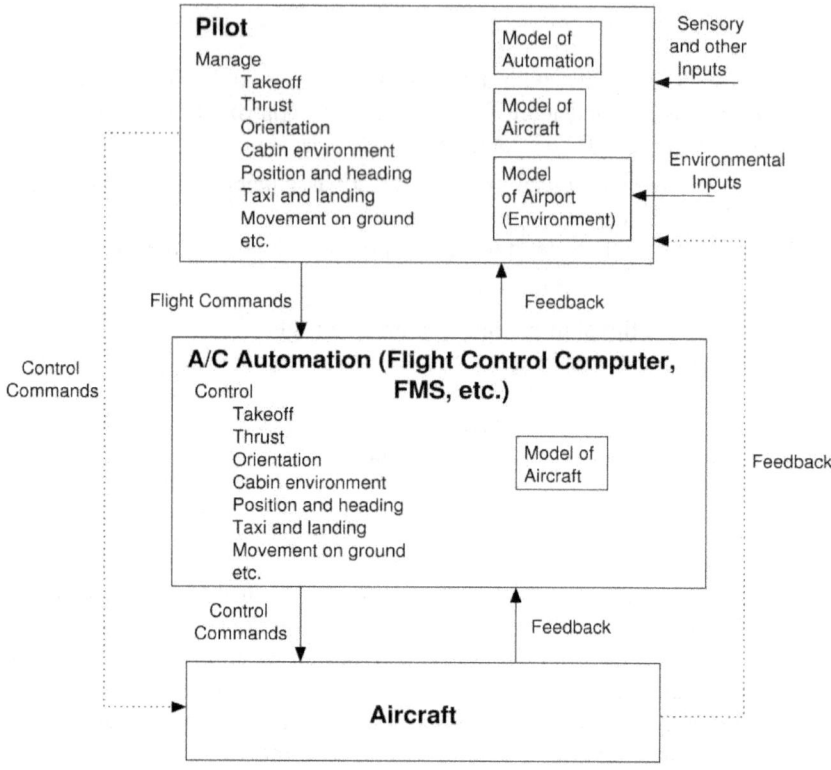

Fig. 10. The high level control structure for an aircraft

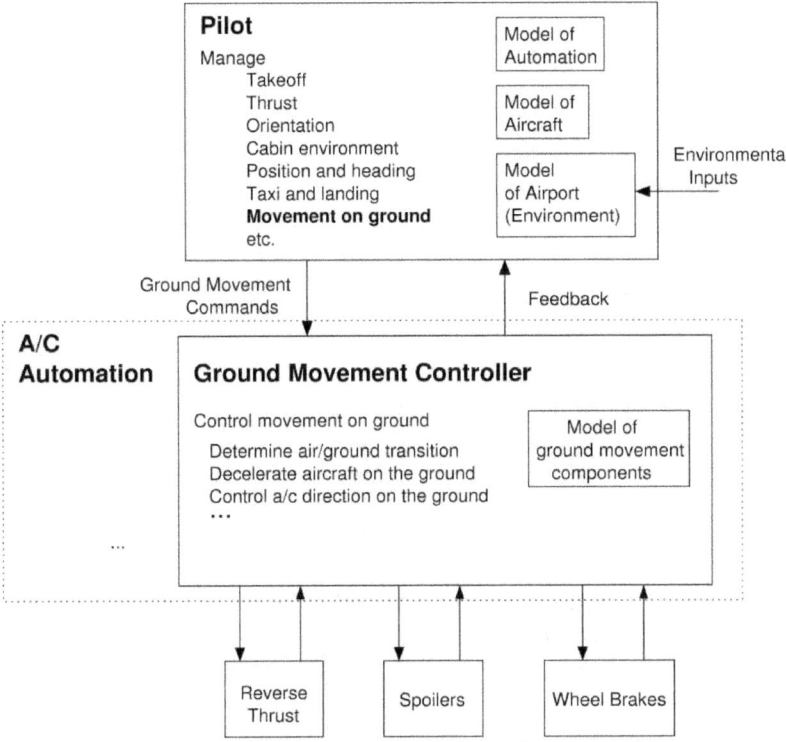

Fig. 11. The portion of the safety control structure related to deceleration on the ground.

A detailed comparison of the process and types of results provided by SAE ARP 4761 and by STPA can be found in (Leveson et.al. 2014). Briefly, Table 1 shows the different types of high-level requirements that are generated by the two approaches. These high-level requirements are then refined into detailed requirements on the system components. While the 4761 component requirements are also stated in terms of failure rates, the STPA requirements are refined into functional requirements on the components, including the software and pilots.

Table 1. System-Level Requirements Generated

ARP 4761 Process (from SAE ARP 4761)	STPA (From Leveson et.al. 2014)
Requirements from the FHA: 1) Loss of all wheel braking during landing or rejected take off shall be less than 5E-7 per flight. 2) Asymmetrical loss of wheel braking coupled with loss of rudder or nose wheel steering during landing or rejected take off shall be less than 5E-7 per flight. 3) Inadvertent wheel braking with all wheels locked during takeoff roll before V1 shall be less than 5E-7 per flight. 4) Inadvertent wheel braking of all wheels during takeoff roll after V1 shall be less than 5E-9 per flight. 5) Undetected inadvertent wheel braking on one wheel w/o locking during takeoff shall be less than 5E-9 per flight Two requirements generated from the Common Cause Analysis: 6) The wheel braking system and thrust reverser system shall be designed to preclude any common threats (tire burst, tire shred, flailing tread, structural deflection, etc.) 7) The wheel braking system and thrust reverser system shall be designed to preclude any common mode failures (hydraulic system, electrical system, maintenance, servicing, operations, design, manufacturing, etc.)	Requirements derived from the STAMP system-level hazards: SC1: Forward motion must be retarded within TBD seconds of a braking command upon landing, rejected takeoff, or taxiing. SC2: The aircraft must not decelerate after V1. SC3: Uncommanded movement must not occur when the aircraft is parked. SC4: Differential braking must not lead to loss of or unintended aircraft directional control SC5: Aircraft must not unintentionally maneuver out of safe regions (taxiways, runways, terminal gates and ramps, etc.) SC6: Main gear rotation must stop when the gear is retracted.

In the ARP 4761 safety assessment process and example, the high-level probabilistic requirements in the left column of Table 1 are refined, usually using fault trees, into more specific (but still probabilistic) requirements on the system components. Software requirements are stated in terms of design assurance levels. Crew requirements are not included.

For example, the brake system control unit (BSCU) requirements generated from the system level requirements are:

1. The probability of "BSCU Fault Causes Loss of Braking Commands" shall be less than 3.3E-5 per flight.
2. The probability of "Loss of a single BSCU shall be less than 5.75E-3 per flight.
3. The probability of "Loss of Normal Brake System Hydraulic Components" shall be less than 3.3E-5 per flight.
4. The probability of "Inadvertent braking due to BSCU" shall be less than 2.5E-9 per flight.
5. No single failure of the BSCU shall lead to "inadvertent braking."
6. The BSCU shall be designed to Development Assurance Level A based on the catastrophic classification of "inadvertent braking due to BSCU."

Additional requirements on other systems are also generated such as

7. The probability of "loss of Green Hydraulic Supply to the Normal brake system shall be less than 3.3E-5 per flight."

Finally, installation requirements and maintenance requirements are generated as well as independence requirements such as:

8. Each BSCU System requires a source of power independent from the source supplied to the other system.

Hardware and software safety requirements generated look like:

1. Each BSCU system will have a target failure rate of less than 1E-4 per hour.
2. The targeted probabilities for the fault tree primary failure events have to be met or approval must be given by the system engineering group before proceeding with the design.
3. There must be no detectable BSCU failures that can cause inadvertent braking.
4. There must be no common mode failures of the command and monitor channels of a BSCU system that could cause them to provide the same incorrect braking command simultaneously.
5. The monitor channel of a BSCU system shall be designed to Development Assurance Level A.

6. The command channel of a BSCU system may be designed to Development Assurance Level B[1]

7. Safety Maintenance Requirements: The switch that selects between system 1 and system 2 must be checked on an interval not to exceed 14,750 hours.

The STPA analysis generates very different types of results. After the first high-level system safety constraints in the right hand column of Table 1 have been identified, they are refined into a more detailed set of functional safety requirements that are associated with specific system components, including the crew, the software, the hardware, and the component interfaces. These safety requirements are generated to prevent the causal scenarios identified by STPA.

Some example requirements on the crew generated from the STPA analysis are:

FC-R1: Crew must not provide manual braking before touchdown [CREW.1c1[2]]

Rationale: Could cause wheel lockup, loss of control, or tire burst.

FC-R2: Crew must not stop manual braking more than TBD seconds before safe taxi speed reached [CREW.1d1]

Rationale: Could result in overspeed or runway overshoot.

FC-R3: The crew must not power off the BSCU during autobraking [CREW.4b1]

Rationale: Autobraking will be disarmed.

etc.

Example requirements that can be generated for the BSCU:

BSCU-R1: A brake command must always be provided during RTO [BSCU.1a1]

Rationale: Could result in not stopping within the available runway length

BSCU-R2: Braking must never be commanded before touchdown [BSCU.1c1]

Rationale: Could result in tire burst, loss of control, injury, or other damage

BSCU-R3: Wheels must be locked after takeoff and before landing gear retraction [BSCU.1a4]

Rationale: Could result in reduced handling margins from wheel rotation in flight.

---

[1] The allocations in 5 and 6 could have been switched, designing the command channel to level A and the monitor channel to level B.

[2] In the examples below, the notation in the square brackets provide traceability information.

Finally, some examples of requirements for the BSCU hydraulic controller (HC) commands to the three individual valves:

HC-R1: The HC must not open the green hydraulics shutoff valve when there is a fault requiring alternate braking [HC.1b1]

Rationale: Both normal and alternate braking would be disabled.

HC-R2: The HC must pulse the anti-skid valve in the event of a skid [HC.2a1]

Rationale: Anti-skid capability is needed to avoid skidding and to achieve full stop in wet or icy conditions.

HC-R3: The HC must not provide a position command that opens the green meter valve when no brake command has been received [HC.3b1]

Rationale: Crew would be unaware that uncommanded braking was being applied.

HC-R4: the HC must always provide a valve position command when braking commands are received.

The next level of more detailed requirements is derived from the causal scenarios generated to identify how each of these requirements could be violated. The specific additional requirements generated will depend on how the design engineers decide to prevent these scenarios, that is, the controls added to the design. If the scenarios are to be prevented by crew actions, then additional crew requirements will be needed. If they are to be eliminated or controlled through software or hardware controls, then those controls must be added to the hardware or design requirements.

For example, consider HC-R4, which requires the hydraulic controller to provide a valve position command when brake commands are received. Why might it not do so? Several scenarios could be identified using STPA. One such scenario might be that the brake commands were sent by the Autobrake controller and a manual braking command was received before or during the Autobrake command. Possible contributors to this scenario are that the manual braking command was intentionally provided by one of the pilots (perhaps because he did not know Autobraking was set or thought that it was not working property), the manual braking command was unintentionally provided by one of the pilots (e.g., foot on pedal during landing or a bump), or a physical disturbance such as a hard landing or sensor failure trips a manual braking command.

The potential for this hazardous scenario occurring might be eliminated or controlled by changes to the hydraulic controller behavior, the autobrake behavior, crew behavior, and/or physical system behavior. These design and requirements decisions must be made by the design engineers, but the STPA analysis will provide them with guidance on the system design decisions needed and why. If changes are made to try to prevent the scenario, there must be further analysis to determine that the changes do not introduce new hazardous scenarios.

Note that redundancy might be an option here to deal with the causal scenario involving the sensor failure. Physical failures are not ignored in the STPA causal analysis; it is just that the connection of the failure to a system hazard is first determined before trying to prevent it. And looking at the impact of a sensor failure in the larger scope of the WBS design might provide insight into a different way of solving the problem that does not involve costly redundancy or monitors.

2.4 A System-Theoretic Approach to Cyber Security

Recently, an Air Force Colonel (Bill Young) working on his Ph.D. research at MIT found that STPA also applies to cyber security (Young and Leveson 2014, Young 2016). The basic argument stems from the observation that it does not matter if, for example, a weapon is fired at a friendly target because of an inadvertent human error, a system design flaw, a component failure, or an intentional enemy action. Hazards need to be eliminated or controlled whatever their cause. In essence, the common goal is "accident prevention" and ensuring that critical functions and services provided by the system and networks are maintained.

The analysis for safety using STPA and that for cyber security starts the same way. The hazards or threats are identified, the control structure is created and the unsafe control actions are identified. The difference occurs only during the development of the causal scenarios. The scenarios identifying inadvertent causes are augmented with those that involve deliberate actions. For example, consider the line on the top right of Figure 10 labelled "inadequate and missing feedback" and "feedback delays." Such missing or inadequate information and delays may have their origins in system design errors or physical failures. They may also result, however, from a cyber threat that purposely inserts unsafe feedback or introduces delays in the system. The controls that need to be introduced to prevent these safety and cyber security problems may be the same or different.

As an example, consider the Iranian Stuxnet case[3], where some hostile actor wanted to delay the development of the Iranian nuclear capability. The loss or accident in this case was damage to the reactor. The point of attack was the centrifuges, which were damaged by spinning too fast and thus had to be frequently replaced, slowing up the progress of the Iranian nuclear program. The hazard, then, was the centrifuges spinning above the maximum speed (to avoid damage to them) and the corresponding constraint was that the centrifuges must never

---

[3] *https://en.wikipedia.org/wiki/Stuxnet*

spin above the maximum speed. The attack here involved a hazardous control action where the centrifuge controller issues an *increase speed* command when the centrifuges are already spinning at the maximum safe speed. One causal scenario involves the controller's process model being incorrect due to incorrect feedback (bad information) about the actual speed of the centrifuges, resulting in the controller thinking that the centrifuges are spinning at less than the maximum speed. There are both deliberate (involving a cyber security threat) and inadvertent events that could lead to this incorrect process model in the centrifuge controller, e.g., information is lost, delayed, or corrupted inadvertently, a purposeful denial of service attack, a man in the middle attack, a virus in the software controller, etc. Design controls created to combat this potential problem may be different for the deliberate and inadvertent cases or, in some cases, can be the same. As an example of the latter, an independent mechanical speed limiter (an interlock) could be used to limit centrifuge speed or an analogue RPM gauge could be introduced into the system design to provide independent, non-digital feedback to the centrifuge controller or a human controller.

The safety and security analyses, in this way, share much of the same effort but extra activities are added to the safety analysis in order to identify the security-related causal scenarios. Requirements and designed controls to handle safety and cyber security may or may not be the same.

2.5 Does it Work? Evaluations in Controlled and Industrial Condiions

STPA has been evaluated and used in most every industry and on very complex, software-intensive systems. For example, the Missile Defense Agency successfully used STPA for a non-advocate safety assessment of inadvertent launch prior to deployment and field testing of the new missile defense system (Pereira 2006). Figure 6 shows a fictional high-level control model for such a system. So many previously unknown paths to inadvertent launch (the hazard analyzed) were identified that deployment had to be delayed six months in order to fix them. These causes included classic component failures but also included such causes as software requirements flaws and potential timing problems.

In more recent commercial and defense system evaluations, where a comparison was made, STPA found the same causes as traditional hazard analysis methods such as FTA, FMECA, Event Tree Analysis (ETA), etc. STPA, however, identified more scenarios, including those involving software and human operators, system design flaws, human factors, and component integration and interactions, for example (Abrecht 2016a. Abrecht et. al 2016b). In several of the evaluations, STPA identified the path to a real mishap in that system (unknown to the analysts) that was not identified by the other techniques.

A final example is a recent dissertation by Col. Kip Johnson (Johnson, 2016), which extends STPA analysis to identify problems in the coordination of multiple controllers. He applied his extended version of STPA to the problem of integrating unmanned aircraft systems (UAS) into military and civilian flight operations to identify causal scenarios involving flawed coordination of flight paths and to generate the safety requirements necessary for safe UAS integration into the national airspace (NAS). RTCA Special Committee (SC) 203 worked for over a decade trying to generate UAS integration requirements using Functional Hazard Analysis (FHA) and the established FAA Safety Risk Management processes (Federal Aviation Administration 2014). Johnson compared his results to the official RTCA SC 203 results published in DO-344. The FHA identified 48 coordination-related scenarios while the extended STPA found 194. Of the 194 scenarios, only 6% related to failures and degradation while 94% were related to design errors (nothing failed). In contrast, all the FHA scenarios were related to failures. A detailed qualitative comparison was performed between the two types of results. Many of the FHA scenarios were assessed as having Minimal Risk or even No Safety Effect while STPA identified catastrophic results for the same scenarios. Similar results have been found in defense system demonstrations.

U.S. Cyber Command (CYBERCOM) has conducted several evaluations of STPA-Sec in comparison to their normal security analysis (using real security experts) on real defense missions, and concluded that STPA-Sec was superior. CYBERCOM has written policy for the DoD to adopt STPA-Sec for Key Terrain Mapping and is currently evaluating it for other cyber security applications. Training of CYBERCOM personnel by Col. Young began two years ago and courses are now being taught by trained Air Force instructors. Other evaluations on different cyber-security problems are being conducted. Col. Young is currently Commander of the Air Force 53rd Electronic Warfare Group.

In all of these comparisons, the STPA analysis was performed before looking at the traditional hazard and risk analysis results. The most surprising result of all the evaluations and comparisons, given that the STPA outputs were more comprehensive, was that in all cases STPA required orders of magnitude less time and effort than the traditional approaches. Data about required resources are currently being collected to validate this anecdotal evidence.

Automated tools are being developed for STPA by universities, research institutes, and private companies and tested in industrial environments. Thomas has created a way to identify potentially unsafe control actions automatically given a list of high-level hazards and a safety control structure (Thomas 2013). The unsafe control actions can, in turn, be converted to executable and analyzable model-based system and component safety requirements. The process is based on an underlying formal mathematical foundation although the users do not have to know or be able to use formal methods to perform the analysis. Research on extending STPA to include more sophisticated human factors analysis and to consider coordination issues in systems with multiple controllers or the need for coordination in decision making between a computer and a human are ongoing.

3 Looking Forward

STAMP represents a paradigm change from what is commonly done today in system safety and cyber-security. That change does not imply that what was done previously was wrong and that the new approach is correct. Einstein suggested that progress in science (moving from one paradigm to another) is like climbing a mountain. As you move further up the mountain you can typically see farther than you could from lower points on the mountain. However, the new perspective does not necessarily invalidate what you saw at the lower perspectives; rather it often extends and enriches your appreciation of the valleys below. A new paradigm does not invalidate the successes or empirical observations made in the old paradigm. In fact, the value of the new paradigm will often depend on its ability to accommodate these. The value of a new paradigm, like STAMP, is that it offers a broader, richer perspective for interpreting the results of the previous approaches. STAMP basically augments the results of the traditional hazard analysis approaches; it does not change them.

Another way to explain this is to use an iceberg analogy (Figure 12). An event and failure-based perspective on mishap causality only illuminates part of the problem, i.e., that part of the iceberg that protrudes above the ocean surface. A system-theoretic perspective allows a larger part of the problem to be examined and handled, namely, the other parts of the problem that are hidden from view by more limited paradigms.

Fig. 12. STAMP provides an enlarged perspective on safety and security problems; that is, the entire iceberg and not just the part that extends above the ocean surface.

So where do we go from here? By introducing a new theoretical foundation for system safety, system theory opens up potential for making progress on problems that have not been easily solved using the old paradigm. Currently research is under way on applying STAMP to other emergent system properties such as manufacturability, quality, engineering optimization, and others. Extensions to STPA are being proposed to include factors that cannot be handled by current hazard analysis methods. And, of course, new types of analysis method based on system theory are possible. A new, more powerful theoretical foundation provides the opportunity to expand our progress in many areas.

Acknowledgments   I would like to acknowledge the hundreds of people I have worked with over the past 36 years, both those who supported what I was doing and those who argued against it. I learned from both groups.

References

Abrecht, Blake (2016a) Systems Theoretic Process Analysis Applied to an Offshore Supply Vessel Dynamic Positioning System, Master's Thesis, Massachusetts Institute of Technology.

Abrecht, B., Arterburn, D., Horney, D., Schneider, J., Abel, B., and Leveson, N. (2016b), A New Approach to Hazard Analysis for Rotorcraft, *American Helicopter Society Specialists Meeting on Development, Affordability, and Qualification*, Huntsville AL, Feb. 9-10.

Ashby, W.R. (1956) *An Introduction to Cybernetics*, London: Chapman and Hall.

Billings, Charles E. (1996) *Aviation Automation: The Search for a Human-Centered Approach,* CRC Press.

Checkland, Peter (1981) *Systems Thinking, Systems Practice*, New York: John Wiley & Sons.

Dekker , Sidney e(2006) *The Field Guide to Understanding Human Error*, London: Ashgate.

Federal Aviation Administration (2014), *Safety Management System Manual*, Version 4.0.

Flach, J., Hancock, P.A., Caird, J., and Vicente, K.J. (1995) *Global Perspectives on the Ecology of Human-Machine Systems*, CRC Press, 1995.

Frola, F. and Miller, C.O. (1985) *System Safety in Aircraft Acquisition*, Washington, D.C.: Logistics Management Institute

Goerlandt, F., Khakzad, N., and Reniers, G. (2016)Validity and Validation of Safety-Related Quantitative Risk Analysis: A Review, Safety Science, in press.

Johnson, Kip (2016), Extending Systems-Theoretic Safety Analyses for Coordination, Ph.D. Dissertation, MIT Aeronautics and Astronautics Dept., September.

Leveson, Nancy (1995) *Safeware: System Safety and Computers*, Waltham, MA: Addison-Wesley.

Leveson, Nancy (2004a). A New Accident Model for Engineering Safer Systems, *Safety Science*, 42(4):237-270, April.

Leveson, Nancy (2004b) The Role of Software in Spacecraft Accidents, AIAA Journal of Spacecraft and Rockets, 2004.

Leveson, Nancy (2005) A Systems-Theoretic Approach to Safety in Software-Intensive Systems, *IEEE Trans. on Dependable and Secure Computing*, Vol. 1, No. 1, January 2005.

Leveson, Nancy (2012). *Engineering a Safer World*, MIT Press, 2012.

Leveson, Nancy (2015) A Systems Approach to Risk Management Through Leading Safety Indicators, *Reliability Engineering and System Safety*, 2015.

Leveson, N., Wilkinson, C., Fleming, C., Thomas, J., and Tracy, I. (2014), *A Comparison of STPA and the ARP 4761 Safety Assessment Process*, MIT PSAS Technical Report.

Lions, J.L. Report by the Inquiry Board, Flight 501 Failure, 1996.

Miller, C.O. (1985) A Comparison of Military and Civil Approaches to Aviation System Safety, *Hazard Prevention*, May/June.

Norman, Don (2002) *The Design of Everyday Things*, New York: Basic Books.

Pereira, Steven J., Grady Lee, and Jeffrey Howard. A System-Theoretic Hazard Analysis Methodology for a Non-Advocate Safety Assessment of the Ballistic Missile Defense System, *Proceedings of the 2006 AIAA Missile Sciences Conference*, Monterey, California, November.

Andrew Rae, John McDermid, and Rob Alexander (2012) "The science and superstition of quantitative risk assessment, *Proceedings of Probabilistic Safety Assessment and Management (PSAM) Conference11*, International Association for Probabilistic Safety Assessment and Management (IAPSAM), Helsinki, June 2012, pp. 2292-2301.

Ramo, Simon (1973) The systems approach. In Ralph F. Miles Jr., editor, *Systems Concepts: Lectures on Contemporary Approaches to Systems*, pages 13–32, New York: John F. Wiley & Sons, New York.

Rasmussen, Jens (1997) Risk Management in a Dynamic Society: A Modelling Problem, *Safety Science*, vol. 27, No. 2/3, Elsevier Science Ltd., pages 183-213.

SAE ARP 4761 (1996), *Guidelines and Methods for Conducting the Safety Assessment Process on Civil Airborne Systems and Equipment*.

Thomas, John (2013) *Extending and Automating a Systems-Theoretic Hazard Analysis for Requirements Generation and Analysis*, Ph.D, Dissertation, Massachusetts Institute of Technology.

Von Bertalanffy (1950) An outline of general system theory, *British Journal for the Philosophy of Science*, Vol. 1, pp. 134-165.

Weinberg, Gerald (1975) *An Introduction to General Systems Thinking*, New York: John Wiley & Sons.

Wiener, Norbert (1948) *Cybernetics: or the Control and Communication in the Animal and the Machine*, Cambridge, MA: MIT Press.

Young, William and Leveson, Nancy (2014) An Integrated Approach to Safety and Security Based on Systems Theory, *Communications of the ACM*, February.

Young, William (2016) Systems-Theoretic Security Engineering Analysis, Ph.D. Dissertation, Massachusetts Institute of Technology, May 2016.

From Safety Cases to Security Cases

R D Alexander, R D Hawkins, T P Kelly

University of York[1]

York, UK

Abstract *Assurance cases are widely used in the safely domain, where they provide a way to justify the safety of a system and render that justification open to review. Assurance cases have not been widely used in security, but there is guidance available and there have been some promising experiments. There are a number of differences between safety and security which have implications for how we create security cases, but they do not appear to be insurmountable. It appears that the process of creating a security case is compatible with typical evaluation processes, and will have additional benefits in terms of training and corporate memory. In this paper we discuss some of the implications and challenges of applying the practice of assurance case construction from the safety domain to the security domain.*

1 Assurance cases are a powerful way to capture arguments about system properties

An assurance case is a structured argument that a system has some properties we desire; that it is safe, or reliable, or secure against attack. Defining *safety* cases in particular, Kelly says *"A safety case should communicate a clear, comprehensive and defensible argument that a system is acceptably safe to operate in a particular context."* (Kelly 1998). We can generalise this easily to properties other than safety.

---

[1] Department of Computer Science, University of York, Deramore Lane, York, YO10 5GH, UK

The argument in an assurance case shows how a high-level claim (e.g. "the system is adequately secure") is ultimately supported by detailed evidence of particular low-level properties (e.g. some statistical testing data targeting a key security requirement on one small software component). An assurance case will refer to a range of evidence items, and will show how different types of evidence (e.g. evidence of good process, results of manual design review, and results of component testing) combine to give us confidence in higher-level properties (Hawkins and Kelly 2010). In other words, the assurance case captures the *rationale* of why the evidence we have produced (the results of the low-level analysis activities we have carried out) gives us reason to believe the high-level claim (which is, ultimately, what we are interested in). Evidence on its own is never enough – even the strongest, most relevant evidence provides only partial support for a given high-level claim, and making that connection between evidence and claims requires an act of subjective judgment. Assurance cases make that subjective reasoning explicit. An assurance case does not replace *any* specific technique for analysis or for generating evidence. What it does do is show the connection between those techniques and the high level claims we want to make. An assurance case also provides a way of capturing assumptions about a system – these may describe a particular context, including a specific version of the system, a description of its environment (including particular threats present in that environment) and a description of how the system will be used.

Security and safety are open problems – the domain is open-ended and ultimately somewhat subjective. Security is worse in this respect, because of increased uncertainty about the capabilities and knowledge of attackers. At any time, new information may arise (about our system or about its environment) and thus it is necessary to review the reasoning about the safety or security of the system. Assurance cases provide a record of what that reasoning was.

In a sense, whenever we honestly claim that a system is acceptably safe or secure, we have an implicit assurance case; we have some mental model behind that claim that we could probably describe if asked. By making an *explicit* assurance case, however, we open up that mental model to review and criticism by others, and we record our reasoning so that others can learn from it. For example, if someone wants to make a change to the system, they can use the assurance case to help assess the impact of that change.

Acceptance of an assurance case is not a mechanical process; it requires subjective assessment by a customer, regulator or evaluator. This does not mean that it is wholly arbitrary and idiosyncratic; we can have rigour in assurance cases by drawing on the work that philosophers – such as Toulmin (Toulmin 1958) – have done on informal logic. See (Kelly and Weaver 2004) and (Bishop et al. 2004) for an explanation of this. Key techniques include assurance case patterns (to capture best practice in argument structure), systematic review processes, and the use of appropriate notations.

It is typical, in safety, for the developer of a system to create a safety case for that system and submit the safety case to the customer or regulator for their assessment. In many safety regimes this is required by standards. If the developer is not required to produce a safety case, it is possible for a third party to produce one as part of a safety evaluation or assessment process. In particular, a high-level assurance case can provide a strong tool for understanding how the safety efforts taken by a developer fit together to create a safe system.

2 Assurance cases have a strong track record in safety

Safety cases have been widely used for over 30 years and are now a mature approach; for example, they are required by the MOD for all equipment acquisitions, under Def Stan 00-056 (MOD 2015).

One benefit of producing a safety case, especially early in development, is that it can help engineers to think about where they need particular arguments, justification and evidence. It can also lead to early recognition of deficiencies in the system design or the process around it. For example, in one case study carried out by researchers at the University of York, creating a safety case revealed a lack of traceability between high-level requirements and evidence about the implementation. Similar results occurred in a security case example produced by Ankrum (see Ankrum 2008 and section 3.1.2 of this paper).

For safety cases to be effective however, they have to be approached with intent to genuinely capture the safety (or not) of the system; failing that, they need to be assessed by evaluators who are competent and willing to find flaws in them. A safety case produced and accepted as part of a box-ticking exercise will not be effective; one of the findings of the report into the Nimrod accident (Haddon-Cave 2009), was that the Nimrod safety case had been such an activity. Kelly, in (Kelly 2008) identifies, through experience of reviewing safety cases, several different ways in which they may be ineffective.

3 There are good reasons to apply assurance cases in the security domain

We explained in sections 1 and 2 that assurance cases are widespread in safety. In this section, we will review the potential for applying them in the security domain. In section 3.1 we look at whether (and how) security assurance cases can be created. In section 3.2 we discuss the differences between safety and security, and in section 3.3 we look at the possible benefits from security cases, given typical security practices and current challenges. In section 3.4 we review some of the practicalities of moving to a security case approach.

3.1 It is practical to create security assurance cases

3.1.1 Methods and guidance are available

There is already some published guidance on creating security cases. Basic advice is provided by Goodenough, Lipson and others – see (Goodenough et al. 2007) and (Lipson et al. 2008) – as part of the "Build Security In" initiative in the USA. The advice there is not particularly detailed, but it is clear and practical. Their motive for proposing security cases is the increasing complexity of systems. They note that many people have responded to increasing system complexity by proposing an empirical, post-hoc approach (treat the developed system as a natural phenomenon, and assess its security after deployment by observing the number of security flaws uncovered). They reject this: they don't see how it can provide the level of confidence needed for high-security systems. Their approach is aimed at vendors; they want vendors to instrument their development processes with evidence-generating activities, and use a security case to capture the result.

A more detailed process is provided by the SAFSEC standard (Dobbing and Lautieri 2006), which was developed by Altran Praxis as a way to unify safety and security cases. In the SAFSEC approach, safety and security risks are treated equivalently – a combined set of mitigations is proposed and a single assurance case is produced that argues that the system will be safe *and* secure. The safety side is based on Def Stan 00-56, and the security side on release 2 of the Common Criteria. It is broadly compatible with similar work on safety-security unification by the Industry Avionics Working Group (IAWG) (IAWG 2007).

The SAFSEC standard provides a strong analogy between safety and security in an assurance case framework – it explains where safety-security commonality exists and provides a specific process to exploit that. As such, it could be followed directly to create a security case. It does not, however, address the cultural, epistemic and economic challenges that we discuss in Section 3.2.

3.1.2 There is some published experience with security cases

There have been a number of security cases created in small-scale case studies. In (Ankrum 2005), Todd Ankrum of MITRE briefly outlines a case study he carried out for the US National Security Agency. The system was a secure enclave project by the NSA's research division, which used several authentication systems and provided an access log. The software had been developed to Common Criteria EAL 5, used formal methods, and extensive documentation was available including a vendor-supplied Security Target document and a NSA-supplied Protection Profile.

When Ankrum and his colleagues produced a security case for this system, they found a number of problems. Despite the obvious rigour with which the project had been developed and documented, it was not possible to fully argue that the system met its security goals. First, although most security threats could be traced to requirements that mitigated them, at least one threat had no such requirements. Second, once the threat-requirements connection was explicitly made, it was not clear that the requirements for each threat sufficiently mitigated the threat; presumably the vendor's process had not made this relationship explicit, so it had not been apparent before. Finally, Ankrum's security case attempted to argue that all security enforcing functions had their dependencies met, and it became apparent that this was not true in all cases.

Lautieri and her colleagues at Altran Praxis describe a case study in (Lautieri et al. 2005) of a Command and Control system, for which they produced a combined safety and security argument. Their paper explains how they create a modular case for this system, and notes that the combination of safety and security was quite acceptable to certifying authorities that were only concerned with one of them; combining the two domains did not cause a communication problem.

Ankrum has also illustrated how an assurance case can capture the implicit argument in a standard that doesn't demand an assurance case. Specifically, in (Ankrum 2005) he briefly outlines an argument that a product has achieved Common Criteria EAL 4, by starting with the claim "Product meets EAL 4" and arguing that it does everything the Common Criteria requires to support that claim. Ankrum's approach has some weaknesses; for example, it is primarily a process rather than product argument, and it is perhaps better described as a compliance case rather than a true security case. The process he gives for creating the case is perhaps over-mechanized; realistic security case creation will involve more subjective judgement than that. It has the value, however, of showing how non-assurance-case standards can be adapted.

The concept of security cases has received increasing attention in the literature in recent years. Knight (Knight 2015) suggests that existing security verification techniques such as proof need to be supported and contextualised within the framework of rigorous (but informal) reasoning that assurance cases offer. He (He and Johnson 2012) present some positive experiences of using security cases (structured using the Goal Structuring Notation) in the Healthcare IT domain. Other papers have continued to explore how best to structure the security case. For example, Yamamoto et al. (Yamamoto et al. 2013) suggest a structure based upon the Common Criteria. There are also some positive experiences in the sub-domain of 'security informed safety cases' (Netkachova et al. 2015) which specifically address the extension of existing safety case practice to include appropriate consideration of security-related safety failings.

3.2 There are significant differences between safety and security

There is no question that safety and security are separated by different goals. The open question is whether the same means (in this paper, assurance cases) can be used to demonstrate achievement of those goals. Cockram and Lautieri say in (Cockram and Lautieri 2007) that the two domains can be served by the same assurance case mechanism: both can use cause-effect models (e.g. fault trees or attack trees), both can derive requirements to mitigate the problems thus identified, and both can argue in an assurance case that those requirements are implemented and that they will achieve the mitigation that is needed. Lautieri et al. (Lautieri et al. 2007) also give an example where separate requirements for safety and security were merged to create a single requirement that served the needs of both domains.

Beyond the basics discussed by Cockram and Lautieri, there are a range of theoretical and practical differences that we need to consider.

3.2.1 Theoretical Differences

The obvious difference between safety and security is the presence of an intelligent adversary; as Anderson (Anderson 2008) puts it, safety deals with Murphy's Law while security deals with Satan's Law. Safety is mostly concerned with the predictable (or random) behaviour of the non-human, non-goal-seeking world, and with adaptive behaviour by humans (e.g. seeking to make their job easier) that is not aimed at reducing safety per se (although it often does as a side effect). Security has to deal with agents whose goal is to compromise systems. These attackers may be systematically probing a system for vulnerabilities (rather than acting randomly) and may realise when they have breached one defence and move to exploit that (whereas non-human phenomena cannot respond to such feedback, and non-malicious humans may try to undo their actions if they realise they have bypassed a defence).

Another contrast is that a lot of effort in traditional system safety has gone into assigning probabilities to basic events (e.g. random failure of a valve) and computing the probabilities of accidents stemming from combinations of those events. Assigning probabilities to the action of unknown intelligent adversaries is dubious – our uncertainty there is *epistemic* (due to lack of knowledge) rather than *aleatory* (due to chance).

Safety does, however, already have to worry about epistemic uncertainty. This has become very apparent over the last 30 years as systemic faults (faults in the design that will cause failures under certain circumstances) have overtaken random failures as the cause of accidents. Partly, this is due to the great strides that system safety has made in handling random failures through component redundancy and architectures that support it. It has also come, however, from increasing use of software. There are methods for assigning failure probabilities to software – e.g. see Bishop (Bishop 2005) – but they do not have great credibility.

Increasingly, software safety is approached in terms of the evidence we can generate and the confidence it gives us that the software's behaviour will be safe in context. Where the link between some evidence and the claims it supports is inadequate, we have an *assurance deficit*. The significance of assurance deficits has been recognised in safety – e.g. see (Hawkins and Kelly 2009) and (Hawkins et al. 2009) – and they may be even more significant for security because of increased uncertainty about attacker's goals, capabilities and actual actions.

Because of this high uncertainty about attacker behaviour, it is common in security-critical development to utilise a variety of security measures that are not responses to specific threats. These measures instead provide a degree of protection against a whole class of attacks. For example, if one process has to be given temporarily elevated privileges, it is common to give it those privileges for only the minimum amount of time. In safety, there is concern about such generic "hardening" – a safety feature added without good rationale may merely be adding to the complexity of the system, and thus increasing our chance of introducing an error and not discovering it.

Well-justified defences against common errors are, however, common in safety. For example, safety-critical programming language subsets (such as MISRA C) often forbid dynamic memory allocation, thereby showing that errors from running out of heap space cannot occur. Similarly, they often forbid recursive calls to functions, allowing the size of the stack to be bounded and thus ruling out all stack overflow errors.

Creating an assurance case can help to capture the rationale for such hardening, and thus distinguish between features that give a clear benefit, and features for which the benefit is uncertain. If it is not possible to justify the inclusion of a security feature (a valid place for it in the argument cannot be found), then you may be spending effort on measures with no actual security impact. Just like safety, all security is a trade-off, and every software feature has a cost (if nothing else, in the complexity of the resulting architecture and hence in the chance for an evaluator to miss a vulnerability).

Security evaluators often use static analysis for vulnerability signature detection – not to detect specific violations of known requirements, but rather to find "holes" in the system that an attacker might be able to exploit (Gutgarts and Temin 2010). This is quite compatible with assurance cases, and can be incorporated into the argument as a complement to claims about dealing with identified threats. In (Hawkins and Kelly 2009), the authors provide a set of argument patterns for software. Although these patterns provide arguments regarding safety, they could be adapted for software security. In such cases the use of static analysis for vulnerability signature detection fits neatly into the part of the argument where it is shown that additional hazardous contributions at the code level have been identified. In other words, it is argued that you have looked well enough for common low-level vulnerabilities that there is adequate confidence in their absence. "Adequate" can be scaled to a level that suits the security criticality of the system.

A final theoretical difference is that security-critical software often has to adapt quickly as attack patterns change (Carter 2010). This has implications for the use of assurance cases, because a case may have to change if the software changes. Even if it turns out that there is no impact on the case, the case maintainer will have to spend time reviewing the case to confirm this. Assurance cases can therefore increase the cost of making changes, and introduce delays. On the other hand, we need to understand the security impact of any changes we make; if we don't, we may fix one vulnerability but create a worse one in the process. If we have an assurance case, then we can use it to trace a low-level change up to its implications in terms of our high-level security claims. The challenge of making assurance cases more dynamic in response to changes to the system or its environment is an area of increasing interest for safety cases, and work in this area will also benefit security cases.

3.2.2 Practical Differences

The previous section talked about the fundamental distinctions between safety and security – those that are likely to endure over time. In this section, we will look at the practical differences, which may be accidents of history; they cannot be ignored, but it may be possible to change them. A number of these distinctions were identified in a meeting between representatives of the safety and security communities in the UK (Carter 2010).

Perhaps the biggest practical concern is the process maturity of security-critical practice. In safety-critical software, even at lower criticality levels, mature development processes are the norm: developers use good processes for requirements management, test planning and configuration control. They monitor their processes for weaknesses, and correct them when they find them. In security-critical projects of an ordinary (commercial-grade) standard, this is not always the case. Requirements may be implicit (requiring evaluators to define their own security targets), testing may be unstructured (making it difficult to relate test schedules to specific requirements) and configuration control may be poor (making it difficult to relate review or test results to specific software versions).

The process problems in security-critical software are not necessarily unique to the domain; more likely, they are present because vendors have not had the economic incentives to do better. Poor process can make it difficult to produce an assurance case. It may be that the system is adequately secure, but because of poor process it is not possible to get adequate evidence of this. Similarly, it may be that we cannot build our case adequately because we cannot understand the security requirements. These are problems, certainly, but they are not problems with assurance cases: they are problems with the system being evaluated. If we cannot understand the system well enough to build an assurance case, then we are not in a position to say that it is secure. Problems with the assurance case may be the "mine canaries" that alert us to problems with the system.

It has been suggested that safety has more problems with requirements, security more with low-level defects in implementation. For example, this view was reported by Carter (Carter 2010) and seems to be an assumption of Lipson (Lipson and Weinstock 2008). This could be a fundamental distinction, but it may just be a side-effect of immature processes in most security-critical software. Once your development process can reliably implement the requirements that you set out to implement, then getting the requirements right is the one place where problems can still manifest (as requirements engineering is ultimately unbounded, there will always be requirements errors). As process maturity improves in security-sensitive software vendors then this difference may vanish.

It has been noted that there are some specific differences in standards and guidance between security and safety in terms of the development practices they demand. For example, King noted in (King 2009) that an avionics system certified to the highest level of DO-178 (RTCA 1992) may not meet Common Criteria EAL 5 or above, because EAL 5 requires that the "developer mathematically prove the security properties of the [software]", whereas DO-178B does not. These are likely incidental, rather than fundamental differences, and agreement could be reached between standards (the inevitable politics aside).

Overall, we suggest that security evaluators remain aware of these practical differences, but experiment with assurance cases based on the safety case model, and see what obstacles they encounter in practice. There is a wealth of experience with assurance cases in the safety domain – where obstacles appear in security, there may already be known solutions.

A final note on practical challenges – safety-critical systems are becoming increasingly software-controlled and increasingly connected. As a consequence, there is an increasing threat of malicious outsiders causing safety-critical failures. If the security-critical software domain does not adopt assurance cases, then the safety-critical software industry will have to extend safety cases to include security. We will, therefore, need to resolve the problems above, and this will be easier if assurance cases are adopted by the security community as well.

3.3 Security cases can help security practice

3.3.1 There are benefits for handling complexity

As noted earlier, the major role of assurance cases is to combine evidence from diverse analyses and show how they complement each other. For example, code review alone provides limited assurance – high assurance will require a range of complementary techniques targeted at fairly specific types of attack. This might include code review, static analysis of code-level requirements (e.g. pre- and post-conditions), static analysis for vulnerability signatures, and statistical or systematic tests.

Assurance cases provide a way to connect system-level properties (e.g. security of certain data) to low-level requirements and analyses. For example, it can be difficult for an evaluator to implicitly maintain the connection between the code in a given source file and the system-level goals to protect certain data or prevent denial of service.

When effective, assurance cases can focus evaluator attention on critical parts of the system at the expense of others. In a world of finite effort and growing software complexity, this is necessary. It is likely that a given evaluator already has some way of doing this – for example, they may concentrate primarily on security enforcing functions, rather than trying to review all of the code. Creating an assurance case allows you to put this in context and see how much assurance such a strategy really gives you.

If a vendor provides a security case, this focussing aspect allows you to treat the case as a summary of the vendor's security thinking – what they have concentrated on, and what they have not. It may be safe to assume that some vulnerabilities will appear in the aspects that they have not addressed, which means that an evaluator can concentrate on attacking those areas. This prioritisation would be difficult to do with only an implicit security case.

Where problems are encountered in assurance because of limited expertise, it may be possible to use an assurance case to "bracket" or encapsulate those problems ready to hand them over to a third-party expert. For example, we may have a portion of a software system that deals with radio communications according to a complex industry standard. As a specialist in software security evaluation, we may not be able to understand that part of the software well enough to assess its security. What we can do is make a number of claims about that component, claims that if true would allow us to support the overall security claims of the system. We can then ask a third-party domain expert to check those claims against the specialist software.

One tool for managing complexity in assurance is modular safety cases (Kelly 2005). For example, if there is an operating system that is often used in multiple systems, a security case module may be created for it. This module would make certain claims about the operating system (for example, that there was no way for a process to elevate its privileges without appropriate authorisation) subject to certain dependencies (for example, that all loaded device drivers could be trusted). This argument module could then be used as part of a security case for any system using that operating system. Similarly, modules might be created for common hardware or for common support applications.

The modular case approach has been explored in other domains (Cockram and Lautieri 2007) with the aim of reducing certification update costs. Both the IAWG (IAWG 2007) and SAFSEC (Dobbing and Lautieri 2006) processes support some degree of modularity. If an evaluator is creating assurance cases, they may be able to create modules for common OSs, hardware and software components, thus simplifying the process and reducing lifetime costs.

The security of a system obviously depends on the context in which it operates. The assurance case for a system may need to change if the system's context changes, for example if the system needs to communicate with a new peer system. Thomas (Thomas 2010) raises this as a potential obstacle to the adoption of assurance cases. It is a genuine problem, although it is one faced in the safety domain as well, particularly as the complexity of safety-critical software increases. Modular assurance cases may help to deal with this challenge.

3.3.2 There are benefits for justifying decisions

When an evaluator believes that a system is insecure, they can demand that the vendor generate additional evidence (e.g. run more tests), or they can demand changes to the design, or they can impose restrictions on how the system is used. If the evaluator is all-powerful then this is not a problem. Realistically, evaluators have to justify their position against claims by vendors that the change will be very expensive. If the problem is an objective vulnerability (something that can be seen e.g. in the code, and is easy to exploit) then this may be straightforward. If the concern is an assurance deficit – perhaps a lack of confidence that a certain requirement is satisfied – the justification needs to be rather more subtle. Assurance cases can help with providing such justification by linking analysis activities and claims about the system design to high-level security properties.

3.3.3 They can enhance typical evaluation processes

As noted in section 1, assurances cases can be created retrospectively. Indeed, if the vendor does not produce an assurance case then the evaluator can. This is similar to when an evaluator creates a security target document because the vendor did not supply one (or supplied an inadequate one). There are potential traps in this activity, particularly if the evaluator is unwilling to reach the conclusion that the system is not secure – see the Nimrod report again (Haddon-Cave 2009) for an analogous example in safety. The cost of creating an assurance case retrospectively may be high, but doing so should also lead to some specific benefits: in creating a retrospective security case an evaluator creates a paper trail for their reasoning – they will capture a justification for why they think the system is adequately secure (or why it is not).

The first benefit of this that if we later need to re-evaluate the product (for example, after a major change or security incident), we know what the original evaluator did and can review only that which has been invalidated by the change.

A second benefit of such records is that they can provide a training tool and a mechanism of corporate memory. Junior evaluators can look at existing security cases to understand what kinds of security processes and evidence are acceptable to the evaluating organisation – they can see what was included and what was emphasized. Similarly, if an experienced evaluator encounters an unfamiliar type of system or an unfamiliar set of security requirements, they can look over past assurance cases for similar systems, and see what their peers did. There are existing techniques, such as assurance case patterns, for distilling this kind of information into a generic form.

The provision of a security case (whether by vendor or evaluator) may help coordinate teams of evaluators. The case could provide a central structure around which activity is organised – one evaluator could be assigned to each abstraction tier, or to a major component, or to a high-level requirement. They can also record their results in terms of the assurance case structure (so, for example, an evaluator working at the source code level may note problems there, and trace their significance back up to security claims made at the system level).

The costs and benefits of assurance cases are likely to vary with the level of security claimed. At lower levels, the assurance being sought may be relatively modest, which will make it easier to create a compelling security case. On the other hand, the product may be off-the-shelf, and when this is combined with low process maturity it may make it hard to create the case at all. As noted above, of course, if the evaluator requires good justification of security then inability to create a case is grounds for rejecting the product.

As we move to higher security levels, the product is more likely to be bespoke and the evaluator is likely to be involved from an early stage. This may make it easier to justify a vendor-created security case, or one produced through vendor-evaluator collaboration. The risk at higher levels is that the case will need to claim a very high level of assurance, and this may be difficult to justify. As ever, if a security case cannot be made compelling, then it may well that the system is not secure.

3.4 There are practical challenges in moving to security cases

As noted in section 3.1.2, there are no public descriptions of major applications of security cases. It is therefore difficult to say what the costs and timescales will be. However, we can draw on experiences of moving from procedural safety standards to safety case approaches. The major need is for expertise: at least the evaluators, and maybe the vendors as well, will need to learn to produce and assess assurance cases. This expertise does not need to be spread uniformly throughout the population, but a sufficient number of staff will need to be skilled at judging assurance cases and fluent in appropriate notations, e.g. GSN. There are published accounts of moving from a prescriptive standard to a safety case approach, and from moving from a textual case to a GSN case. For example, see Chinneck et al (Chinneck et al. 2004).

The move to assurance cases does not always have a major disruptive effect on the products being assured, or on the techniques used to evaluate them. In the first instance, assurance cases merely provide a way to relate what is already done to the goals that are already held (although those goals may previously have been implicit). Their impact will not, for example, be comparable to moving a vendor's software development approach from procedural to object-oriented. What they *may* do, in terms of disruptive influence, is reveal that there are weaknesses in the products or in the evaluation of them. That is, after all, the point of an assurance case: it should allow a third party to assess whether the system has the properties that are claimed for it. When an assurance case does reveal problems, then evaluators and vendors may, of course, decide that they need to change their processes and tools. *That* may be disruptive.

If an evaluator already produces a security target document, then an assurance case is partly an extension of that - it maps threats onto requirements, and requirements onto justification of their adequacy and evidence that they have been implemented. If the vendor already follows the requirements of the Common Criteria, particularly the refinement structures required by the higher EALs, then there may be a clear mapping onto the existing software safety case patterns, (Hawkins and Kelly 2009), which use a similar structure of refinement through 'tiers' of development.

4 Conclusion – there is potential, cost, and risk

Adopting assurance cases can offer benefits to security assurance practice, potentially increasing the rigour and transparency of the security evaluations.
However, there will be costs to adopting assurance cases (e.g. associated with
training), and there is an element of financial risk here. Assurance cases are
widely used in safety, but there has been limited use in security thus far. To
gain the benefits from assurance cases, both developers and regulators will need
to develop significant in-house expertise, and address some of the challenges
specific to security that we have highlighted in this paper. Finally, it is worth
recognising that some have expressed concerns that the preparation and presentation of a 'full and frank' security case (that highlights, for example, contextual
assumptions) could itself present a security risk, i.e. it could aid an attacker
greatly in the formulation of their attack vectors. These concerns bring us back
to long-standing debates concerning the effectiveness of *security by obscurity*.

Acknowledgments   We acknowledge the financial support provided by the UK CESG for
the original study underlying this paper.

References

Anderson R. Security Engineering: A Guide to Building Dependable Distributed Systems.
2nd ed: John Wiley & Sons; 2008.
Ankrum TS, Kromholz AH. Structured Assurance Cases: Three Common Standards. 2005.
www.asq509.org/ht/action/GetDocumentAction/id/2132
Bishop P, Bloomfield R, Guerra S. The future of goal-based assurance cases. Proceedings of
the Workshop on Assurance Cases - supplemental Volume of the 2004 International Con-
ference on Dependable Systems and Networks, 2004.
Bishop P. SILs and Software. Safety Critical Systems Club Newsletter. 2005;14:2.
Carter A-L. Safety-Critical Versus Security-Critical Software. British Computer Society;
2010.
Chinneck P, Pumfrey D, Kelly T. Turning Up the HEAT on Safety Case Construction. Pro-
ceedings of the Twelfth Safety-critical Systems Symposium, 2004.
Cockram TJ, Lautieri SR. Combining Security and Safety Principles in Practice. 2nd IET
International Conference on System Safety, London, 2007.
Dobbing B, Lautieri S. SafSec Methodology: Standard. 3.1 ed., Altran Praxis; 2006.
Goodenough J, Lipson H, Weinstock C. Arguing Security - Creating Security Assurance Cas-
es. 2007. https://buildsecurityin.us-cert.gov/bsi/articles/knowledge/assurance/643-
BSI.html
Gutgarts PB, Temin A. Security-critical versus safety-critical software. IEEE International
Conference on Technologies for Homeland Security (HST), Waltham, MA 2010.
Haddon-Cave C. The Nimrod Review. 2009.
Hawkins R, Kelly T, Knight J, Graydon P. A New Approach to Creating Clear Safety Argu-
ments. Proceedings of the Safety-critical Systems Symposium, 2011.
Hawkins R, Kelly T. A Structured Approach to Selecting and Justifying Software Safety Evi-
dence. Proceedings of the 5th IET System Safety Conference, Manchester, 2010.
Hawkins R, Kelly T. A Systematic Approach for Developing Software Safety Arguments.
Proceedings of the 27th International Systems Safety Conference, 2009.

Hawkins R, Kelly T. Software Safety Assurance - What Is Sufficient? Proceedings of the 4th IET System Safety Conference, London, UK, 2009.

He, Ying, and C. W. Johnson. "Generic security cases for information system security in healthcare systems." *System Safety, incorporating the Cyber Security Conference 2012, 7th IET International Conference on.* IET, 2012.

IAWG. Short Study: SafSec Coherence. Industrial Avionics Working Group; 2007.

Kelly T, Weaver R. The Goal Structuring Notation - A Safety Argument Notation. Proceedings of the Dependable Systems and Networks 2004 Workshop on Assurance Cases, 2004.

Kelly T. Are 'Safety Cases' Working? Safety Critical Systems Club Newsletter. 2008;17:2

Kelly T. Arguing Safety - A Systematic Approach to Managing Safety Cases [PhD Thesis]: University of York; 1998.

Kelly T. Using Software Architecture Techniques to Support the Modular Certification of Safety-Critical Systems. Proceedings of the Eleventh Australian Workshop on Safety-Related Programmable Systems, Melbourne, Australia, 2005.

King T. Two different realms: RTOS support for safety-critical vs. security-critical systems 2009. http://www.vmecritical.com/articles/id/?4031

Knight, J., 2015. The importance of security cases: Proof is good, but not enough. *IEEE Security & Privacy, 13*(4), pp.73-75.

Lautieri S, Cooper D, Jackson D. SafSec: Commonalities Between Safety and Security Assurance. Thirteenth Safety Critical Systems Symposium, Southampton, 2005.

Lipson H, Weinstock C. Evidence of Assurance: Laying the Foundation for a Credible Security Case. 2008. https://buildsecurityin.us-cert.gov/bsi/articles/knowledge/assurance/973-BSI.html

MoD Defence Standard 00-056 Issue 6 - Safety Management Requirements for Defence Systems. Ministry of Defence; 2015.

Netkachova, K., Müller, K., Paulitsch, M. and Bloomfield, R., 2015, September. A layered approach to architecting security-informed safety cases (applied to an avionics case study). In 2015 IEEE/AIAA 34th Digital Avionics Systems Conference (DASC) (pp. 1-36). IEEE.

RTCA. DO-178B: Software Considerations in Airborne Systems and Equipment Certification. 1992

Thomas R. Software Assurance Using Structured Assurance Case Models. Journal of Research of the National Institute of Standards and Technology. 2010.

Toulmin S. The Uses of Argument: Cambridge University Press; 1958.

Yamamoto, S., Kaneko, T. and Tanaka, H., 2013, March. A proposal on security case based on common criteria. In Information and Communication Technology-EurAsia Conference (pp. 331-336). Springer Berlin Heidelberg.

Cyber Safety and Security for Reduced Crew Operations (RCO)

Kevin R. Driscoll

Honeywell International, Inc.

Abstract *The civil aviation industry is looking into reduced crew operations (RCO) that would cut today's two-person flight crews down to single-person crews with support from ground-based crews. Shared responsibility across air and ground personnel will require highly reliable and secure data communication with supporting automation, which will be safety-critical for passenger and cargo aircraft. This paper looks at the different types and degrees of authority delegation given from the air to the ground and the ramifications of each, including the safety and security hazards introduced, the mitigation mechanisms for these hazards, and other demands on an RCO system architecture, which would be highly invasive into (almost) all safety-critical avionics. The related areas of unmanned aerial systems and autonomous ground vehicles are reviewed to find problems that RCO may face, and related aviation accident scenarios are described. Potential problems with RCO data communication encryption are identified. This paper concludes with questioning the economic viability of RCO in the light of the expense of overcoming the safety and security hazards it would introduce.*

1 Introduction

During the latter half of the past century, advances in avionics and related technologies have: (1) reduced the workload of aircraft flight crews and (2) allowed for the reduction of aircraft crew from five in the 1940s to two. Today, there are research efforts underway for Reduced Crew Operations (RCO) or Single-Pilot Operations (SPO), which usually relate to US Federal Aviation Regulations (FAR) Part 121 (US 2016) operations or equivalents. Some FAR Part 135 operations already are approved for single-pilot operations. The "reduced crew operations" phrase can be read in one of two ways: reduced crew-operations[1] or reduced-crew operations[2], relating respectively to the reduction trends of the past half-century.

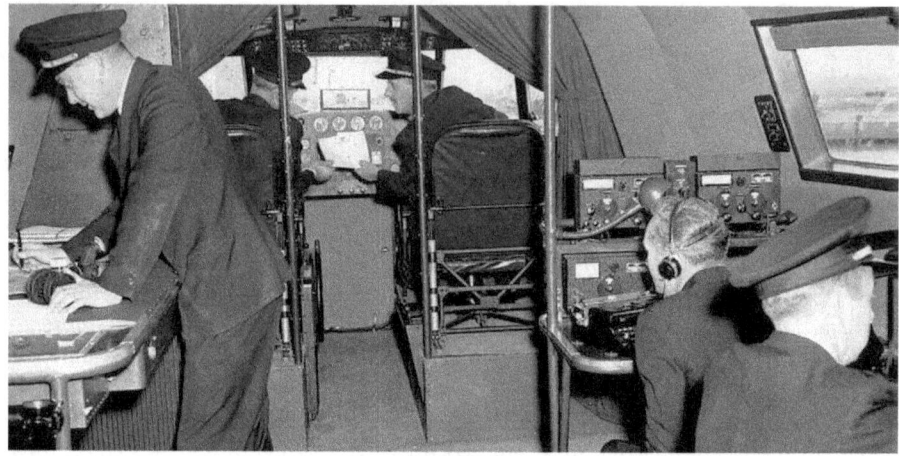

Fig. 1. Typical Civil Aviation Flight Deck Crew from the 1940s

The Advanced Cockpit for Reduction of Stress and Workload (ACROSS) study, which is funded, in part, by the European Commission under its Seventh Framework Program, is typical of efforts in this area (EU 2013). It includes objectives covering both interpretations of the "reduced crew operations" phrase. Similarly, this paper covers both interpretations of the "reduced crew operations" phrase, but with emphasis on the latter interpretation due to its greater safety and security issues.

As an apparently logical extension to the crew reduction trend, the RCO concept would reduce today's two-person flight crew down to a single-person crew, with support from crew on the ground that is shared among multiple aircraft. However, significant safety and security hazards could be introduced. Shared responsibility across on-air and on-ground personnel will require a highly reliable data communication system that offers very low latency and jitter, as well as high data integrity. In addition, effective protection of the end-to-end information system will be critical to ensure the safety of passengers for passenger aircraft and the survival of the aircraft, crew, and cargo for cargo aircraft.

---

1 Reducing the number of operations that must be performed by the crew
2 Reducing the number of crew members

2. Levels of Authority Delegation

The types and degree of safety and security hazards introduced by an RCO system depends on the degree of authority that an airborne crew relinquishes to the ground and any of its supporting automation. The following subsections describe different degrees of this authority delegation, the ramifications of each (including hazards introduced), the mitigation mechanisms for these hazards, and other demands on an RCO system architecture.

Much of the existing RCO research has been aimed at the human part of potential RCO systems (e.g. determining the meaning of cockpit resource management (CRM) when some of the cockpit resources are not in the cockpit or anywhere near the cockpit). This paper will not revisit this previous research, examples of which can be found in the ACROSS work mentioned above and some NASA SPO work (Comerford et al 2013). This paper will focus mainly on the safety and security aspects of hardware and software; it will cover only the human parts of RCO that have not been well explored in other research and are tightly tied to the hardware and software. Something that falls into the latter category is handover effects during changes of authority delegation, particularly changes to whether the airborne crew (AC) or the ground crew (GC) is the pilot-in-command. What exactly happens when more authority is shifted toward the AC or toward the GC? Is there a time during the handover when neither have control of the aircraft?

In these subsections, it should be understood that the terms 'ground crew' and 'GC' refer not only to the totality of the ground component of an RCO system, but also to any airborne automation that supports the ground crew interface.

A possible variation for each of those authority delegation degrees in which the GC is in command includes the addition of an untrained or lesser-trained person in the cockpit who just carries out commands from the GC. This does not represent any difference in the level of authority allocation. Such a variation is best viewed as a 'biology-based actuator' for the GC.

2.1. AC is pilot-in-command, GC is just standby redundancy

This is the minimal level of authority delegation. The GC actually has no immediate authority over the aircraft, but the GC has the capability of elevating its authority to one of the following levels of delegation. An issue here is how that elevation is performed. For RCO operations with a single AC member, this elevation must be able to be performed after the single AC member becomes incapacitated. It is clear that some form of automation would have to be used to detect that the AC has become incapacitated and to elevate the GC authority to take over from the loss of AC capability. This automated ability to detect inca-pacitated crew and elevate GC authority must be a full-time capability. It is incorrect to assume that the dependability requirements for an RCO system can be reduced based on an argument that it is called into play only after the AC has been incapacitated. Even at this lowest level of authority delegation, an RCO system must have the full-time capability of assuming authority over the air-craft. Thus, while an RCO system can be argued to have lower availability re-quirements due to this argument, the integrity (commission failures) require-ments for an RCO system are just as stringent as for any other full-time safety-critical aircraft system.

2.2. AC is pilot-in-command, GC is active second pilot

At this level of authority delegation, the GC is another pair of eyes, sharing the see-and-avoid responsibility with the AC. In addition, the AC may delegate some specific sub-duties to the GC. Again, this calls into question the meaning of RCO CRM. What are the GC's 'eyes'? Adding multiple video cameras would require high-bandwidth and potentially safety-critical communication from the aircraft to the ground. There is a question of whether there will be sufficient available bandwidth to support this video traffic. Papers have been written showing that L band communication has insufficient capacity and that C-band would be required. Given that many of the arguments for RCO envi-sion its greatest use in transoceanic flight, this would mean C-band satellites. However, there are no C-band satellites for commercial aviation. Moreover, there are no plans to create any such satellites.

2.3. GC is pilot-in-command, AC is active second pilot

At this level of authority delegation, the GC has immediate full authority over the control of aircraft. In addition to the communication requirements of the previous section, the communication for this level of delegation has an addi-tional requirement for low round-trip latency and jitter (the variation in laten-cy). This communication is also now fully safety-critical.

This level of authority delegation begins to raise the AC recovery time issue (the amount of time it takes an AC to resume control of the aircraft if GC com-munications or onboard systems fail). During this time, neither crew is in con-trol of the aircraft.

2.4. GC is pilot-in-command, AC is just standby

At this level of authority delegation, the AC is further out of the loop. The AC would require more recovery time if GC communications or onboard systems fail. There are varying degrees of the AC being out of the loop. These include being in the cockpit: eating, doing logbook, working on schedule, napping, etc. or being out of the cockpit: lavatory, sleeping, checking on abnormalities, etc.

2.5. GC is pilot-in-command, AC is incapacitated/unavailable

At this level of authority delegation, the GC has full authority over the aircraft, with the AC being incapacitated or otherwise unavailable to share in any cockpit duties. Sometimes overlooked when considering AC incapacitation is the fact that the AC's incapacitation may be of a type (for example, seizure or dementia) that would cause them to perform some action(s) that are indistinguishable, at least in part, from the cases described in the next subsection.

2.6. GC is pilot-in-command, AC is an adversary or is suicidal

At this level of authority delegation, the GC has full authority over the aircraft and has to deal with an AC that may be an adversary, such as a hijacker, or an authorized AC member that has become suicidal. After the suicidal hijackings of 9/11, the German wings suicide, and the potential Malaysian MH370 suicide[3], there have been calls to prevent these kinds of aircraft loss by creating some mechanism for the control of aircraft from the ground. However, there are huge (probably insurmountable) problems trying to do this. One of these is the fact that the system would have to prevent all the possible ways that an adversarial or suicidal AC could prevent an aircraft safely completing its flight; and, there a lot of ways that this could be done. Another problem is that any solution to this scenario creates a new, and probably more dangerous, scenario described in the next subsection.

---

[3] There have been several other such incidents: EgyptAir Flight 990, LAM Mozambique Airlines Flight 470, Royal Air Maroc Flight 630, SilkAir Flight 185, Japan Airlines Flight 350, Pacific Air Lines Flight 773, Pacific Southwest Airlines flight 1771, Federal Express Flight 705, Air France Flight 8969, Air Botswana, …

2.7. AC is pilot-in-command, GC is an adversary

At this level of authority delegation, one can argue that the GC has been given too much authority. If a GC can wrest complete control of an aircraft away from an AC, such a capability could be subverted by someone (inside or outside the system) or a component failure. This could lead to an AC that wants to continue safe flight, but would not be able to do so.

Such a design would violate the "do thy patient no harm" principle by creating a new cyber attack pathway into the aircraft and another source of natural failures that could adversely affect *all* safety-critical systems on an aircraft!

3. RCO Authority Questions

3.1. Should the GC to be able to override a 'rogue' AC?

The term 'rogue AC' is used here to mean adversarial crew, suicidal crew, or crew with incapacitation indistinguishable from the latter. An important design decision for an RCO system is to determine if that system should have the ability for the GC to override a rogue AC.

Providing the ability for the GC to override the AC leads to a troubling question: "Who has the ultimate authority, the AC or the GC?" The answer to this must be the same for all situations. Otherwise, who has the authority to decide what the situation is? Whoever/whatever has that decision authority is the ultimate authority. The decision to establish who/what has the ultimate authority must be made at the RCO system design time and is fixed for the life of the design. This means that any RCO system design that has a solution for the 'GC is pilot-in-command, AC is an adversary or is suicidal' scenario is mutually exclusive to any RCO system design that has a solution for the 'AC is pilot-in-command, GC is an adversary' scenario. There are no exceptions to this mutual exclusion. One cannot design an RCO system that can handle both of these scenarios; it has to be one or the other.

Obviously, the ultimate authority would have to be the GC if we want RCO to have the capability for the GC to override the AC. However, why should a GC be any less prone to being rogue than an AC? One can argue that there is a greater probability for a GC going rogue. They do not have to face certain death and they can crash more than one aircraft.

One could envision redundant GCs. However, redundant GCs would have to be fully independent of each other, including independent communication channels from and to the aircraft.

For this level of GC authority to be viable, the RCO system would have to prevent/mitigate:

- all the possible ways that a rogue AC could make the flight unsafe
- all the failure modes described the section 'RCO Interface to Onboard Safety-Critical Systems' below
- all the security intrusions that could have a severe safety impact

3.2. Should the GC be able to take over for an incapacitated AC?

It is highly probable that this will be required for single-person AC. The occurrence of having an incapacitated crewmember is not that rare. In the UK, this happens about once every 10 days (e.g. 36 in 2004 and 32 in 2009).

There are many ways that an incapacitation can cause, and has caused, an AC to inadvertently activate some control that is adverse to safety. These are more than just the Hollywood cliché of an AC having a heart attack and falling onto the controls. There have been a number of instances of seizures with limb extension. This can cause, for example, a hard-over push on the rudder pedal and then having the foot slip off the rudder pedal and jam it in the hard-over position. There have also been instances of dementia where an aircrew member was unaware of what they were doing.

Given these types of active incapacitation, solving the AC incapacitation problem is not any easier than the adversarial AC problem. We then have the following implication chain for the design of an RCO system:

single person AC → tolerate incapacitation → assume adversarial AC action(s)

4. RCO Interface to Onboard Safety-Critical Systems

4.1. Murphy and Satan

The points at which an RCO system integrates with traditional aircraft systems and the control paths that this integration needs to intercept will depend on the types and degree RCO authority. A failure of a component within an RCO system or a successful external attack into the RCO system can be coupled into traditional safety-critical aircraft systems via this RCO integration. These two sources of RCO-introduced safety hazards can be characterized as 'Murphy and Satan' (random naturally occurring failures and failures induced by humans with malicious intent, respectively). Protections must be provided for both; and, these protections must provide dependability commensurate with the highest criticality level for all aircraft functions that could be adversely affected.

When dependability requirements restrict the probability of failure to be less than 10^{-7} for a one hour exposure (approximately the failure probability of a single integrated circuit), Murphy is indistinguishable from Satan. That is, the worst possible human adversary attack also could be produced by Murphy with help from Mother Nature, with one major exception. This exception is that we assume independent components of a system will fail independently from natural sources, but humans can mount co-ordinated attacks against multiple components. However, the RCO system interface into existing aircraft safety-critical systems is an exception to this exception. That is, a failure of this interface can appear as a co-ordinated attack against multiple aircraft safety-critical systems, which had been independent until coupled through the RCO system interface! Thus, the RCO system interface not only would have to be Level A, the highest level defined by DO-178 (RTCA 2011), it would have to be what is euphemistically called 'Level A+', because it would control a very large number of Level A functions.

Those who are not well versed in the way that things can fail usually assume that failures are somewhat benign, often consisting only of omission failures. However, at the low levels of failure probability allowed for safety-critical aviation functions, failure modes can happen that are unbelievable ... until we find out that they actually do occur. Examples of these can be found in the Real System Failures area (Driscoll 2012) of the Nasa DASHlink web pages.

The design of the RCO system interface into the rest of the aircraft safety-critical systems must be able to tolerate failures of commission (an integrity issue) as well as failures of omission (an availability issue). The same consideration also must be given to the communication link from the GC to the aircraft. The proper balance between integrity and availability must be struck. The reason for this is that fault-tolerance mechanisms that promote one of these characteristics typically are detrimental to the other. To illustrate this, we can look at a simple dual-redundant communication link. Two versions of this link are shown in Figure 2 below, one designed for availability and the other designed for integrity.

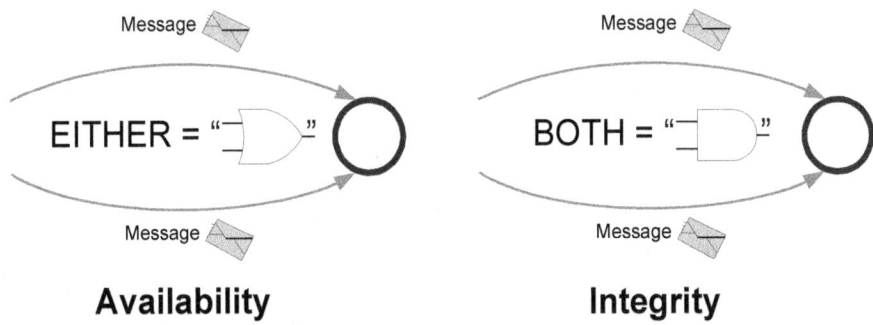

Fig. 2. Availability versus Integrity

For the highest levels of safety-criticality, in-line integrity mechanisms such as checksums and CRCs are insufficient in themselves (Koopman et al. 2015) (Morris et al. 2005). For these dual communication link examples, the only fault detection mechanism with sufficient coverage for full GC authority is the comparison of the two messages arriving via the two independent communication paths. What the receiver does with the messages when the messages do not match depends if the system is designed for availability or integrity. If it needs availability (strive to continue operation), the receiver will arbitrarily select one of the two messages as its input. If it needs integrity (only do correct operation), it will reject both messages. Thus, a dual system designed for availability will accept either message (an OR function) and a system designed for integrity needs to have both messages (an AND function). Note that integrity does not imply safety. For this to be the case, taking no action must be safe. In general, it is not always possible to design a system that has a failsafe state and no type of dual-redundant architecture could be designed that would be safe for such systems. Simple dual redundancy can give you availability or integrity, but not both. If both of these characteristics are needed, the system needs to be at least triplex.

This availability vs. integrity observation holds for RCO communications from the ground to the aircraft and its interface into aircraft safety critical systems. First, we need to determine the degree of availability and/or the degree of integrity needed. With sufficient onboard automation, the availability requirements for communication would not be very stringent. The need for this communication is conditional on those events with sufficiently high workload or for AC incapacitation. The probability of this on-demand availability working correctly when called upon need not be very high if the probability of it being called upon is low enough. On the other hand, the integrity requirements are not conditional. The ability of the RCO system to contain integrity failures must be full-time. One cannot make an on-demand argument for RCO system integrity, similar to arguments that can be made for systems like autoland. We cannot rely on the AC to turn off the RCO system interface (thus preventing integrity failures) for all times except when they become incapacitated. Then, turn the RCO system on when they are (becoming) incapacitated. If an RCO system is designed to detect AC incapacitation, then it must be on all the time.

Even with just these high-level qualitative observations of availability and integrity requirements, we can make some statements about redundancy for RCO communication from the ground.

For availability, simplex (no redundancy) may be sufficient, except for placement of redundant antennas on the aircraft to prevent shadowing of the antennas during certain manoeuvres where parts of the aircraft may block the RF signal. The final determination of whether simplex is sufficient will depend on the specific demands placed on the RCO system. If communication redundancy is required for availability, each of the redundant communication paths must have sufficient bandwidth to carry the entire RCO communication demand.

On the other hand, integrity requirements would demand at least dual communication redundancy, with the redundant paths possibly being asymmetric. That is, one of the redundant paths would have to carry the entire RCO communication demand, but other paths could be just the equivalent of an 'enable'. This latter capability would be particularly useful when using redundant GC. Note that for asymmetric communication paths, the path(s) with lower demand possibly could be accommodated within existing communication equipment.

Of course, if the system needs redundancy for both availability and integrity, the communication path would have to be triplex. When the shadowing requirement is added to this, we are faced with the extreme demand of finding locations for six antennas on the aircraft.

4.2. Traditional Three Layers of Aircraft Control Automation

When looking at suitable locations for where an RCO system would connect into existing safety-critical onboard systems, one probably would begin with the traditional three layers of aircraft control. These can be roughly broken into:

- Flight Management System (planning, source-to-destination profile)
- Auto Pilot (altitude, heading, speed)
- Flight Control (stick and rudder – attitude control, stability) and Engine Control

This list is in top-down order, in which each of the upper items in the list provide inputs into the next item lower down the list. Providing RCO inputs into systems near the top of this list requires less stringent communication latency and jitter requirements than for those closer to the bottom. However, RCO inputs intercepting signals in systems toward the bottom of the list have greater authority for taking control away from a rogue AC. Any design for an RCO system will have to deal with this authority vs. latency trade-off.

4.3. Other Potentially Safety-Critical Systems

While the three layers of aircraft control described in the previous section are the most often studied and cited locations for an RCO system to interface with other aircraft systems, there are many other controls typically used by an AC that an RCO system may have to control and many of them are safety-critical. Here is a partial list of such controls:

- Power
  - Conversion (AC/DC, DC/AC) and distribution (tie relays and switches)
  - Circuit breakers
- Fuel distribution (centre of gravity control, jettison)
- Flight-control surface trim
- Landing gear
- Spoiler, thrust reverse, and braking systems
- De-icing and pitot heat
- Radio tuning and audio?
- . . .

Each of the items in this list has been implicated as a contributing factor to incidents in which their misuse has led to a catastrophic event. Thus, it should be clear that an RCO system must be able to control all of these. However, these levels of pervasiveness and invasiveness of the RCO interface have not been adequately addressed by previous RCO and SPO studies (for Part 121), which typically have concentrated on the traditional three levels of aircraft control.

5. RCO Airborne System Architecture

Depending on the requirements for handling rogue pilots, an RCO system may need to intercept all signals/systems that could possibly cause an aircraft to not continue safe flight, including systems not in the three traditional control layers of Flight Management System (FMS), autopilot, and flight control. Even without a rogue pilot (i.e. just 'benign' loss of AC) many signals/systems will need to be intercepted to provide a ground override. A couple of possible on-aircraft RCO system architectures are shown in Figures 3 and 4 below. Both are expensive, safety-critical, and highly disruptive to many current aircraft systems.

These characteristics are unavoidable; they would be true of any RCO airborne system architecture.

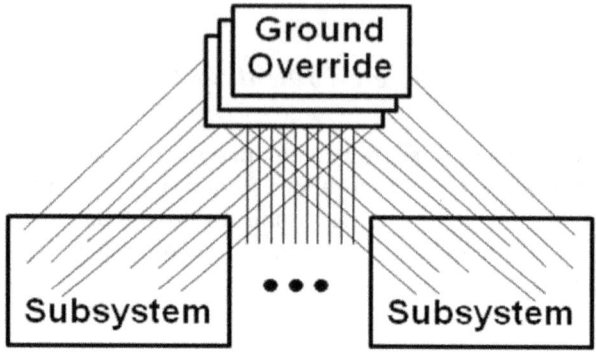

Fig. 3. Centralized 'Porcupine'

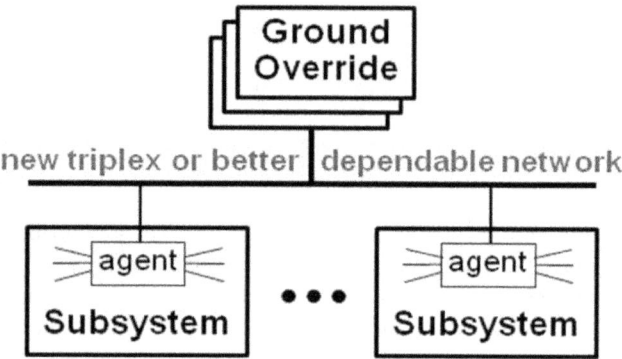

Fig. 4. Remote Agents

In the 'Porcupine' architecture, individual signals go out from the RCO interface boxes that provide the GC ground override capability to all the points in all the other subsystems where the RCO must intercept some existing signal. The name 'Porcupine' comes from the fact that there might not be enough surface area on the Ground Override boxes to accommodate all the possible signal wires.

In the Remote Agents architecture, the mass of Porcupine wires are replaced by a new high-dependability network that connects the Ground Override boxes to remote agents within each of the other aircraft subsystems. The only difference between this Remote Agents architecture and the Porcupine architecture is the structure of the signal interconnects.

The Ground Override boxes are shown as triplex. This is because many of the places where the other subsystems must be intercepted have no always-safe state. Therefore, a dual-redundant control is insufficient. One example of this is the landing gear. The gear must be up for high-speed flight and down for landing. Neither state (up/down) is safe in the other situation. It should be clear that many of the other safety-critical controls have no universally safe state.

One set of safety-critical controls might not be so obviously lacking of universally safe states, but actually is very probably the set of controls that will make it prohibitively expensive to retrofit an RCO system into an existing aircraft. This set of controls is the circuit breakers. There are a lot of them. If some downstream electrical malfunction could cause a fire, the safe state for the circuit breaker is off. This is the reason that circuit breakers exist. On the other hand, there are a number of electrical subsystems on the aircraft for which the safe state is having one or more circuit breakers on. Therefore, dual redundancy is not enough. In addition, it is not feasible to make circuit breakers, switches, and relays fail-operational (providing availability and integrity, simultaneously) by using triplex redundancy. The common way of providing fail-operational capability in circuit breakers, switches, and relays is the quad configuration shown below.

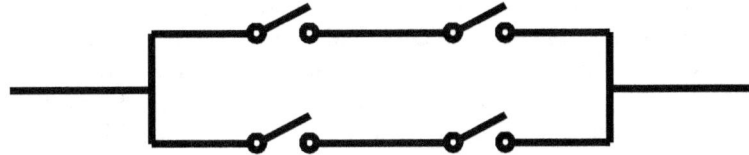

Fig. 5. Quad Circuit Breaker, Switch, or Relay

This configuration is single fault tolerant to any single stuck-closed or stuck-open component.

However, this configuration presents the problem of how to connect a triplex controller to these quad components. The easiest solution is to make the controller quad instead of triplex. Then, each member of the quad controller would independently control one component of the quad circuit breaker. In the Porcupine architecture, the Ground Override boxes would have to be quad and there would have to be four independent control signals from these boxes to each of the quad circuit breakers. In the Remote Agents architecture, the agents would have to be quad and have a triplex-to-quad conversion voting plane between it and the triplex dependable network, or the entire system would have to be quad.

In addition, it is quite probable that many of these intercept points will need to have their actions co-ordinated. As soon as any type of co-ordination is required among redundant elements, the possibility of a Byzantine fault is created (Driscoll et al. 2004). To tolerate one Byzantine fault, a minimum of four fault sets as needed (Lampart et al. 1982). Thus, the Ground Override boxes may need to be a minimum of quad to cover these faults.

6. Related Research and Development

6.1. Previous RCO and SPO R&D

While safety, security, and certification issues for the hardware and software that make up the nonhuman parts of an RCO or SPO system have not been totally neglected in previous studies, these issues certainly have taken a backseat to studying the human parts of these systems. This could be an example of 'design procrastination' where the difficult and uninteresting parts of a design problem are delayed to the end of the process. What little publication has been created for these areas of RCO and SPO was consulted for our research.

To "not reinvent the wheel" and "not look under rocks that have already been examined before", previous R&D in fields adjacent to RCO and SPO were also examined. The two most applicable adjacent fields are unmanned aerial systems (UASs) and autonomous ground vehicles. Both of these areas are currently hotbeds of R&D activity looking at some issues that could be applicable to RCO.

6.2. R&D Done In Adjacent Fields

6.2.1. UASs

While unmanned aerial systems have some issues with availability, safety, and security for remote control of aircraft, there is no AC to share control responsibility and the dependability requirements are much less stringent than for civil transport operations. A few pieces of information from this field are incorporated in subsequent sections of this paper.

6.2.2. (Semi-) Autonomous Ground Vehicles

While ground vehicles don't fly, aren't remotely controlled (yet), and their safety requirements are less than for aircraft (largely because ground vehicle crashes don't cause people to avoid these vehicles as much as plane crashes cause them to avoid flying), research and development (R&D) in (semi-)autonomously-driven ground vehicles covers an important issue not present in unmanned aerial systems. This issue is the control authority that an in-vehicle person relinquishes, which is much larger than previous systems (e.g. aircraft autopilots and ground vehicle adaptive cruise control). In fact, we are now looking at situations where the automation/remote control authority supersedes that of the in-vehicle person and that person may not even have the ability to take back some portion of that control.

A hot topic in (semi-)autonomous vehicle R&D is the issue of full autonomy vs. shared responsibility. Ford says that the possible interim step to fully autonomous vehicles, where driving responsibility is shared between digital driving systems and human drivers, cannot be done safely. The problem is the handoff from the digital system back to the human when something unexpected happens. Designers cannot anticipate every possible situation a vehicle can encounter. Dr. Ken Washington, Ford's VP of research and advanced engineering, stated:

> Right now, there's no good answer, which is why we're kind of avoiding that space.

This problem of control being handed back to a human when automation fails is already an emerging problem for cockpit automation systems. Introduction of RCO and/or SPO will exacerbate this problem.

7. Control Hand-Back Problems

Nasa's Paul Schutte, in his "How to Make the Most of Your Human" presentation for HCI International 2015, had a slide that said:

- One reason why computers are so reliable at what they are programmed to do is because they *give up* at the first sign of trouble.
- When the autopilot reaches its maximum authority, it throws up its hands and tosses control back to the human, whether the human is ready for it or not.
- Pilots routinely must intervene whether it's simply resetting a circuit breaker or turning off the automation.
- The main reason why humans are still on the flight deck is to manage risk by dealing with or avoiding the unexpected, unanticipated, or complex situations

The same things can be said about the RCO communication path from a GC to the aircraft, the path into the aircraft's safety-critical systems, and any of its onboard supporting automation. The issue is that there may be nobody/nothing in control of the aircraft between the time that the communication or automation fails and the time that the AC can retake control of the aircraft. The duration of this time depends on how out of the loop the AC is at the time of the failure. The following subsections describe these varying degrees of being out of the loop and include illustrative events from actual aviation incidents and accidents.

7.1. Time to get to the controls, when out of the cockpit

On a Delta (Chautauqua) flight 6132 (Hradecky 2011), the captain got stuck in the lavatory due to a broken door latch. The one cabin crewmember of this commuter flight had to go into the cockpit when the captain left it. The captain yelled for a passenger to tell the cockpit what was going on. The passenger banged on the cockpit door and yelled through it, trying to explain the situation. The problem with this was…he had a thick Middle East accent. No one in the cockpit was going to open the door in that circumstance. The captain had to breakdown the lavatory door. The flight continued without further incident after the AC rescinded their radio message that they were potentially in a hijack situation.

A common reason for leaving the cockpit is to investigate an abnormal situation (e.g. smoke). One can argue this is precisely the wrong time to leave the cockpit unattended. The abnormality being investigated could be something that would cause the loss of RCO communication or its interface to critical systems. The temporal corollary to Murphy's Law is: *When things do go wrong, they will go wrong at the most inopportune time.* For example, a half-hour into a scheduled 12-hour flight from Seattle to Beijing, a cockpit crew member rushed to the rear of the airplane to investigate the smell of smoke, which is never a good sign. On an RCO flight, this would have been the entire crew (!), away from the cockpit for a significant amount of time.

Fig. 6. Fire Trucks Surround an Aircraft at the Seattle-Tacoma Airport

The airplane returned to Seattle for over-night repairs, which replaced a cabin air recirculation fan and one third of the cabin seats. Therefore, this had to be a nontrivial fire. Such events are not that rare. Some have said (Learmount 2010) that SPO would require automation that has no hand-back mode (e.g. no autopilot trip) if the crew has to leave the cockpit.

7.2. Time to get to the controls, when in the cockpit

On an Aeroflot Flight 593 (an A310), the captain allowed his two children to sit in the front two cockpit seats. The son accidently disengaged the autopilot lateral control. While there were two members of the cockpit crew in the cockpit, having the children in the way plus the g-forces caused by the lack of the autopilot lateral control prevented crew getting back into their seats and at the controls in time. All 63 passengers and 12 crewmembers died in the crash (Megill S 1994).

7.3. Once at the controls, time to regain situational awareness under normal conditions

An Air Canada Flight 878 (a B767) incident report said, 'Under the effects of significant sleep inertia (when performance and situational awareness are degraded immediately after waking up)' a pilot mistook the planet Venus as lights of another airplane on a collision course and he dove to avoid it. While this manoeuvre managed to avoid a collision with the planet, 14 passengers and 2 crewmembers were injured because they were not wearing seatbelts (CA 2011). Figure 7 shows some crewmembers examining damage caused by passengers hitting the ceiling.

Fig. 7. Flight 878's Crew Examining Ceiling Damage (Pickering 2011)

The cognitive delay due to the 'startle effect' is present even when the crew is fully awake. Audi says its tests show it takes an average of three to seven seconds, and as long as 10, for a driver to snap to attention and take control, even with flashing lights and verbal warnings (Davis 2015).

> … anyone who gets behind the wheel [of a semi-autonomous car] must be properly trained. For Audi, this means learning to be a better than average driver. […] if you need to grab the wheel, the odds are something's gone terribly amiss.

The Air France Flight 447 crash is now well known (BEA 2012). It was a scheduled passenger flight from Rio de Janeiro to Paris, which crashed in 2009. The Airbus A330 entered an aerodynamic stall from which it did not recover and crashed into the Atlantic Ocean, killing all 228 persons aboard the aircraft. When the airspeed indicators failed, the autopilot sounded the caution alarm (startle effect) and threw the control immediately to pilots (who were unprepared).

Another crash in which the startle effect was cited as a significant contributing factor (NTSB 2009) was Colgan Air Flight 3407, marketed as Continental Connection under a codeshare agreement with Continental Airlines. It was a scheduled passenger flight from Newark, NJ, to Buffalo, NY, which crashed in 2009. The Bombardier Dash-8 Q400 aircraft entered an aerodynamic stall from which it did not recover and crashed into a house in Clarence Center, NY, killing all 49 passengers and crew on board, as well as one person inside the house.

7.4. Recovery time can be even longer if diagnosis is required

The crew of Qantas Flight 32, in which an A380 engine disintegrated, needed 50 minutes to sort out all the ECAM[1] warning messages (the crew had no time for ACARS[2]) and assess the aircraft damage (ATSB 2010). It was lucky that this flight had five cockpit crewmembers (three normal crew plus a Check Captain and a Supervising Check Captain). Therefore, they had the luxury of having an extra person they could send aft to look out the windows and assess damage. Dealing with abnormal situations may require additional AC, versus a reduction in crew. Richard Woodward (a Qantas A380 pilot and deputy president of the Australian and International Pilots Association) said that the 'number of failures is unprecedented [...] There is probably a one in 100 million chance to have all that go wrong.' (Schneider 2010) However, there have been over a half-dozen previous similar incidents. The Sioux City DC-10 crash is well known (NTSB 1989). Again, they were lucky to have additional crew on board, which prevented the crash from being worse than it was.

8. Are There Real Communication Threats?

When a capability is created to remotely control an aircraft, the security of the communication used for this control is an obvious concern. However, would someone really try to interfere with the flight of an RCO aircraft or is this just a 'Hollywood' fantasy? It turns out that the answer is, "Just because you're paranoid, that doesn't mean that they are not out to get you." We have to assume there will be bad actors that are out to get us because there have been a number of instances in the past where radio communications to aircraft have been attacked. Some of these instances are described below in subsections grouped by the type of perpetrators.

---

[1] 'electronic centralized aircraft monitor' is a status display system, developed by Airbus

[2] 'aircraft communications addressing and reporting system' is an aviation radio text service

8.1. Individuals

Officially called a 'phantom controller' (a.k.a. 'bogus', 'fake', or 'phony' controller), there are individuals who like to pretend that they are air-traffic controllers. In the UK, there were 18 in 1999 (Morgan 2000). The US will only say that it happens 'several times a year'. It has been said that these instances have been underreported in order to prevent copycats. This is hard to verify, but from reasonable sources. Jim Epik has written a novel, called "Phantom Controller", based on his investigations into these incidents. He also has created a petition to encrypt ATC communications.

8.2. Ad Hoc / Transitory Groups

During the 1981 Professional Air Traffic Controllers Organization (PATCO) strike, some of the striking members became phantom controllers (Factor 2014).

Opposing factions in civil wars would like to wrest control of the airspace over their contested country from others involved in the war. Thus, they will interfere with the other factions' aviation communication.

8.3. Nation-State Sponsored

An Air France captain said[3] that his aircraft received bogus air-traffic control instructions during a flight back to France from Japan. He believed that his aircraft was targeted because he had transmitted a PAN-PAN[4] message indicating that an electrical problem had caused half of his cockpit avionics to be inoperative and the crew would be under a heavy workload. The attacker (indications were that it was North Korea) made six attempts to cause the aircraft to fly into an unsafe situation. The captain suggested that encrypting the PAN-PAN message for secrecy might have prevented this attack.

---

[3] in conversation with this author

[4] The international radio-telephony urgency signal, similar to, but less severe than, the well known MAYDAY international radiotelephony distress signal.

9. Communication Encryption

The only viable current method to protect aircraft communications is the use of encryption. However, there are a number of problems to overcome when employing encryption to protect RCO communications. These problems include each nation's laws governing cryptography, the latency introduced by encryption, and other ways that current encryption algorithms are ill suited for use in a real-time cyber-physical system (CPS).

9.1. Cryptography Laws

Most countries have some restrictions on the export, import, and/or domestic use of encryption technology. They may prohibit, limit, and/or require licensing for encryption use within its territories. Tables listing many of these restrictions can be found on the web, including the Crypto Law Survey (Koops 2013).

Underlying the use of encryption is a cryptographic key management infrastructure. There are two aspects to key management: trust and logistics. Trust involves three questions:

1. Who do you trust?
2. With what?
3. To do what?

For example:

4. Can an airline trust the UK government?
5. With its cryptographic keys?
6. Not to reveal these keys (To North Korea? To the US? To Israel?)

Specifics are important. Some airlines would be more upset than others to find that the UK had revealed their cryptographic keys to Israel.

Key management logistics are the mechanisms to enforce the trust. This includes the creation of keys with their ownership association, key distribution, and revocations. Key management allows only authorized users to have possession of private or secret keys, often only for a set period of time.

These cryptographic keys can have an ordinary use (e.g. RCO communications) and an extraordinary use (e.g. a government investigation). The laws governing cryptography in many countries require that some arm of the country's government have access to the plaintext that has been encrypted. (Plaintext is anything that will be encrypted or has been decrypted.) Usually, the easiest way to give a government access to the plaintext is to allow them access to the cryptographic keys used for the encryption. This allows the government to decrypt what they want. However, this still can complicate the key management infrastructure.

Much of the literature covering future encryption systems for aircraft communications assumes that just saying an X.509-based public key infrastructure (PKI) will be used for key management is a sufficient explanation for how the key management problems will be solved. However, full PKI is heavy weight and does not solve all the problems by itself. PKI does not answer the trust questions. The trust questions include the question of whose keys will be used. As a complex example to illustrate the point: say that an aircraft manufacturer includes some cryptographic equipment made by an avionics supplier; the aircraft manufacturer sells the aircraft to a leasing company; the leasing company leases the aircraft to a scheduled airline; the airline rents the aircraft to a charter company at times when the airline isn't using that aircraft; and, the charter company hires a crew that normally works for a rival airline. Whose keys should be used for the encryption? Should the cryptographic equipment have a dedicated link to some key management infrastructure owned by the avionics company? the aircraft manufacturer? the leasing company? the airline? Alternatively, should the crew load keys as part of pre-flight? If so, what keys should be used? the charter airline's keys? the crew's personal keys? the keys they use as employees of the rival airline? Should some aircraft systems have keys (or use a key infrastructure) that is different to other systems? For example, an engine manufacturer with a "power by the hour" arrangement might like to have engine performance data transmitted to them using their own key. There are some current uses of PKI for aircraft communication. However, it is unknown whether this PKI can be used for RCO.

Encryption can be used to provide secrecy and/or authentication. These two properties do not need to be tied together. Often glossed over in discussions of key management is the fact that key distribution needs secrecy protection for private and secret keys, even if these keys are used only for authentication (not secrecy). Authentication schemes need either private keys (for public-key system signatures) or secret keys (for message authentication codes). The need for secrecy in the distribution of these keys complicates the key management infrastructure and can cause problems with national laws that restrict encryption used for secrecy, when an encrypted channel is used to provide secrecy for key distribution rather than using physically secure communication path for the key distribution. This invalidates the sometimes-heard argument that the infrastructure for authentication-only encryption is simpler than for secrecy encryption.

9.2. Encryption Latency

One problem encountered by UAS operations is communication latency. The sum of the communication latencies can be on the order of a couple seconds, which can make closed-loop remote control of an aircraft difficult. Encryption of this communication can be an aggregating factor in these latencies. The communication for each iteration around the closed loop incurs the latency of two encrypts and two decrypts (the four arrows in Figure 8).

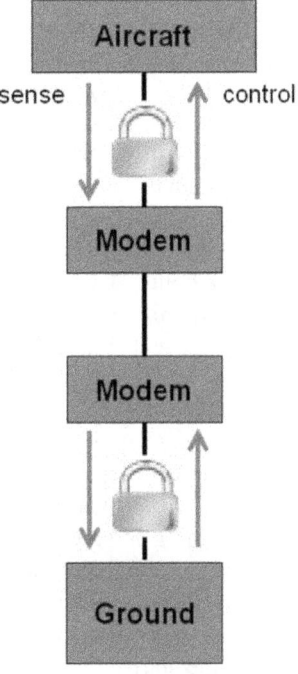

Fig. 8. Encryption Latency

If AES (or similar block cipher) is used to provide secrecy and integrity, a block (e.g. 128 bits) of store-and-forward latency has to be added, plus the latency for any added initialization vector (IV) and/or integrity data (e.g. 32 bits each). These latencies depend on communication speed (the slower the link, the longer these latencies) and they have to be added to the cryptographic computation latencies. The sum of these latencies doubles if handshakes (e.g. ACK/NAK) are used and are encrypted.

UASs mitigate this cryptographic latency problem by using very high-speed (e.g. 10 Gbps) communication links and special low-latency hardware encryption, e.g. KG-340 encryptors (Ballard 2014) and Single-Chip Crypto field programmable gate arrays. RCO communications probably will not be able to find such a wide bandwidth for its use and adding additional high-speed encryption hardware can be expensive.

9.3. General-purpose cryptography ill suited for a CPS

An RCO system is an example of a cyber-physical system with real-time and other constraints not seen in general-purpose processing. While latency and jitter may be the main differences in requirements/constraints between general-purpose processing and cyber-physical system processing, there are a number of other problems with employing general-purpose cryptographic algorithms in cyber-physical systems. Many of these problems compound the latency and jitter problems.

The remainder of this subsection deals with symmetric encryption algorithms implemented in software, possibly with hardware support in the form of instructions in the processor's instruction set architecture (ISA) or an adjunct crypto field programmable gate array. Asymmetric (public key) algorithms are not discussed because their use can be restricted to key exchanges that can be performed during pre-flight or other times when temporal performance is not important. Also not discussed are stand-alone encryption 'boxes', because their added costs in terms of cash outlay, size, weight, power, and latency makes them much less desirable.

One problem with using general-purpose cryptographic algorithms in cyberphysical systems is the slow start-up for each key change due to key scheduling being done. Key scheduling is the conversion of the cryptographic key into data that the encryption algorithm uses internally. When the encryption is not used for real-time cyber-physical systems, it makes sense for key scheduling to be made expensive. The rationale is that legitimate users incur these start-up delay costs a relatively small number of times. On the other hand, an attacker that uses some form of brute force key-related search would have to try a huge number of different keys and incur the start-up delay cost a vastly greater number of times. Another reason that general-purpose cryptographic algorithm key scheduling is slow is that they have been optimized for peak throughput performance, which is usually measured in clock-cycles-per-byte. In the "my algorithm is faster than your algorithm" speed propaganda wars, start-up does not count. Therefore, to game the system, algorithm designers can put more work into the start-up to make the rest of the algorithm run faster. With the cryptography speed propaganda focused on peak throughput, worst-case throughput is ignored. However, in cyber-physical systems, typically only the worst-case timing counts, peak is not important. A missed real-time deadline cannot be helped by finishing early at other times. For cyber-physical systems, latency and jitter are both usually more important than throughput. Often, jitter is more important than latency because control algorithms can better deal with a known latency rather than with instances of unknown jitter.

A cryptographic algorithm characteristic related to low-latency is 'key agility'. This is the ability of an algorithm to easily and quickly change from one key (and/or IV) to another. Good key agility may be required if different avionics subsystems and/or applications within the aircraft need different keys and there is a centralized provider of encryption services for them. Good key agility also may be required as an aircraft crosses geographic boundaries that delineate jurisdictions where different keys must be used and the keys must be quickly changed at the boundary in order to avoid a communication 'dead zone' where encrypted communication cannot be performed.

In order to provide the properties of continuing authentication, secrecy, and integrity, most existing cryptographic systems use separate secrecy and integrity algorithms or use an added integrity mode that is wrapped around a secrecy algorithm. Compared to an authenticating encryption (AE) (a.k.a. integrity-aware) algorithm, which intrinsically provides continuing authentication and integrity along with secrecy, these approaches consume additional bandwidth (which may expensive or not even available) and exacerbate the temporal problems.

An AE algorithm called BeepBeep, designed specifically for real-time cyber-physical and/or retrofit applications, solves the problems discussed in this section and additional problems that general-purpose cryptography has when used in cyber-physical systems (Driscoll 2002). Beepbeep has a tiny code size, zero working data memory, low latency, good key agility, and provides continuing authentication, secrecy, and integrity in a single pass. In the last couple of decades, there have been several competitions and initiatives to create new encryption algorithms, e.g. AES (NIST 2001), CRYPTREC (CRYPTREC 2014), eSTREAM (ECRYPT 2012), and NESSIE (EU 2004). However, none of these had explicit goals addressing the problems discussed here.

10. Conclusions

The RCO concept does not seem to be economically viable for Part 121 operations, at least in the short term where existing aircraft would have to be retrofitted for RCO capability. This is due to the very high cost of implementing an RCO system that can safely and securely provide the capability of controlling an aircraft in which some actions by an incapacitated crew could be similar to that of an adversary.

The cost replaced by RCO is the salary and benefits of a First Officer. However, 100% of that cost cannot be eliminated. There are GC labour and maintenance costs. If there is a 1:1 replacement of AC with GC, obviously, there is no labour cost savings. There have been estimates that a GC can handle five aircraft simultaneously (Comerford D et al 2013). However, that must be for benign conditions and the GC intercepting the aircraft systems at the FMS level (possibly, the Autopilot level). If the GC has to intercept the aircraft systems at the Flight Control level (required for adversarial or incapacitated AC), it is hard to imagine that the aircraft:GC ratio can be better than 1:1. For there to be any labour savings, an RCO system on an aircraft must have interception points at the lowest level (for full control) and at a higher level (to reduce the number of crew actions that need to be taken and to allow a greater aircraft:GC ratio). The number of such aircraft that could be handled by one GC would be some number between one and five, depending on how many simultaneous incapacitations that would have to be handled.

The cost of completely redesigning and replacing most of the cockpit avionics and adding a quad-redundant (or better) Ground Override system with highly invasive tentacles into most systems (many also being quad), will be more than the cost for the original avionics and will have fewer aircraft over which to spread the development cost. This development cost also will be higher than the original development cost due to the Ground Override system needing to be, euphemistically, "Level A+" because of its potential to be a single point of failure for *all* of the critical avionics. In addition to these aircraft costs, we must add the development, deployment, and operation cost for the ground segment. These ground costs could be very large when they include the development, deployment, and operation costs of a C-band satellite system. The amortization of all these large costs (including time value of money) must be less than the crew labour cost saved.

There have been some public statements of belief for potential cost benefits. For example, the account of Dr. R. Mike Norman's presentation at NASA's Single-Pilot Operations Technical Interchange Meeting (Comerford D et al 2013) includes, 'SPO may have economic benefit, but once again, new costs associated with SPO were not addressed'. Given the arguments presented here, these unaddressed RCO/SPO costs probably would overwhelm any AC labour cost savings.

The inclusion of RCO within future aircraft designs would cost less than for retrofit. There are two reasons for this: the first is that some avionics developments will make it easier to add RCO functionality, just as a by-product of their creation for other reasons. A good example of this is the replacement of individual circuit breakers with an integrated 'electronic fuse box'. This will make it a lot easier for an RCO Ground Override interface to control the equivalent of circuit breakers. The second reason is that future avionics can anticipate the possible addition of RCO. However, the degree to which creators of avionics would be willing to add 'hooks' for an RCO option is unknown, given that these 'hooks' would add some cost for all same-type aircraft, including aircraft that do not use RCO. It is unclear if the reduced cost for RCO in future aircraft would make RCO economically viable.

References

ATSB (2013) Final Investigation Report, AO-2010-089 -- In-flight uncontained engine failure overhead Batam Island, Indonesia 4 November 2010 VH-OQA Airbus A380-842

Ballard M (2014) NSA encryption no smoking gun, says drone net contractor http://www.computerweekly.com/blog/Public-Sector-IT/NSA-encryption-no-smoking-gun-says-drone-net-contractor Accessed October 30, 2016

BEA - Bureau d'Enquêtes et d'Analyses pour la sécurité de l'aviation civile (2012) Final Report on the accident on 1st June 2009 to the Airbus A330-203 registered F-GZCP operated by Air France flight AF 447 Rio de Janeiro – Paris. Ministère de l'Écologie, du Développement durable, des Transports et du Logement. France

CA (2011) Aviation Investigation Report A11F0012 http://www.tsb.gc.ca/eng/rapports-reports/aviation/2011/a11f0012/a11f0012.pdf Accessed October 30, 2016

Comerford D et al (2013) NASA/CP—2013-216513 Single-Pilot Operations Technical Interchange Meeting: Proceedings and Findings.

CRYPTREC (2014) http://www.cryptrec.go.jp/english Accessed October 30, 2016

Davis A (2015) www.wired.com/2015/01/rode-500-miles-self-driving-car-saw-future-boring Accessed 30 September 2016

Driscoll K (2002) BeepBeep: Embedded Real-Time Encryption. Fast Software Encryption 2002 / Lecture Notes in Computer Science Vol. 2365 pp 164-178

Driscoll K et al (2004) The Real Byzantine Generals. Proceedings of the 23rd Digital Avionics Systems Conference (DASC). pp 6.D.4-1 - 6.D.4-11

Driscoll K (2012) http://c3.nasa.gov/dashlink/resources/624 Accessed 30 September 2016

ECRYPT (2012) eSTREAM: the ECRYPT Stream Cipher Project http://www.ecrypt.eu.org/stream Accessed October 30, 2016

EU - European Union (2004) NESSIE https://www.cosic.esat.kuleuven.be/nessie Accessed 30 September 2016

EU - European Union (2013) http://www.across-fp7.eu Accessed 30 September 2016

Factor N [pseudonym] (2014) http://jobstr.com/threads/show/4473-air-traffic-controller/6 Accessed October 30, 2016

Hradecky S (2011) http://avherald.com/h?article=4464222f Accessed October 30, 2016

Koopman P et al (2015) FAA report DOT/FAA/TC-14/49 Selection of Cyclic Redundancy Code and Checksum Algorithms to Ensure Critical Data Integrity

Koops B (2013) http://www.cryptolaw.org Accessed 30 September 2016

Lamport L et al (1982) The Byzantine Generals Problem. ACM Transactions on Programming Languages and Systems, Vol.4, No.3, July 1982, pp 382-401

Learmount   D   (2010) https://web.archive.org/web/20100715144540/http://www.flightglobal.com/blogs/learmount/2010/06/the-lonely-airline-pilot.html Accessed 30 September 2016

Megill S (1994) Report on the investigation into the crash of A310-308, registration F-OGQS, on 22 March 1994 near the city of Mezhdurechensk http://aviation-safety.net/get.php?http://asndata.aviation-safety.net/reports/1994/19940323-0_A310_F-OGQS.pdf Accessed October 30, 2016

Morgan   D   (2000)   Hackers   Attack   Air   Traffic   Control http://abcnews.go.com/US/story?id=95993&page=1 Accessed October 30, 2016

Morris J et al (2005) Coverage and the Use of CRCs in Ultra-Dependable Systems. Proceedings of the 2005 International Conference on Dependable Systems and Networks

NIST - US National Institute of Standards and Technology (2001) Overview of the AES Development Effort http://csrc.nist.gov/archive/aes Accessed October 30, 2016

NTSB (1989) AAR-SO/06 Aircraft Accident Report, United Airlines Flight 232, McDonnell Douglas DC-1040, Sioux Gateway Airport, Sioux City, Iowa, July 19, 1989

NTSB (2009) AAR-10/01 Loss of Control on Approach Colgan Air, Inc. Operating as Continental Connection Flight 3407, Bombardier DHC-8-400, N200WQ Clarence Center, New York February 12, 2009 pp 87, 89, 127, 152

Pickering L (2011) https://www.youtube.com/watch?v=Bkigne0S70I Accessed October 30, 2016

RTCA (2011) DO-178C Software Considerations in Airborne Systems and Equipment Certification.

Schneider K (2010) http://www.news.com.au/travel/travel-updates/qantas-jet-could-have-exploded/story-e6frfq80-1225956388231 Accessed 30 September 2016

US   (2016)   http://www.ecfr.gov/cgi-bin/text-idx?tpl=/ecfrbrowse/Title14/14cfr121_main_02.tpl Accessed October 30, 2016

Waking up to The Insider as a Safety-Critical Threat

Ryan Meeks and Robert Dickie

Frazer-Nash Consultancy Ltd

Bristol, UK

Abstract   *The Insider threat is rarely considered as part of functional safety to inform design, process and procedure. Worryingly, it is often neglected as part of safety and risk management practices entirely. This must change in light of high profile cases in recent years where Insiders have been seen to pose a severe threat. Industry must attempt to analyse and understand Insider threat risk and build this into integral processes, which will require close collaboration across diverse technical areas and specialisms. Government policy may even be developed in the coming years, similar to that of US Executive Order 13587, which necessitates a more comprehensive consideration of these risks. Now is the time for safety-critical industries to wake up to the Insider threat as one of the most real and present dangers to organisations in the modern age.*

This paper is a thought-piece about how Insider threat could be dealt with as part of normal engineering practice, and proposes a concept methodology for the formal assessment of Insider threat risk to systems and organisations. The paper deals only with deliberate and malicious acts (intended to do harm in some way), rather than the unintentional insider threat.

1 Introduction

Insider threats are defined as attacks to an organisation or system from people within it. These people could be permanent or temporary employees or even part of the supply chain, but who generally have access to the organisation's critical systems, assets and information. They have deeper knowledge of the organisation than external attackers and so are better placed to exploit known weaknesses and vulnerabilities by acting 'under the radar' for longer periods of time.

Insider threat attacks are often categorised into five main types (CERT, 2012):

1. **Sabotage** – a deliberate attack to destroy, damage or obstruct the organisation;
2. **Intellectual Property (IP) theft** – the theft of material that the attacker does not have the legal right to access or take for their own use;
3. **Fraud** – deceptive actions that lead to personal financial gain;
4. **Espionage** – infiltration into an organisation with the intention of obtaining information that is of use to others (typically competitors or adversaries);
5. **Unintentional** – non-malicious actions by individuals that inadvertently act to facilitate an attack with unforeseen consequences[1].

It is important to understand what attack type an organisation is most vulnerable to, as this helps to understand how insiders will behave and also what may influence their attack motivations. For example, an insider trying to siphon funds from the company bank accounts (fraud) will have different motivations and will exhibit different threat indicators than an insider attempting to insert malicious code onto the server (sabotage), who will have different motivations to someone willing to engage in interpersonal violence (personal/ ideological).

1.2 Safety-Critical Consequences

Insiders are mostly intelligent adversaries that do not reside within the lower echelons of the organisation as is often thought. Their knowledge of, and access to, critical assets and information allows them to conduct more complex and concealed attacks, moving laterally across the organisation to maximise the attack consequences and reduce the chance of being caught. There is no other type of threat that resides so closely within the 'trusted inner circle', and so in the modern age the malicious Insider must be considered one of the most dangerous safety-critical threats. Indeed, some research suggests that 25% of all security breaches are as a result of malicious insiders (Forester Research, 2013).

The impact of Insider threat only worsens when you consider the severity of consequences within safety-critical industries. These often form part of the Critical National Infrastructure (CNI) and so safety and security practices are often robust and comprehensive. However, there remains a glaring oversight of Insid-

---

[1] This paper does not deal with unintentional insider threat, only the four malicious attack types above. Unintentional attacks should be dealt with as a human reliability/ human error issue, rather than being linked to the malicious drivers that this paper deals with

er threat risk within many of these practices, which perpetuates the systematic and organisational vulnerabilities that malicious insiders are happy to exploit.

The prevalence of this threat in safety-critical industries does not just pose financial, operational and reputational risks to the organisation concerned, but it may also present a real risk to life and public safety. Insider threat has potential life-threatening consequences given the nature of the systems and equipment that the malicious insider may have access to.

2 Background

2.1 Evolution of Safety

Safety engineering as we know it today began in its infancy in the first part of the 20th century, with professional societies and numerous codes, standards and laws being passed in the wake of high profile disasters; one of the incidents being the sinking of RMS Titanic. Following the Titanic tragedy and the identification of a number of engineering failures (e.g. low bulkheads and the brittle fracture of the hull steel) much stricter standards for safety regulation were implemented for ships at sea. A similar pattern followed in many other industries as the consequences of engineering failures became increasingly apparent as the discipline attempted to conquer ever-more complex challenges.

Fig. 1. The Loss of RMS Titanic led to significant developments in engineering safety (image: History Extra, 2015)

Throughout the second half of the 20th century the safety engineering discipline grew immeasurably as the importance of safety was realised and various laws and legislation mandated a safety-based approach within complex engineering programmes. Engineers became better at understanding the inherent weaknesses in the materials they selected, the manufacturing techniques they used and the wider socio-technical support mechanisms that they were better at identifying risks and hazards and mitigating these as part of a structured design process. Inevitably however, engineering failures did still occur due to competing technical and business objectives but on the whole the prevalence of major engineering disasters has reduced significantly across most industries.

At around the same time industries also started to realise that there was another major risk to the success of their systems; the human user. While it is well understood that humans are often the most integral component of any complex socio-technical system, they can also be the most unreliable, unpredictable and susceptible to negative influences. This requires systems to be designed in a way that treads the fine line between exploiting the strengths of human users (e.g. decision making, contextual knowledge, semantic appreciation) while mitigating their inherent weaknesses (e.g. fatigue, processing speed). This is not to say that the human component is an inherent weakness of the system design however. In fact, when integrated effectively the human is the most intelligent, robust, flexible and innovative part of any system and an absolutely integral part of Safety management.

In the middle half of the 20th century the disciplines of Human Factors (HF), Ergonomics and Engineering Psychology rose to prominence due to their ability to apply scientific expertise of human psychology and physiology to adopt a user-centred approach to engineering problems. In the same way as the Titanic did for engineering safety, similar disasters such as Chernobyl, Bhopal and Three Mile Island were attributed to a large degree to human error and the failures of the system design (Meshkati, 1991). This led to the development of analysis methods that were well suited to ensuring that engineering design was able to take account of human reliability. It is encouraging that industries are becoming increasingly better at ensuring that HF is built into the core design process and that human-related considerations are raised at the project onset.

Fig. 2. The Chernobyl disaster in 1986 highlighted the importance of controlling human error within system design (image: Nuclear News, 2015)

The engineering world has become much better at understanding and controlling the risks associated with system design and human reliability. At the same time however the world is experiencing a technology revolution. The pace of technological development is so rapid that engineering industries are struggling to keep up. The rise of areas such as artificial intelligence, the Internet of Things and cyber environments pose a significant risk to engineering safety, although the exact nature of this risk remains under-appreciated. How engineering programmes embrace technological development while ensuring that they pose no critical risk to system safety is a challenge that engineers face daily in the modern world.

The risk area that forms the main subject of this paper is the Insider threat. This is differentiated from human reliability outlined earlier in this section as it deals with malicious activity, rather than the inherent unintentional human fallibilities that HF analysis mostly considers. Human error implies by its very definition that the error is an unintentional consequence of user activity, whereas Insider threat considers the scenario that a user may intentionally attack a system or organisation as the result of a direct motivation to do so, usually to cause harm. Historically this has typically not been considered a viable threat source, however a number of modern examples have emphasised the severity and prevalence of the Insider as a safety-critical threat. Insider threat risks have increased in recent years (Crowd Research Partners, 2015) whilst other threat sources, such as human reliability and engineering design, have been better controlled. The ability of Insiders to conduct more harmful attacks has become greater given the increased complexity of cyber environments and technological advances. This critically under-considered threat necessitates that the next evolution of safety-critical engineering must seek to address the malicious insider.

2.2 An Increasing and Persistent Threat

In recent years there has been a number of Insiders who have conducted complex and damaging attacks to systems and organisations, including:

- **Edward Snowden** – stole vast amounts of classified government-owned national and defence information and revealed them to the media and published them on the internet. It is thought that Snowden stole over 900,000 Department of Defence files (Leopold 2015) and the financial and security consequences of his actions are still not entirely known;
- **Bradley Manning** – stole nearly 750,000 sensitive documents, including military reports from the Afghan and Iraq wars, and disclosed them on the internet site Wikileaks (Guardian, 2013). In a similar vein to the Snowden case, the consequences of Manning's actions are still being fully determined;
- **Robert Duronio** – a former IBM computer programmer who inserted malicious 'logic bomb' code onto the network, which brought down 2000 computers and cost the company over \$3million to repair, due to a disgruntlement over a bonus payment (HBR, 2014). Duronio can be considered one of the first 'modern insiders' who used a cyber environment to conduct his attack.

Fig. 3. Cases such as that of Edward Snowden (pictured) have highlighted the risk posed by modern insiders to large and complex organisations (image: Financial Times, 2013)

These cases provide the catalyst to spur industry into developing new policy and approaches to Insider threat. However, many have been slow to react, even in the face of severe attacks that have caused loss of life, perhaps due to the need to consider the psychological and socio-cultural influences to Insider threat that fall outside of typical expertise boundaries. These knowledge gaps lead to systematic weaknesses that the malicious insider is acutely aware of and provides an environment where they can move freely with little concern of their actions being identified.

It is also important to consider tragic cases such as that of German Wings flight 9525 (BEA, 2016), which demonstrate that insiders can pose a real threat to life. This poses the questions; how far do we go in the definition of Insider threat, and, do we expand this definition to consider risks associated with mental health and addiction issues? Alternatively, are these issues addressed by other means?

Fig. 4. German Wings flight 9525 demonstrated the potential life-threatening risk that malicious insiders can pose (image: CNN (2015)

Insider threat is also not a sector-specific problem. While it is seen to be more prominent in some industries than others (e.g. the 'Green-on-Blue' issue for the military in Afghanistan[1]) it can occur within any organisation. Fundamentally, insider threat deals with human psychology and behaviour and therefore any organisation that employs humans is at risk of malicious activity. Understanding of insider threat in some industries has risen to prominence in recent years, such as:

- **Banking and Financial Services** – this is the sector at most risk of insider attacks over any other (Raytheon, 2015) either conducted entirely by insiders or by external attackers being intentionally or unintentionally facilitated by an insider. This has typically involved fraudulent activity for financial gain or the theft of personally identifiable information (PII) from internal systems;
- **Nuclear** – as a safety-critical component of the CNI the nuclear industry will always be an attractive target for the malicious insider (e.g. the Doel and Tihange power station incidents in Belgium (The Ecologist, 2016));

---

[1] 'Green-on-blue' is a phrase used to describe an attack on NATO Coalition forces by members of the Afghan security forces, who they are working alongside.

- **Aviation** – the aviation security environment was required to become much more robust in the post-9/11 environment, which included a better understanding of the risk posed by insiders. However, insiders are seen as the best way of bypassing aviation security measures and have been used to devastating effect in recent cases in Egypt and Somalia (CBS, 2016).

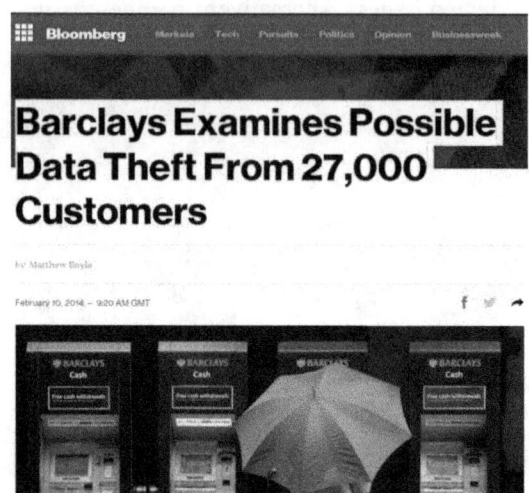

Fig. 5. Insiders are increasingly being seen as the perpetrators and facilitators of attacks in various industries; including Banking, Nuclear and Aviation (image: Bloomberg, 2014)

Perhaps as a result of specific incidents and the nature of their operations these industries are starting to 'wake up' to the Insider as a safety-critical threat. Contrary to popular belief, programmes that are seen to be overtly active in countering an organisation's insider threat are supported by employees rather than met with scepticism and rejection. However, many industries are lagging behind and are perhaps unwilling to be seen to be questioning the integrity of those in their 'inner circle'. This is a mind-set that needs to change. However, this is not to say that insider threat is necessarily the biggest threat that organisations face, especially in relation to a number of external threats, but it is certainly prevalent enough to warrant more consideration than it currently receives.

2.3 Lack of Current Approaches

In addition to a general lack of appreciation of the level of Insider threat risk, our research suggests that there is also an absence of any validated approaches for assessing it. The way in which different engineering disciplines communicate an understanding of risks is generally through quantitative or qualitative assessment that can be used to determine the level of risk associated with the system. While there are a number of technical approaches that allow aspects such as safety risk and human reliability to be assessed based on expert judgement, there appears to be no such current methods for the assessment of Insider threat risk.

While there is a clear need to consider Insider threat risk as a core component of system design, it will never be effectively integrated into current risk-based safety and security practices until an agreed method of analysing the risk, validated by academic understanding, is developed. As the Safety world evolves to address Insider threat as a key consideration, it must also ensure that robust technical analysis approaches are developed to empower practitioners 'on the ground' with usable tools. These tools must also be 'intelligent' enough to be able to distinguish between the malicious and the unintentional insider; probably by focussing on behavioural patterns, rather than behavioural anomalies. This is an absolute necessary development, as the probability of a given risk occurring is entirely different when the human component is acting maliciously, rather than unintentionally. In order to truly understand the risk probability there is a need to understand both intentional and unintentional human actions.

3 Learning from Others

We believe that there are methods within certain technical areas that could be adopted as a general concept framework for the development of an Insider threat risk assessment method.

3.1 Risk Analysis

Conventional risk analysis is a well understand process that hinges on the product of probability / likelihood and consequence assessments. Whilst assessing the probability of insider threats may not be possible (hence the concept of susceptibility is introduced in Section 4), the use of a consequence multiplier is an important tool which reflects the fact that the consequences of insider threats can have a range of severities (e.g. minor disruption / injury though to loss of company / major loss of life). The use of a consequence multiplier is therefore taken forward into the concept methodology in section 4.

3.1 Human Reliability Assessment - HEART

Human Reliability Assessment (HRA) is the approach to evaluating the probability of human error occurring throughout the completion of a given task or in a specific condition of operation. There are a number of different analysis methods; including the Human Error Assessment and Reduction Technique (HEART) (Williams, 1986). This method analyses the tasks that users have to complete and the conditions of the operation and attribute a numerical rating to indicate the probably of human error based on an informed subjective judgement. The numerical ratings given to certain tasks and conditions are defined by a number of validated assumptions related to the 'nominal human unreliability'. There are a number of guidelines as to how to apply the numerical ratings that provide a framework for practitioners to work to and to provide consistency in their application.

Many HRA methods, including HEART, provide an output that expresses human error probability in terms of its likelihood to occur in a given task. Table 1 presents some generic tasks and nominal human unreliability figures from the HEART methodology.

Table 1. HEART Task Categories

Generic task	Proposed nominal human unreliability
Totally unfamiliar, performed at speed with no real idea of the likely consequences	0.55
Complex task requiring a high level of comprehension and skill	0.16
Fairly simple task performed rapidly or given scant attention	0.09
Routine, highly practiced, rapid task involving a relatively low level of skill	0.02
Restore or shift a system to original or new state following procedures, with some checking	0.003

This nominal figure is then multiplied by a figure associated with given error-producing conditions (EPCs), which ultimately represent the assessed proportion of effect. For every task that includes a human-system interface the human error risk is assessed by multiplying the nominal human unreliability by the EPC, to determine whether the risk is acceptable. Table 2 outlines some of the EPCs from HEART.

Table 2. HEART EPCs

EPC	Max. predicted amount by which unreliability might change, going from good conditions to bad
A shortage of time available for error detection and correction	X17
Operator inexperience	X3
High levels of emotional stress	X1.4
A poor or hostile environment	X1.15
An incentive to use other more dangerous procedures	X2
No obvious means of reversing an unintended action	X8

This two-stage method is the main take away from HRA that has been used as a framework for the concept methodology in section 4.

4 A Proposed Concept for Assessing Insider Threat Risk

4.1 High Level Methodology Concept

This section outlines a high level methodology concept that it is believed could eventually be integrated with existing safety, security and risk practices. Whilst there are still some challenges to overcome, it is presented here as a potentially useful concept to develop a targeted Insider threat assessment methodology.

Please note that the figures provided in the examples below (tables 3, 4, 5, 6 and 7) are merely example assumptions and have not been validated by research or analysis. They act as notional figures to demonstrate the concept methodology, and therefore must not be taken from this paper for formal application.

Stage 1 – Define Critical Assets

The first stage in determining the Insider threat risk within an organisation is to determine what they actually consider to be their critical assets; those systems, equipment, locations and information that the organisation cannot function without, would be severely degraded should they be lost, or would cause a threat to life should they fail to function. Practitioners need to liaise with key contacts within the organisation to identify these critical assets, determine their importance and the risk appetite related to each of them. For example, certain information may be critical to the organisation but if it were over-protected it would significantly impede the ability of staff to do their job, and so the organisation is therefore willing to accept a higher level of inherent Insider threat risk.

Stage 2 – Identify Insider Threat Scenarios

Once the critical assets are identified it is important that the analysis identifies who has access and responsibility in relation to each. Our belief is that the largest single influence to Insider threat motivation is exposure, and so the identification of access paths and 'silos of responsibility' is essential. It is also important at this stage to define the relationships between people who have similar levels of access and responsibility for the critical assets. Determining the types of relationships at this stage is important as it allows the identification of social engineering possibilities and the ability for malicious Insiders to collaborate, much in the same way that Social Network Analysis (Scott, 2013) methods work.

The purpose of this stage is to help identify how Insiders ('lone wolves' or groups) may behave within the organisation, what assets they may target and how they may access these for the purposes of an attack. This is achieved through the conceptualisation of Insider threat scenarios that could occur within the organisation. A workshop could be held with key subject matter experts (SMEs) to define the 'step-by-step' insider actions, and define consequence definitions (Negligible through to Catastrophic) based on the functions of the organisation in question. These will be used later in the method to determine the ultimate risk rating once their consequence has been discussed.

Stage 3.1 – Susceptibility Rating of Insider Threat Conditions

In a similar vein to how the HEART EPCs are allocated nominal ratings and subsequent multipliers, there are a number of conditions within an organisation, technical systems and people (which can be identified during stages 1 and 2) that affect the inherent Insider threat risk. The overall susceptibility has two main inputs; system design (incorporating both technical and organisational aspects) and user behaviour; which are dealt with in different ways to determine the overall susceptibility of the system or organisation.

The first stage in the rating process involves identifying the system design susceptibilities (SDS) within a system or organisation, in a similar way to how 'nominal human unreliability' is defined within HEART. Table 3 provides an overview of how some Insider threat technical and organisational conditions may be attributed susceptibility ratings based on their level of perceived severity.

Table 3. Insider threat conditions susceptibility rating examples (system design)

Insider Threat Risk	Rating
Excessive user system privileges	8
Poor password policy	5
Large number of active technical access paths	9
Lack of network monitoring	4
Unprotected sensitive server folder locations	6
Regular use of removable media	6

Once the technical/organisation conditions have been identified and rated, the next stage is to identify the behavioural susceptibility conditions (BSC). These are the perceptions, motivations, ideologies and behaviours of the people that could affect Insider threat risk. Table 4 provides an overview of how some Insider threat behavioural conditions may be attributed multiplier ratings (much in the same way as HEART EPCs) based on their level of perceived severity.

Table 4. Insider threat conditions susceptibility multiplier rating examples (behavioural)

Insider Threat Risk	Rating
Overlooked for promotion	X6
Pattern of unusual working hours	X5
Serving notice period	X8
Perceived mistreatment	X4
High degree of technical work ownership	X4
Workplace conflict/ violence	X9

The inherent system design susceptibilities in Table 3 are multiplied by those multiplier ratings allocated to the identified behavioural conditions in Table 4. Based on the scenarios outlined in Stage 2, susceptibility scores can then be allocated to individual threat conditions, or the Insider threat scenario as a whole.

Stage 3.2 – Defining Insider Threat Consequence

The outputs from stage 3.1 will be a number of susceptibility scores, attributed to conditions that are deemed to occur in given Insider threat scenarios (defined in stage 2). However, analysis of the risk to the organisation should also consider the severity of the consequences of the insider threat. Therefore, there may be a need to allocate a further escalation (or de-escalation) to the susceptibility ratings in stage 3.1 should the consequence of a certain scenario be deemed more or less severe. This stage acts in a similar manner to the Consequence multiplier in conventional risk assessments.

It is recommended that during this stage a workshop is held with qualified and knowledgeable people within the organisation in question (preferably senior management) to run through the defined scenarios from stage 2 and rate their perceived severity, based on an understanding of the specific consequences. The scenarios could be categorised into the following, as outlined in Table 5.

Table 5. Insider threat Scenario Consequence Categories

Consequence Category	(De)escalation Factor
Negligible	0.9
Marginal	0
Critical	1.2
Catastrophic	1.5

The (de)escalation factors are then applied to the susceptibility ratings from stage 3.1, based on the defined scenarios from stage 2. Combining the susceptibility score by an (de)escalation value creates the ultimate Insider threat risk rating. For the purposes of demonstrating this method assumed (de)escalation figures have been applied between the four consequence categories. However, this poses the question; what is the escalation between the consequence categories, and does the escalation factor increase exponentially as the consequences categories become worse?

For example, a basic scenario might include the theft of critical design information by an employee. The regular use of removable media would rate this as 6 (see Table 3), and with a high degree of technical work ownership escalating this susceptibility score to 24 (see Table 4). If the consequences related to that scenario are then deemed in the stage 3.2 workshop as being 'catastrophic' to the organisation, then this score then gets escalated by a factor of 1.5, producing a final Insider threat risk rating of 36.

Stage 3.3 – Insider Threat Mitigation Rating

Much in the same way that ratings and multipliers are allocated to system design and behavioural susceptibilities in stage 3.1, this stage deals with mitigations; those protective measures in place to counter the Insider threat risk. The mitigations are split into the same two categories and allocated mitigation ratings and multiplier scores in the same way. Tables 6 and 7 provide example mitigations in the given format.

Table 6. Insider threat mitigation rating examples (technical/ organisational)

Insider Threat Risk	Rating
Regular access privilege audit reviews	9
Implementation of UBA software tools	6
Use of digital watermarking	3
Restricted use of personal email	4
Separation of duty technical controls	5
Backup of critical information	3

Table 7. Insider threat mitigation multiplier rating examples (behavioural)

Insider Threat Risk	Rating
Clear definition of job/role responsibilities	X2
Clear acceptable use policy	X3
Insider threat training/awareness programme	X6
Formal employee assistance programme	X6
Regular performance reviews and supervision	X4
Strict notice period policy and processes	X7

The system design mitigations (SDM) in Table 6 are multiplied by those multiplier ratings allocated to the identified behavioural mitigations (BM) in Table 7. Based on the scenarios outlined in Stage 2, mitigation scores can then be allocated to individual threat conditions, or the Insider threat scenario as a whole (but must be consistent with the scope used in stage 3.1).

For example, if the organisation conducted regular access privilege audit reviews (rated as 9), with a clear acceptable use policy (multiplier of 3) this would produce a mitigation score of 27 (see Table 7).

Stage 4 – Residual Insider Threat Risk

The final stage in this method is to subtract the Insider threat risk rating (incorporating the susceptibility score and the (de)escalation figure) from the mitigation scored to determine the residual risk score for a given scenario. Essentially, if the final figure is a positive value then the Insider threat residual risk is deemed to be intolerable due to inadequate mitigations. On the other hand, if the figure is a negative value then the Insider threat residual risk is deemed to be tolerable due to adequate mitigations. This activity also serves to provide a 'hierarchy' of Insider threat risk scenarios, which allows the organisation to prioritise and focus their counter-Insider threat efforts in the future. Table 8 presents a typical output from an example scenario; in this case the Insider threat residual risk is intolerable.

Table 8. Insider threat rating summary – *'Scenario 1'*

Measure	Rating
Insider threat risk rating	36
Insider threat mitigation rating	27
Insider threat Residual Risk	**9**

Stages 3 and 4 of the above methodology can be described by the formula below:

Insider threat Risk rating – Insider threat Mitigation rating = Residual Risk

$$\left(\left(\frac{(SDS1 \times SDS2...)}{(BC1 \times BC2...)}\right)(de)escalation\ factor\right) - \left(\frac{(SDM1 \times SDM2...)}{(BM1 \times BM2...)}\right) = Insider\ threat\ Residual\ Risk$$

4.2 Challenges

Challenges associated with developing the methodology are summarised below:

4.2.1 Validation

A key dependency for this method is that there is a validated list of assumptions with rating and multiplier figures for both susceptibility conditions and mitigation measures, as per Tables 3, 4, 5, 6 and 7. For example, an agreement that the susceptibility rating for 'excessive user system privileges' is greater than 'poor password policy', based on an informed assumption that it poses a higher level of risk. Similarly, there are some mitigation measures that provide a much greater level of protection than others, and therefore the ratings can be allocated to reflect this. As outlined above, the relationship between the consequence categories and the (de)escalation factors in stage 3.2 also requires analysis and validation.

There is a need to turn the notional numbers outlined in section 4 into validated numbers; based on relevant research, suitably qualified and experienced person (SQEP) input and the analysis of case studies. It is our belief that the knowledge is already exists, but needs to be synthesised and distilled into a format that is appropriate for a methodology such as this.

4.2.2 Tolerability and SFAIRP

The method outlined above attempts to deal with tolerability by assessing the residual risk score; however the setting of this tolerable / intolerable boundary warrants further consideration. For safety related consequences this may be guided by societal perceptions of tolerability, but for non-safety related consequences, there may need to be flexibility within the methodology, such that an organisation can set its own tolerance threshold level based on its own risk appetite. In addition another aspect that needs further consideration within the methodology, but that has not been discussed in this paper, is the legal requirement to reduce risks to life 'So Far As Is Reasonably Practicable' (SFAIRP).

4.2.3 SQEP

The influences associated with Insider threat are broad and varied, and so it is important that a SQEP practitioner is employed to define the specific conditions and characteristics associated with the technical, organisational and behavioural aspects. A current question for debate is; who is this SQEP practitioner and what skills, qualifications and experience must they possess in order to make the expert judgements? We believe that it is important that this is achieved by integrating HF and Security expertise into Safety approaches, not by attempting to turn Safety professionals into HF and Security experts.

5 Conclusion and a Way Forward

This paper discusses the need for Insider threat risk assessment. It highlights the current lack of validated approaches for defining Insider threat risk and demonstrates how existing methodologies, such as HEART combined with conventional risk assessment, could provide the concept framework for the development of a similar method for the assessment of Insider threat risk.

A high level concept method for assessing Insider threat vulnerability is proposed. This provides a basic framework to enable the main risk areas to be identified, although the concept still requires further development and validation before it can be effectively integrated into mature safety, security and risk practices.

Insider threats will become more severe as cyber environments become increasingly complex, something that is recognised in the wake of policy such as the US Executive Order 13587 (US Government 2012). Organisations need to think about how they counter Insider threat risk with more robust solutions, policy and dedicated core programmes. One thing that is certain is that we must wake up to the malicious Insider as a safety-critical threat in the modern age, or risk becoming the victim of a concealed, intelligent and dangerous internal adversary.

References

Bureau d'Enquêtes at d'Analyses (BEA) (2016). Accident to the Airbus A320-211, registered D-AIPX and operated by Germanwings, flight GWI18G, on 03/24/15 at Prads-Haute-Bléone. BAE2015-0125

Bloomberg (2014). Barclays Examines Possible Data Theft from 27,000 Customers. http://www.bloomberg.com/news/articles/2014-02-09/barclays-probes-possible-theft-of-data-from-27-000-customers. Accessed 18 August 2016

CBS (2016). "Insider Threat" highlights security loopholes at U.S airports. http://www.cbsnews.com/news/tsa-insider-threat-could-pose-security-loophole-at-u-s-airports/. Accessed 3 October 2016

CNN (2015). Germanwings co-pilot – an accident waiting to happen? http://edition.cnn.com/2015/04/07/opinions/abend-co-pilot-warning-signs/. Accessed 20 October 2016

Computer Emergency Response Team (CERT), Software Engineering Instite, Carnegie Mellon (2012). The CERT Guide to Insider Threats: How to Prevent, Detect, and Respond to Information Technology Crimes (Theft, Sabotage, Fraud). Pearson Education, Inc.

Crowd Research Partners (2015). Insider Threat Spotlight Report. http://www.crowdresearchpartners.com/wp-content/uploads/2015/02/Insider-Threat-Report-2015.pdf. Accessed 3 October 2016

The Ecologist (2016). 'Dirty Bomb' security risk at Belgian nuclear power plants. http://www.theecologist.org/blogs_and_comments/Blogs/2987502/dirty_bomb_security_risk_at_belgian_nuclear_power_plants.html. Accessed 3 October 2016

Financial Times (2013). Edward Snowden has a new job in Moscow, says lawyer. https://www.ft.com/content/a5c13c14-423a-11e3-bb85-00144feabdc0. Accessed 20 October 2016

Forrester Research (2013). Understand the State of Data Security and Privacy: 2015 to 2016

The Guardian (2013). Bradley Manning Trial: what we know from the leaked Wikileaks documents. https://www.theguardian.com/world/2013/jul/30/bradley-manning-wikileaks-revelations. Accessed 3 October 2016

Harvard Business Review (HBR) (2014). The Danger from Within. September 2014

History Extra (2015). 4 Revelations about the Titanic Disaster. http://www.historyextra.com/article/maritime/four-revelations-about-titanic-disaster. Accessed 18 August 2016

Leopold, J (2015). Exclusive: Inside Washington's Quest to Bring down Edward Snowden. Vice News.

Meshkati, N (1991). Human factors in large-scale technological systems' accidents: Three Mile Island, Bhopal, Chernobyl. Organization Environment, Volume 5, No. 2. Institute of Safety and Systems Management

Nuclear News (2015). Chernobly nuclear reactor 1986 and today. https://nuclear-news.net/2015/10/26/chernobyl-nuclear-reactor-1986-and-today/. Accessed 20 October 2016

Ponemon Institute LLC (2015). 2015 Cost of Cyber Crime Study: Global

PWC (2015). Managing Insider Threats (MW-15-1443 LL)

Raytheon (2015). The Financial Industry and the Insider Threat: total awareness leads to secured enterprise (white paper). 300132.0415

Scott (2013). Social Network Analysis. 3rd edition. Sage publications Ltd. ISBN 978-1-4462-0903-5

United States Government (2012). Executive Order 13587: National Insider Threat Policy

Vormetric (2015). Vormetric Insider Threat Report: Trends and Future Directions in Data Security, Global Edition

Williams, J. C (1986). HEART – a proposed method for assessing and reducing human error. 9th Advances in Reliability Technology Symposium, University of Bradford.

From the IBM 29 Card Punch to the Boeing 787 Dreamliner (and Beyond)

Dewi Daniels

Software Safety Limited and Aeronautique Associates Limited

Abstract   *This paper is a reflection on the author's career in software engineering over the last 35 years, with an emphasis on what he has learned along the way and its relevance to safety-critical avionic software development.*

1 Learning How to Program (1978-1981)

Many years ago, my wife and I visited the Science Museum in South Kensington, London. My wife pointed at an IBM 29 card punch and said, 'Look at what they used in the days before even you were involved in computing'. I rather sheepishly admitted that I suspected that the card punch had been donated by Imperial College, which is next door to the Science Museum, and that I had almost certainly used that very card punch.

Fig 1. IBM 029 Card Punch[1]

I was in the last undergraduate intake to be taught programming using punched cards at the Department of Computing at Imperial College. The following year's intake were taught to program using interactive IBM 3270 terminals. We used to submit our batch jobs just before a lecture, then at the end of the lecture, visit the computer room to pick up our line printer output, which often contained an error message such as 'missing semi-colon at line 5'.

Fig 2. IBM 3270 Terminal[2]

---

[1] By waelder - Own work, CC BY 2.5,
https://commons.wikimedia.org/w/index.php?curid=1962578

Learning to program using punched cards had a lasting effect on my approach to programming. I still review my code very carefully before I compile it. If a test fails, my instinct is to reread the code very carefully to see if I can figure out the reason for the unexpected behaviour. I have observed that those programmers who learned how to program later using interactive terminals or personal computers tend to use the compiler to find the syntax errors, and that when a test fails, their first instinct is to single-step through the code in the debugger to see where it goes wrong. Both approaches have advantages and disadvantages, but they are two very different approaches.

I don't think I could have wished for a better foundation for my career in software engineering than the education I received at Imperial College. The teaching emphasised general principles that have stood me in good stead throughout my career, instead of specific technologies.

We were taught to program in Pascal (Wirth 1976), FORTRAN, COBOL and Simula 67 (Birtwistle 1974), none of which are in common use today, yet the principles I was taught are as valid today as they were in 1981. Which leads me to a story: many years ago, my wife asked that we switch from On Digital[3] to Sky[4] because our video cassette recorder recorded a blue screen frequently when our Nokia set top box had crashed. Shortly afterwards, I spotted a recruitment advert for Nokia. I half-seriously submitted my CV, recounting our poor experience with the Nokia set top box, explaining that I had no idea what any of the buzzwords or acronyms in their advert meant, but that I could introduce software engineering techniques into their organisation that would cost very little to implement, but which would greatly improve the reliability of their set top boxes. Nokia didn't even bother to reply. Given Nokia's current woes, I most definitely feel it was their loss and not mine!

A major benefit of my education at Imperial College was the ability to approach problem-solving in an analytical and methodical manner and to express my ideas clearly and concisely, which are skills that I believe would have benefited me greatly in any career I had chosen to follow.

The course at Imperial College emphasised the principles of good software design, and the importance of abstraction and information hiding. I owe a great debt to my programming tutor, Iain Stinson, whose ideas on software design have influenced me throughout my career. He taught the principles expounded by Dijkstra (Diskstra 1976), Wirth (Wirth 1976) and Parnas (Parnas 1972).

I find the main difference between average programmers and great programmers (not that I consider myself one of the latter) is the latter's ability to find the right abstraction, eliminating unnecessary complexity. The best pro-

---

[2] By Retro-Computing Society of Rhode Island - Own work, CC BY-SA 3.0, https://commons.wikimedia.org/w/index.php?curid=7354001

[3] An early terrestrial digital television provider, long since defunct.

[4] A satellite digital television provider that now dominates the UK market.

grams look so simple that they give the misleading impression of looking as if they were easy to write. Conversely, the worst programs look so complicated that they look as if they were very difficult to write. As Blaise Pascal wrote:

> I would have written a shorter letter, but I did not have the time.

I also like the quotation from Sir Tony Hoare:

> There are two ways of constructing a software design: one way is to make it so simple there are obviously no deficiencies, and the other is to make it so complicated that there are no obvious deficiencies. The first method is far more difficult.

The course at Imperial College was very much a software engineering course rather than a computer science course. Imperial College was rather unusual at that time in teaching project management as part of the undergraduate curriculum. The teaching included practical programming projects where we were given the opportunity to lead a project team. The main textbook used was (Brooks 1974).

2 Learning How to Program For a Living (1981-1984)

After Imperial College, I was offered a place to study for a PhD at the Computer Laboratory at the University of Cambridge, where I had the enjoyable privilege of being interviewed by Roger Needham. I turned down their kind offer, which in hindsight is probably just as well, as I had little idea what I wanted to research at that time.

I felt that I still had very little experience of computing, and was keen to put the skills I had learned at Imperial College into practice. I therefore joined Logica Limited[5], where I became a member of a software team implementing a Cambridge Ring computer network.

Logica had a very long-lasting influence on the UK software industry. Many of the companies I later worked for were founded by logibods (the term for someone who has ever been employed by Logica). What I most remember was that Logica had a strong commitment to their software quality system, and I think that has remained true of the companies founded by logibods.

It certainly felt good to put my new programming skills into practice.

---

[5] Logica was purchased by CGI Group in 2012.

3 Learning About High Integrity Software (1984-1989)

After Logica, I worked for Novus Systems Technology Limited, a company founded by three people who had left Logica. I worked in a small team developing a software product that allowed Digital Equipment Corporation (DEC) VAX minicomputers to interface to airline reservation systems. Applications developed using this product allowed travel agencies, tour operators and car rental companies to place reservations through the computer systems operated by airlines such as British Airways and American Airlines. These applications had very stringent availability requirements, as downtime would result in loss of sales. The product was launched at DECworld in Boston in 1987, where a kindly old gentleman asked me for a demonstration and seemed to be quite impressed. The kindly old gentleman turned out to be Ken Olsen, the founder of DEC.

Fig 3. VAX 11/780 Minicomputer[6]

At Novus Systems Technology, I had the pleasure of working for Dave Thomas. Dave is one of the smartest programmers I've ever worked with. I

---

[6] By Emiliano Russo, Associazione Culturale VerdeBinario –
http://it.wikipedia.org/wiki/Immagine:VAX_11-780_all.jpg, Public Domain,
https://commons.wikimedia.org/w/index.php?curid=4807141

learned a lot from him. Dave went on to be one of the authors of the Agile Manifesto (Agile Manifesto 2001) and co-author of (Hunt and Thomas 1999).

One of the things I learned at Novus Systems Technology was that it's possible to achieve a great deal with a very small software team, provided that team is skilled, motivated and work well together.

4 Learning About Formal Methods (1990-1995)

In 1990, I joined Program Validation Limited (PVL) in Southampton, which had been founded by Prof. Bernard Carré. At PVL, I helped develop the SPARK Examiner, a tool intended for the development of high integrity software.

I also carried out program proof of a number of full authority digital engine controllers (FADECs) developed by Lucas Aerospace (now part of Rolls-Royce Controls and Data Services). The first engine on which we worked was the Rolls-Royce RB211-535. We conducted program proof after the software had already been verified to DO-178B (RTCA 1992) Level A.

Fig 4. Rolls-Royce RB211-535 Engine[7]

It was very exciting to conduct industrial-scale program proof, as we had been taught about program proof at Imperial College, using (Dijkstra 1976).

A surprising benefit was that this experience turned out to be invaluable even on projects that make no use of formal methods. I believe that my experience of writing pre- and post-conditions has greatly improved my ability to write clear and concise English-language requirements that do not constrain the implemen-

---

[7] Rolls_royce_engine_on_boeing_757-300_-_original.jpg: TobiasK at en.wikipedia (Tobias Kierk)

tation. Throughout my career, I have been surprised by the exceedingly poor quality of requirements produced by many organisations. In particular, I have come across many very smart and experienced programmers who seem totally unable to write detailed and precise requirements without resorting to writing pseudo-code.

Another surprising benefit of learning how to do program proof was that I believe it made me a better tester. I found that most of the defects found by program proof related to boundary values, emphasising the importance of boundary value analysis as described in (Myers 1979).

5 Learning About Correctness by Construction (1998-2004)

5.1 Lean

When Bernard Carré retired from PVL, the company was purchased by Praxis Critical Systems. At Praxis, I continued to work closely with my friend and colleague from PVL, Peter Amey.

Peter and I became interested in Lean, as practiced by the Japanese car industry. Our interest resulted from a chance conversation between myself, Peter and Anthony Hall. I suggested that cost, schedule and scope did not necessarily have to be traded off against each other; rather, good engineering practice reduces cost, schedule and effort. Anthony responded, 'Like the Japanese car industry, you mean?'

Peter and I read (Womack et al. 1990) and we attended a day school on Lean at the Cardiff Business School. One of the things that most impressed me about (Womack et al. 1990) was that the authors were able to obtain production data that allowed comparison between different methods of car production, and therefore demonstrate the efficacy or otherwise of alternate methods.

More recently, I have been impressed by Ben Goldacre's writings on evidence-based medicine (EBM), such as (Goldacre 2013). EBM is an approach to practising medicine that selects treatments based on objective, clinical evidence rather than on expert opinion. EBM practitioners go to great lengths to construct very clever clinical experiments that demonstrate the effect of a particular treatment compared to alternative treatments, taking into account the placebo effect.

It's regrettable that no such data is available for the efficacy (or otherwise) of programming techniques and technologies. For example, 25 years after the introduction of DO-178B, there is no published evidence as to the defect rate achieved by the various DO-178B software levels, or the effectiveness of any of the individual DO-178B objectives. Although DO-178B was based on the consensus opinion of a large group of industry and regulatory experts; it is not evidence-based.

It's not just DO-178B; there is little or no objective evidence for the effectiveness or otherwise of any of the major software development tools or techniques. Most of the software industry seems to be led by fashion, rather than by evidence. I would love to see the development of a new field of evidence-based software engineering.

Lean is concerned with maximising customer value while minimizing waste, where waste is defined as anything that is not of value to the customer. Waste could involve a part that is needlessly expensive to manufacture, remedial work to fix a defective part, or having to wait for a part to come in stock.

There are five key principles as defined by the Lean Enterprise Institute (LEI 2000):

1. **Value.** Find out what matters to your customers.
2. **Value stream.** List all the steps that are carried out. Eliminate steps that do not add value. Ensure that someone 'owns' the entire value stream.
3. **Flow.** Ensure the remaining steps flow smoothly, in a logical order.
4. **Pull.** Reduce time to market by 'pulling' the product as needed.
5. **Perfection.** Whenever a defect is found, stop the production line, find the root cause, and fix the process so the defect can't occur again.

I attended a day school on Lean by Tom and Mary Poppendieck, where they gave the following example of an application of the value stream principle.

Traditional companies typically organise themselves along functional lines. There will be someone in charge of research and development, another in charge of manufacturing, another in charges of sales, and so on. Dell is organised by value stream. There is one person in charge of the entire value stream associated with each product line, from R&D, through sales and manufacturing to post-sales support.

Many years ago, there was a shipping strike in the Pacific. Most laptops sold in the USA were manufactured in South East Asia and shipped by container ship to the west coast. When the value stream manager for Dell laptops heard of the shipping strike, he chartered a fleet of cargo aircraft to transport the laptops from the factories to the USA. This increased Dell's costs, but it meant they were able to carry on delivering laptops during the strike. Dell's rivals were organised along traditional lines. The person in charge of sales had to ask the board to direct the person in charge of shipping to charter aircraft, which required an increase to the shipping budget. By the time the board had made their decision, Dell had chartered all the aircraft available. Dell were able to increase their prices during the strike because they were the only vendor able to deliver laptops, while their competitors lost sales because they were unable to deliver any laptops. This was therefore a successful application of the value stream principle.

Despite Toyota's recent fall from grace, I am still a fan of Lean. Eliminating waste lies at the heart of engineering. As the American civil engineer, Arthur Wellington, wrote in 1887:

> It would be well if engineering were less generally thought of, and even defined, as the art of constructing. In a certain important sense it is rather the art of not constructing; or to define it rudely but not inaptly, it is the art of doing that well with one dollar, which any bungler can do with two after a fashion.

The novelist and aeronautical engineer, Nevil Shute, put it more succinctly in 1960:

> An engineer is a man who can do for five bob what any bloody fool can do for a quid[8].

Lean has been widely adopted in other engineering disciplines. For example, most of the engineers at my local branch of the IET seem to be very familiar with Lean. Lean seems to have gained surprisingly little traction in software engineering. (Poppendieck and Poppendieck 2003) and (Middleton and Sutton 2005) are two rare examples of the application of Lean to software development.

5.2 Agile

Our interest in Lean led to an interest in Agile, which shares many aims with Lean, but originated from within the software industry rather than the car industry. The Agile Manifesto lists twelve principles:

1. Our highest priority is to satisfy the customer through early and continuous delivery of valuable software.
2. Welcome changing requirements, even late in development. Agile processes harness change for the customer's competitive advantage.
3. Deliver working software frequently, from a couple of weeks to a couple of months, with a preference to the shorter timescale.
4. Business people and developers must work together daily throughout the project.
5. Build projects around motivated individuals. Give them the environment and support they need, and trust them to get the job done.
6. The most efficient and effective method of conveying information to and within a development team is face-to-face conversation.
7. Working software is the primary measure of progress.

---

[8] For readers who don't remember pre-decimal British coinage, a 'bob' was 5p. A 'quid' is £1.

8. Agile processes promote sustainable development. The sponsors, developers, and users should be able to maintain a constant pace indefinitely.
9. Continuous attention to technical excellence and good design enhances agility.
10. Simplicity — the art of maximizing the amount of work not done — is essential.
11. The best architectures, requirements, and designs emerge from self-organizing teams.
12. At regular intervals, the team reflects on how to become more effective, then tunes and adjusts its behaviour accordingly.

There are clearly some parallels between Agile and Lean. However, I have come across a number of instances where we found that working software failed to deliver value to our team. I'm concerned that Agile approaches can over-emphasise the importance of writing code, while ignoring or placing insufficient emphasis on what Brooks called a 'programming system product' (Brooks 1974). I suppose it depends on your definition of *working software*.

For example, we recently downloaded an open source real-time operating system (RTOS). At first, we were impressed with how quickly we were able to develop our embedded real-time application. However, during integration testing, we found that our application would stop after anything from a few minutes to a few hours. I found that I could construct a short test program that exhibited the same behaviour, demonstrating that the problem was in the RTOS, not in our application. We contacted the developer, who stated that we must have configured the RTOS incorrectly. He claimed there was no defect in the RTOS, because no one else had reported such a problem. The configuration tool for this RTOS had a very complex interface and was very poorly documented. The developer was not interested in fixing our problem, even though we were willing to pay him. We ported our application to another RTOS, which has worked extremely well.

The developer would argue that he delivered working software, because we did not find any defects, but I would argue that the software failed to deliver customer value because the RTOS configuration tool was unintuitive and poorly documented.

In the case of safety-critical systems, working software is certainly important, but is not sufficient. It is also necessary to develop compelling evidence that the software-intensive system is safe for its intended function *before* it is deployed in service. Safety cannot easily be added after the fact, but needs to be designed in from the beginning.

One of the strengths of DO-178B (RTCA 1992) and DO-178C (RTCA 2011) is that it they are objective-based. They don't prescribe a specific software process, but they do define objectives that the selected software process must satisfy. An avionics software project still has to meet all the DO-178B/DO-178C objectives, whether it follows the waterfall model or an Agile life cycle.

5.3 Correctness by Construction

While I was the Software Engineering Service Line Leader at Praxis, we formalised an approach to software development that we called Correctness by Construction (CbyC). CbyC is a method in which the use of formal methods is central, but incorporating some aspects of Lean and Agile. The seven key principles of Correctness by Construction (CERT 2006a) are:

1. **Expect requirements to change.** Changing requirements are managed by adopting an incremental approach and paying increased attention to design to accommodate change. Apply more rigour, rather than less, to avoid costly and unnecessary rework.

2. **Know why you're testing.** Recognize that there are two distinct forms of testing, one to build correct software (debugging) and another to show that the software built is correct (verification). These two forms of testing require two very different approaches.

3. **Eliminate errors before testing.** Better yet, deploy techniques that make it difficult to introduce errors in the first place. Testing is the second most expensive way of finding errors. The most expensive is to let your customers find them for you.

4. **Write software that is easy to verify.** If you don't, verification and validation (including testing) can take up to 60% of the total effort. Coding typically takes only 10%. Even doubling the effort on coding will be worthwhile if it reduces the burden of verification by as little as 20%.

5. **Develop incrementally.** Make very small changes, incrementally. After each change, verify that the updated system behaves according to its updated specification. Making small changes makes the software much easier to verify.

6. **Some aspects of software development are just plain hard.** There is no silver bullet. Don't expect any tool or method to make everything easy. The best tools and methods take care of the easy problems, allowing you to focus on the difficult problems.

7. **Software is not useful by itself.** The executable software is only part of the picture. It is of no use without user manuals, business processes, design documentation, well-commented source code, and test cases. These should be produced as an intrinsic part of the development, not added at the end. In particular, recognize that design documentation serves two distinct purposes:

     i.   To allow the developers to get from a set of requirements to an implementation. Much of this type of documentation outlives its usefulness after implementation.

     ii.   To allow the maintainers to understand how the implementation satisfies the requirements. A document aimed at maintainers is much shorter, cheaper to produce and more useful than a traditional design document

An article by Peter Amey on the official website of the US Department of Homeland Security (CERT 2006b) states:

> Praxis has evolved CbyC over the last 12 years and used the approach to produce software in an industrial environment with extremely low defect rates; rates are fewer than 0.05 defects per 1,000 lines of code, with good productivity, up to around 30 lines of code per person day averaged over the development lifecycle.

6 Learning About CMMI and Offshoring (2004-2010)

After Praxis, I joined Silver Software, another company specializing in safety-critical software development, especially in the aviation and rail industries. Silver Software had about 400 engineers located in the UK, India and Spain. I eventually became the Chief Engineer for the company.

During my time at Silver Software, it became the first CMMI-SW Level 5 company in the UK. At CMMI Level 5, organizations are expected to seek continuous improvement.

My colleague, Dinos Appla, started a lunchtime Agile Club, which was very well attended. The company sponsored a number of people to become certified ScrumMasters (CSMs). We applied Agile techniques successfully on a number of projects, including a safety-critical railway system developed to EN 50128 SIL 4 (Nicoll 2014).

We found that having CMMI Level 5 made it easier, not harder, to implement Agile. Because we collected standardized metrics on every project, we were able to get immediate feedback as to whether a new technique was more or less effective. CMMI Level 5 fits very well with Scrum, since it makes it easy to measure the project's trajectory.

We did a lot of work on the Boeing 787 and acquired a good reputation with Boeing. This meant we were asked to support a high-level Boeing team visiting suppliers who were late delivering (their subsystems were described as 'the last box on the aircraft'), to determine the cause for the delay and to give the suppliers practical support to help them get back on schedule. I found this very interesting and rewarding, as I felt we were making a difference and the suppliers were glad of our help.

Fig 5. First Flight of the Boeing 787 Dreamliner, 15 December 2009[9]

While I was at Silver Software, I was able to get involved in the committee that was updating DO-178B to DO-178C. I found this a very positive experience and I made a number of very good friends on the committee. I was initially a bit skeptical about 'design by committee', but I came to see the value of a consensus-based approach. The involvement of a large and diverse group of people ensured that all aspects were considered, while the participation of so many companies and regulatory authorities ensured that the published document had industry and regulatory authority buy-in.

7 Some Surprising Lessons From Gliding (2009-present)

In 2009, I achieved a lifelong ambition by flying a glider solo. I now hold a pilot's licence and am a member of the Black Mountains Gliding Club (BMGC 2016), where I fly my own glider.

I've learned a surprising amount about decision-making from gliding. I'm naturally a deep, analytical thinker. I like to collect lots of evidence and take my time to consider as many options as possible. I dislike making quick decisions. I often find myself dwelling on past decisions that I've made.

In gliding, it's important to be decisive, and to make each decision at the right time. Decisions are made with the aim of achieving a goal. For example, I might choose to fly a task, which means turning around a number of declared waypoints. At the beginning of the task, you don't worry about decisions you don't need to make until later, such as on which runway to land. Instead, you focus on the next decision. For example, which cloud do I go to next to find lift? It's also important to have a backup plan. For example, if that cloud

---

[9] By Dave Sizer from Seattle, WA, USA (787 First Flight Uploaded by Altair78) [CC BY 2.0 (http://creativecommons.org/licenses/by/2.0)], via Wikimedia Commons

doesn't work, I can go to that ridge. It's important to be decisive. If you vacillate between two clouds, you'll guarantee that you stay in the sinking air and end up on the ground. Finally, it's best to defer a decision until it needs to be made (but no longer). That means that you have the best and most up-to-date information on which to base that decision.

I'm surprised how many of these principles I've been able to apply to my professional life. I feel much more comfortable than before about deferring decisions. I find that avoiding making premature decisions keeps more options open, and that the final decision is based on better information. Also, being decisive but always having a backup plan means that I make better progress.

8 Learning How to Run a Company (2010-2015)

I was invited to form the UK subsidiary of an American company in 2010. I learned a lot about running a company in my time at Verocel. We started with just me working from home in 2010 and ended up with a turnover of over £2 million in 2013 and 2014.

I was influenced by Joel Spolsky (Spolsky 2007). He writes that his goal in starting his company, Fog Creek, was 'to create a software company where we would want to work'. I felt the key to a successful company was a) generating a steady stream of sales and b) employing people I could trust to do a good job. I found the most effective way of selling was to sell to people who knew me and who trusted me. I tended to recruit either people I knew or people who came recommended by people I knew.

I had a great team working for me at Verocel Limited. I'd like to think that I treated my staff well, and that they would all like to work for me again.

9 Learning about Unmanned Air Systems (2015-present)

I'm currently acting as Chief Software Engineer for Callen-Lenz Associates Limited, where I'm leading a team that is developing a flight control system for an optionally manned rotorcraft to DO-178C Level A. I think we're about to see some very exciting developments in unmanned and optionally manned air systems.

10 Conclusion

While the programming languages and tools have changed, the basic principles remain unchanged. I would like to leave you with the following principles to consider:

1. Decompose the problem into small steps. Ideally, each step should result in working software.
2. Write clear requirements that don't constrain the implementation.
3. Find the right abstraction to avoid unnecessary complexity and hide implementation detail.
4. Keep the software simple and easy to understand.
5. Verify that all the requirements have been satisfied.

Acknowledgments   I would like to thank my programming tutor at Imperial College, Iain Stinson, who taught me most of what I know about programming, and who has influenced my programming style to this day. I would also like to thank my good friend, Prof. Matt Jaffe of Embry-Riddle Aeronautical University, for reviewing this paper.

References

Agile Manifesto (2001), http://agilemanifesto.org, accessed 26 September 2016.
BMGC (2016), http://blackmountainsgliding.co.uk, accessed 26 September 2016.
Birtwistle G M, Dahl O-J, Myrhaug B, Nygaard K (1974) Simula BEGIN. Auerbach/Studentlitteratur.
Brooks F (1974) The Mythical Man Month and Other Essays on Software Engineering. Addison Wesley Longman.
CERT (2006a) https://www.us-cert.gov/bsi/articles/knowledge/sdlc-process/secure-software-development-life-cycle-processes#correctness, accessed 17 October 2016.
CERT (2006b) https://www.us-cert.gov/bsi/articles/knowledge/sdlc-process/correctness-by-construction, accessed 27 September 2016.
Dijkstra E (1976) A Discipline of Programming, Prentice Hall.
Myers G (1979) The Art of Software Testing, John Wiley & Sons.
Womack J, Jones D, Roos D (1990) The Machine That Changed the World, Simon & Schuster.
Goldacre B (2013) Bad Pharma: How Medicine is Broken, and How We Can Fix It, Fourth Estate.
LEI (2000) http://www.lean.org/WhatsLean/Principles.cfm, accessed 17 October 2016.
Middleton P and Sutton J (2005) Lean Software Strategies: Proven Techniques for Managers and Developers, Productivity Press.
Nicoll (2014)
http://www.slideshare.net/AdaCore/david-nicoll-experienceofagileforscdevelopment,
accessed 26 September 2016
Parnas D L (1972) On the Criteria To Be Used in Decomposing Systems into Modules, Communications of the ACM 15 (12), 1053-58.
Poppendieck M and Poppendieck T (2003) Lean Software Development: An Agile Toolkit, Addison Wesley.
RTCA (1992) DO-178B Software Considerations in Airborne Systems and Equipment Certification.

RTCA (2011) DO-178C Software Considerations in Airborne Systems and Equipment Certification.

Spolsky J (2007) Smart and Gets Things Done, Springer-Verlag.

Wirth N (1976) Algorithms + Data Structures = Programs, Prentice-Hall.

Analysis of Effects induced by EM disturbances on COTS Devices, from an EM Security and Functional Safety perspective

José Lopes Esteves[1], Chaouki Kasmi[1], Andy Degraeve[2], Davy Pissoort[2], Keith Armstrong[3]

[1] Wireless Security Lab, FNISA-ANSSI, France
[2] KU Leuven, Technology Campus Ostend, Belgium
[3] Cherry Clough Consultants Ltd, UK

Abstract *Electromagnetic Security refers to the compliance of electronic devices with Information Security requirements with regards to electromagnetic disturbances, and has important implications where there are Functional Safety or other risks to be managed. In this study, the resilience and the integrity of electronic devices are the topics of interest. Many studies have been devoted to the detection, the analysis and the classification of failures and damage by electromagnetic interference induced on commercial off-the-shelf devices, which mostly require external monitoring and measurement equipment. More recently, an approach based on the exploitation of existing internal resources of the tested devices has been proposed. Monitoring of effects due to exposure to EM fields is important for both EM Security and Functional Safety. This paper shows how this can be done by measuring the existing internal parameters of computers and smartphones during exposure to unintentional or intentional electromagnetic fields.*

1 Information Security, Functional Safety and EMC

The field of information security (INFOSEC 2016) is based on the following three main principles: *confidentiality*, *availability* and *integrity*. The main purpose is to prevent the access, the use, the disclosure and the modification of sensitive information by unauthorized third parties while still providing authorized entities with these abilities.

The discipline of ElectroMagnetic Compatibility (EMC), however, focuses instead on the emissions (IEC 61000-6-3, IEC61000-6-4) and the immunity of electronic devices (IEC 61000-4-1, IEC 61000-4-3). The main goal is to ensure the operation and robustness of electronic devices by reducing their electromagnetic (EM) emissions and improving sufficiently their immunity to the EM disturbances expected to occur in their operational EM environments.

When taken into account for an INFOSEC risk analysis, information leakage due to compromising EM emissions generated by electronic devices (Van Eck and Laborato 1985), as well as their resilience to exposure to EM disturbances (Sabath and Römer 2008, Savage et al 2008, STRUCUTRES 2016, HiPOW 2016) are important concerns. This topic is known as Electromagnetic Security, which may have implications for the assessment of threats to the correct operation of computers and other information technologies, including (but not limited to) Functional Safety risks, see (Pissoort and Armstrong 2016).

Where errors, malfunctions or failures in electronic systems could increase safety risks, Functional Safety standards and/or Safety or Product Liability laws/regulations might not be complied with. Indeed, EM disturbances can cause EM interference (EMI) with all types of electronics, so must be taken into account for Functional Safety and/or Product Liability. However, it is unrealistic to perform enough EM immunity testing to ensure that a digital processor or its software is functionally safe enough having regard to electromagnetic disturbances, see (Pissoort and Armstrong 2016) for more on this subject.

Unless physical damage is caused by EMI, there is usually no evidence that it has occurred, transient malfunctions may not even be noticed at the time, and any data indicating them will generally be erased by the next power-off/on cycle. Event data recording can be used to improve the possibility of establishing that a malfunction has occurred that could have been caused by EMI. Whenever an anomaly is detected in the operation of safety-related electronics (such as a data value out of range, a checksum failure, or a sequencing error) relevant data can be recorded. This data can then be analysed statistically in real time or at some later time to detect and diagnose trends due to sporadic failures and to propose remedial action. Hence; it would be helpful to know if the electronic devices in common use record data which could be useful for detecting whether EMI had probably occurred, and when.

Manufacturers of Commercial-Off-The-Shelf (COTS) electronic devices make more and more information accessible to users about the running state of the equipment in the system logs of the operating system or the embedded firmware. Additionally, modern devices contain many hardware sensors that can be polled directly by low-level communication buses (e.g. I^2C (NXP 2014)) or via high level application programming interfaces (APIs) provided by the system (e.g. kernel, drivers). Modern CPUs can also retrieve information about the voltage levels of some components. The operating system viewpoint of abnormal behaviour caused by sources of EM disturbances has been reported (Kasmi et al 2014, Kasmi and Lopes-Esteves 2015) and analysed in detail (Kasmi et al 2015).

It has been concluded that "system events" logs and external interfaces statistics, combined with available hardware sensors' data, can be a good source of information to diagnose critical malfunctions in hardware and software, as well as the probable cause of system reboots or shutdowns. As this information is available to the operating system, it is possible to design real-time remote monitoring software. In this paper, we propose to rely on the recorded events log files and embedded sensors measurements to detect and record functional failures when devices are exposed to radio-frequency (RF) fields. It will be demonstrated that the proposed health monitoring software allows for overcoming both functional safety and information security challenges.

The paper is organized as follows. First, in Section 2 after presenting the main aspects of EMC for Functional Safety, the proposed real-time recording approach is described. In Section 3 the results obtained for two types of devices, computers and embedded devices, are summarized. Additionally, the outcomes of the presented approach for a Functional Safety analysis with regards to EM disturbances are discussed.

2 EMC and EMI Susceptibility testing

In practical EMC experiments, operators would like to be able to detect and correlate the effects of intentional or unintentional electromagnetic interference (EMI) with hardware and software faults that can be recorded on an equipment under test (EUT). Already applied in the industry in immunity testing, it has been reported that the effects of high-power electromagnetic (HPEM) disturbances on the CAN-Bus networks (Bosch 1991) used in automobiles can be estimated (Mirschberger 2012). Unfortunately, for our purposes, the hardware and the software internal resources of these automotive electronic devices cannot be monitored, due to access to these resources being restricted to manufacturers only.

Estimating the effects during exposure to EM disturbances requires the test engineer to be able to monitor the electronic device health status, which is generally impossible to do directly for safety reasons due to the effects of certain EM disturbances on human health. Moreover, short temporary failures may have disappeared at the time the test engineer is allowed to enter the experiment facility.

Helpfully, more and more accurate information about the CPU (e.g. load, temperature), the internal memories (e.g. integrity) and the software crashes (e.g. operating system malfunctions, firmware errors and drivers errors) can be obtained using documented commands provided by electronic system manufacturers.

By collecting this data in real time, it would be possible to record and to analyze its response to electromagnetic disturbances on the EUT in a finer grain and to trace the malfunctions from the hardware to the software level. Experimental tests, reported in (Choi et al 2014), were conducted in order to have an idea of the effects induced by electromagnetic attacks against a large infrastructure containing a complex IT network.

In order to analyze a potential failure of the involved computers, the authors reported a handmade test conducted on each of those devices. Nevertheless, this approach is known to be time-consuming and leads to the missing of the temporary failures recovered by the operating system running on the tested COTS IT systems. Real-time monitoring of the faults occurring during the tests would overcome this limitation. In the following study, we focus on COTS devices to show that more clear-cut details can be obtained directly by the software running on the EUT itself.

2.1 Local software analysis of effects

In order to use a software approach to assess the functional behaviour of an EUT whilst it is exposed to EM disturbances, the fundamental step is to identify suitable observables (Astrom and Murray 2016) which can be accessed by software. To achieve this, a decomposition of the EUT based on the different EM coupling paths is proposed, as described below. This decomposition allows for determining a set of hardware coupling interfaces for EMI roughly related to the characteristics of the EM disturbances. Then, an identification of software interfaces, accessible by the operating system or overlying applications and linked to the hardware coupling interfaces is necessary. In this section, an overview of the systematic decomposition and of the software interfaces identified is proposed in what follows.

Generally, an electronic device deployed in its working environment is composed of Printed Circuit Boards (PCBs) interconnected by wires and cables, all packaged in an enclosure which has exposed external interfaces intended for connecting peripherals (e.g. networking devices, human interface devices) or power supply sockets. From an EMC point of view, several coupling paths can be identified which will be more or less prone to propagating EM disturbances. For example, peripheral cables have lengths and shielding characteristics which significantly differ from shorter PCB lines contained in a shielded enclosure. Moreover, when wireless communication controllers are embedded in the EUT, antennas are straightforward front-door coupling interfaces (see item 3 in the list below) located in areas where the shielding strategy allows for maximizing the reception of EM signals. Thus, an identification of the hardware coupling interfaces based on the coupling path seems to be an appropriate approach. Three categories have been considered in this study and are described in the following:

Field to cable (back-door coupling). In this category, all the external cables or wirings are considered as possible hardware coupling interfaces. USB and PS/2 peripherals, such as keyboards, mice, external webcams and other human interface devices (HID) are commonly connected to the EUT with cables with varying lengths and shielding characteristics. Display devices also fall into this category. Those peripherals are managed by the EUT with specific controllers, which interface with the operating system through device drivers. The data reported through the system logs by the drivers was a useful observable for monitoring hardware malfunction during exposure to EM disturbances.

Field to PCB (back-door coupling). This category of hardware coupling interfaces regroups all the EUTs internal busses involving conductive communication tracks on the PCBs. From a functional viewpoint, these tracks generally support communications between integrated circuits, either digital (e.g. microprocessor to flash memory chip) or analog (e.g. microprocessor to temperature sensor), and low voltage power supply. Again, the operating system has access to relevant data for monitoring failures on those busses. The system logs provide useful information about communication disruption, the measurements from the sensors can be accessed by software and integrity check routines on storage components (e.g. random access memory) can also be implemented.

Field to Antenna (front-door coupling). The wireless communication interfaces (e.g. 2G/3G modem card, Wi-Fi), act as interfaces for direct coupling. Depending on the hardware components and the software drivers, different information about the EM environment can be gathered. The noise floor (NF), the signal to noise ratio (SNR) and the received power (RP) can be available. The data rate and integrity of ongoing communication can also be monitored.

2.2 Application to large scale functional safety testing

In the previous section, a coupling path based approach for identifying hardware coupling interfaces and their related software accessible observables has been described. As a result, the conception of a local functional susceptibility testing software has been derived, which would access efficient observables, allowing a real-time monitoring of the effects of EMI on the EUT. As discussed in (Choi et al 2014), a distributed architecture for the software facilitates its deployment on several networking capable EUTs. In this section, a similar distributed architecture of the monitoring software adapted to functional susceptibility testing is proposed. The benefits of this approach from a functional safety viewpoint are then discussed, both for a large scale functional testing and for in-field testing.

2.2.1 Distributed real-time monitoring software architecture

We propose a distributed software architecture providing a real-time monitoring of the effects resulting from unintentional and intentional EMI on a set of EUTs. The software architecture is depicted in Figure 1. The implemented architecture is composed of several software subsystems. The first subsystem is the Symptom Observer Subsystem, which gathers data from a set of sensors and is specific to the host machine hardware and software configuration.

The data is then processed by the Analysis Subsystem, which parses the collected data, correlates the symptoms and determines the immunity level of the EUT machine. The Monitoring Subsystem aggregates the data and handles the user interface. It gives an overview of the evolution of the immunity level and raises messages to the operator. It provides great flexibility and the possibility to deploy each subsystem both locally or distributed over a network. Moreover, the Monitoring subsystem could be synchronized with the software driving the EM sources of the testing facility in order to better correlate temporally the effects with the stimuli.

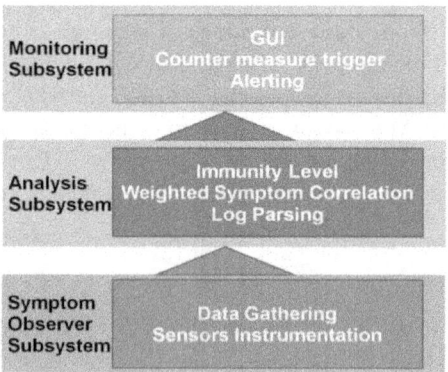

Fig. 1. General architecture of the health monitoring software

2.2.2 Benefits for functional safety testing

The presented approach relies on the main idea of exploiting the capabilities of the EUT to perform an autonomous self-monitoring of a set of observables which are expected to react during EM exposure. This strategy offers the possibility to obtain information on the EUT during the tests without the need of additional hardware, which is usefully cost-effective.

Furthermore, the distributed architecture of the proposed software solution provides many benefits. First, the Symptom Observer only performs the collection of the relevant data, which reduces the computation needed to perform the monitoring to the minimum on the EUT. The analysis can be done on dedicated machines which are more suited for this kind of tasks. Moreover, it enables the testing of several EUTs in parallel while centralizing the results to a small set of monitoring computers. This allows performing large-scale testing campaigns by testing several instances of the same device. Also, the Symptom Observer is the only part which is device-specific and must be adapted to each new kind of EUT. This facilitates in-field testing, where different EUTs can be analyzed simultaneously and in their operational environment.

2.3 Limitations

The main idea of the proposed approach is relying on the capability of the EUT of performing a kind of self-monitoring during the tests. Obviously, this implies that the integrity of the measurement unit cannot be guaranteed, which results in some limitations regarding its resilience and the reliability of the measurements (Kasmi et al 2015). These limitations have to be taken into account when analyzing the reported behaviour of the EUT during the tests.

First, this approach is highly dependent of the observables that are identified and monitored. This basically means that an exhaustive coverage of all the possible disturbances is not guaranteed. The impact of this limitation could be reduced by involving electronic devices manufacturers, which should provide a lower-level access to observables on their products along with specific software stacks dedicated to testing, and standardization bodies which could give a significant impulse for the generalization of these requirements.

Second, the EUT is likely to be disturbed during EM exposure in such a way that the observables and the hardware and software parts involved in the collection of the data could be corrupted. If a corruption is noticed, it should lead to a finer-grain analysis of the underlying symptoms resulting in the identification of a new observable. But there could be some cases where the disturbance inhibits the capacity of detecting its impacts on the already monitored observables.

This limitation can also be extended to the whole monitoring network infrastructure (routers, cables, and monitoring computer) which can be corrupted during the tests.

3 Experimental results

The proposed approach has been applied to two types of targets: desktop computers and smartphones. These two categories are interesting as they provide different resources that can be used as observables. More specifically, smartphones have more analog sensors whereas desktop computers possess wired IO interfaces (e.g. keyboard, mouse…) and a more verbose logging system. Thus, according to the target, different sets of observables can be considered. The most common observables are listed in the following subsections.

The experiments were performed in a shielded room (Faraday Cage). The devices under test have been equipped with the described custom software in charge of gathering data from the observables and sending it to a monitoring computer outside the shielded room. The networking chain involved a wired (Ethernet) router for desktop computers and a wireless (Wi-Fi) router for smartphones which routed the data stream through an optical fibre link outside the shielded room (Figure 2).

During the experiments, the targets were illuminated with continuous waves in the 100 MHz – 20 GHz frequency band modulated in amplitude with a repetition rate in the range 1 Hz – 50 kHz (50 % duty cycle). In what follows, the most responsive observables and the observed effects of HPEM exposure are summarized for each type of targets.

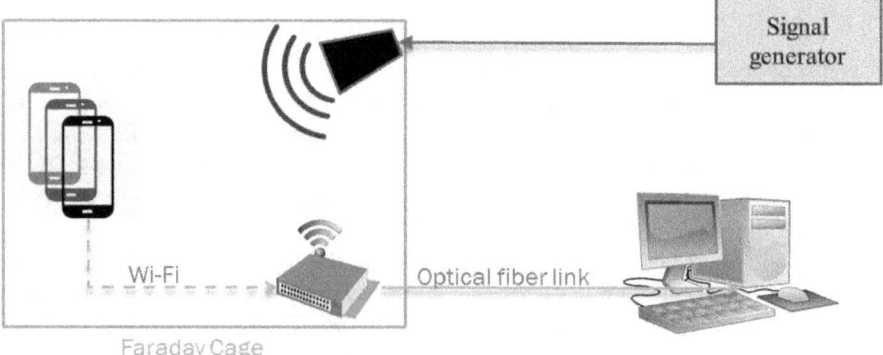

Fig. 2. Experimental test setup; the wireless link has been replaced by an Ethernet wired link, depending on the EUT

3.1 COTS computers

COTS desktop computers usually enclose several analog sensors and system diagnostics and management chips which provide low-level information about the internal state of critical components on the motherboard and peripherals. In (Kasmi et al 2014), the definition of some observables and their reaction to HPEM fields has been studied, as reported below:

- Temperature sensors (in the CPU and HDD) show a high susceptibility and report unrealistic temperatures to the operating system;

- The CPU cooling fan activity is affected by the erroneous temperature readings;

- Wired peripherals device drivers (PS/2, USB) report several errors, from disconnect/reconnect sequences to random input interpretation;

- Wired and wireless networking interfaces statistics show a drastic increase of errors and a decrease of the link throughput;

- Analog input interfaces, such as RF communication interfaces or the audio interface allow for gathering raw signal data of the front-door and back-door, respectively, coupling effects (Figure 3).

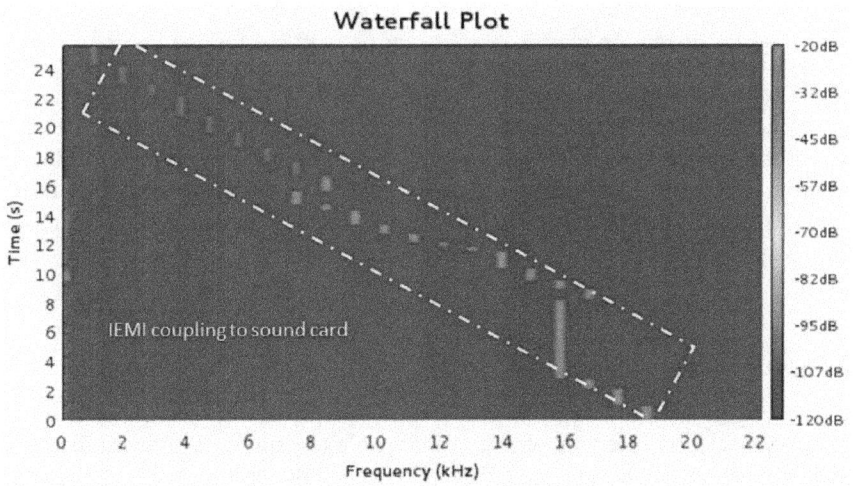

Fig. 3. Back-door coupling effects observed on the audio front-end

As mentioned earlier, the effects range from disabling the system from interacting with the user or other systems, to causing errors in the reported temperatures or fan speeds. With this approach, the functional criticality of these effects can be studied with a fine grain and acceptance thresholds for each observable can be defined (Kasmi et al 2015).

3.2 COTS Smartphones

As mentioned in (Kasmi and Lopes-Esteves 2015), modern smartphones include many sensors (e.g. triaxial magnetic field, accelerometer gyroscope and light sensors). Observables can be gathered from these embedded sensors enclosed in real-time in order to correlate the disturbances during exposure to unintentionally or intentionally-generated radio-frequency fields. The following effects were detected during the HPEM exposure experiments:

- GPS module: unexpected GPS receiver activity;

- Wi-Fi interface: data rate decreasing, packet errors increasing, communication stopped;

- Audio interface (with/without headphones): voice command interface launch;

- Touch screen effects: unusable, applications stopped;

- Battery charge: Errors on the reported charge capacity.

From a Functional Safety perspective, many of the listed effects can be seen as critical ones as users may lose the usability of the wireless connectivity as well as the accessibility to the devices as the touchscreen. As a consequence, it is necessary to harden these interfaces against HPEM to improve their resilience. Appropriate health monitoring software would have provided EMC/EMI testers with information to feedback to manufacturers that would have pointed-out these weaknesses.

4 Conclusion

In this paper, it has been shown that we are able to exploit the internal resources of COTS devices and their operating systems to collect information about their health status during exposure to unintentionally or intentionally-generated RF fields.

An effective set of observables has been identified and it has been shown that these can be responsive to electromagnetic disturbances. A software-based monitoring agent has been designed by automating the instrumentation of the identified observables and correlating the observed symptoms due to the unintentionally or intentionally-generated fields. Its architecture was designed to easily allow a flexible deployment on several kinds of devices and different network topologies. We show that this distributed agent network provides a novel way to monitor and record functional failures induced by lack of electromagnetic compatibility, both temporally and spatially over a whole infrastructure.

The application and the testing of the proposed software during electromagnetic compatibility testing and along its life cycle opens a new way for accessing the functional safety of critical infrastructures as well as providing valuable knowledge for their hardening against electromagnetic security threats.

References

Astrom K J, Murray R M (2016) Observability: see Chapter 8 of Output Feedback, v3.0h, Princeton University Press. http://www.cds.caltech.edu/~murray/FBSwiki. Accessed 25 October 2016

Bosch R (1991) CANbus Specification, Version 2.0. http://esd.cs.ucr.edu/webres/can20.pdf. Accessed 25 October 2016

Choi J S, Lee J, Ryu J et al (2014) Evaluation of Effects of Electronic Equipments in Actual Environments. In Proc. of AMEREM

IEC 61000-4-1 Electromagnetic compatibility (EMC) - Part 4-1: Testing and measurement techniques - Overview of immunity tests

IEC 61000-4-3 Electromagnetic compatibility (EMC) - Part 4-3: Testing and measurement techniques - Radiated, radio-frequency, electromagnetic field immunity test

IEC 61000-6-3 Electromagnetic compatibility (EMC) - Part 6-3: Generic standards - Emission standard for residential, commercial and light-industrial environments

IEC 61000-6-4 Electromagnetic compatibility (EMC) - Part 6-4: Generic standards Emission standard for industrial environments

IET (2013) Overview of techniques and measures related to EMC and Functional Safety. Institution of Engineering and Technology. http://www.theiet.org/factfiles/emc/emc-overview.cfm?type=pdf. Accessed 25 October 2016

INFOSEC (2016) Information Security. https://en.wikipedia.org/w/index.php?title=Information_security&oldid=743986660. Wikipedia contributors. Accessed 25 October 2016

Kasmi C, Lopes-Esteves J (2015) Automated analysis of the effects induced by radio-frequency pulses on embedded systems for EMC Functional Safety. Radio Science Conference (URSI AT-RASC)

Kasmi C, Lopes-Esteves J, Renard M (2015). Automation of the Immunity Testing of COTS Computers by the Instrumentation of the Internal Sensors and Involving the OS Logs. System Design and Assessment Notes, SDAN 44. http://www.ece.unm.edu/summa/notes/index.html. Accessed 25 October 2016

Kasmi C, Lopes-Esteves J, Picard N et al (2014) Event Logs Generated by an Operating System Running on a COTS Computer During IEMI Exposure. IEEE Transactions on Electromagnetic Compatibility 56:1723-1726

Mirschberger J, Sonnemann F, Urban J et al (2012) High-Power Electromagnetic effects on Distributed and Automotive CAN-bus systems. In Proc. of European Electromagnetics Conf. (EUROEM)

NXP (2014) UM10204: I²C-bus specification and user manual, Rev. 6. NXP Semiconductors. http://www.nxp.com/documents/user_manual/UM10204.pdf. Accessed 25 October 2016

Pissoort D, Armstrong K (2016) Why is the IEEE developing a standard on managing EMI risks. IEEE 2016 International Symposium on EMC

Project HiPOW (2016) Protection of Critical Infrastructure against High Power Microwave Threat. European Research Project FP7. http://www.hipow-project.eu/hipow/project. Accessed 25 October 2016

Project SECRET (2016) SECurity of Railways against Electromagnetic aTtacks. European Research Project FP7. http://www.secret-project.eu. Accessed 25 October 2016

Project STRUCTURES (2016) Strategies for The impRovement of critical infrastrUCTUre Resilience to Electromagnetic attackS. European Research Project FP7. http://www.structures-project.eu/overview. Accessed 25 October 2016

Sabath F, Römer B (2008) Susceptibility of IT-Networks to HPM and UWB Threats. In Proc. of European Electromagnetics Conf. (EUROEM)

Savage E B, Radasky W A, Smith K S et al (2008) Susceptibility of Network Interface Cards to High-Level Conducted Pulses. In Proc. of European Electromagnetics Conf. (EUROEM)

Van Eck W, Laborato N (1985) Electromagnetic radiation from video display units: An eavesdropping risk. Computers and Security 4:269–286

Sneak Path Analysis: Realising the Potential

Steve Gregory

AWE plc

Aldermaston

Abstract   *Sneak Analysis (SA) is a technique originating from work carried out by Boeing in the 1960s. Sneak Path Analysis (SPA) is a derivative of SA that focuses on the identification of latent, unintended paths that, under specific circumstances, can cause otherwise functional systems to exhibit undesired behaviours. SPA has not, however, been globally embraced by the safety engineering community. This is unfortunate, as there are some unique concepts and features of SPA (such as the explicit identification of systematic design flaws) that cannot be substituted by any other safety analysis techniques. This paper summarises a project that resulted in the development of a revised SPA procedure that can be implemented as an extension of the widely utilised Hazard and Operability (HAZOP) study. The procedure was applied to practical examples and was shown to be repeatable, efficient and applicable to a range of technologies.*

1 The Origins of Sneak Analysis

Sneak Circuit Analysis (SCA) was originally developed by The Boeing Company in the 1960s, under contract to the National Aeronautics and Space Administration (NASA), Houston, Texas, USA. The work began with a detailed review of a number of recorded incidents including accidental missile launches, accidental arming and dropping of bombs and unannounced failures of aircraft electrical systems. However, the primary driver for the investigation was the notorious Mercury Redstone MR-1 incident (Rankin 1984).

The Redstone launch appeared to be going successfully when, just a fraction after take-off, the engine cut-out and the booster fell a few centimetres back to the launch platform. Rankin describes how, following this event, "the payload, a Mercury capsule, jettisoned and came to rest about 400m away" (Rankin 1984). This was obviously not the desired outcome of the launch and the subsequent investigation determined that there had been no component failures.

The circuit was designed such that activation of the on-board firing switch would energise and latch the ignition coil, starting the engine. An on-board abort switch was provided to allow the engine cut-off coil to energise in the event of an abort command; a similar functionality was also provided by an externally-supplied abort switch located at the remote launch pad, enabling a ground operator to abort the launch if required. The ground operations and indications were connected to the on-board circuit via two umbilical cables, which were intended to break away from the missile as it elevated from the launch pad (preventing a ground-commanded abort once the launch was in progress).

Fig. 1. Mercury Redstone MR-1 during capsule jettison (image courtesy of NASA)

This incident was caused by a "sneak circuit" that was formed when the ground umbilical plug disconnected 29ms before the main control umbilical cable (Rankin 1984). During this short time period, the current flowing through the ground umbilical back to the missile skin was diverted via the abort indicator coil suppressor diode and, unfortunately, the engine cut-off coil. The time period was sufficient to allow the engine cut-off coil to energise and latch, causing the launch to abort. This is shown in Figure 2.

This case was of particular interest to NASA as neither component failures nor operator errors had been necessary to result in the incident. The fault was correctly assigned as a systematic design error, in that the design had not been sufficiently constrained.

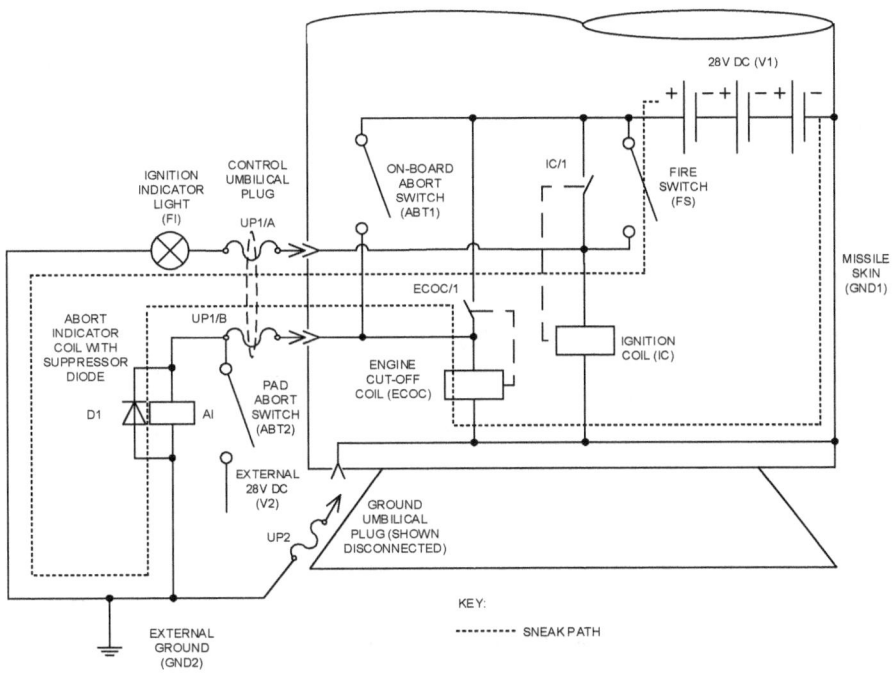

Fig. 2. Mercury Redstone MR-1 sneak circuit

Miller describes how Boeing's original approach to identifying such sneak paths involved the production of an equivalent topographic representation of the entire circuit and then applying specific "clues" to each topograph. These clues were intended to prompt the analyst to systematically consider factors with the aim of identifying sneak paths or conditions. The list of candidate sneak paths would then be screened by taking account of factors such as the consequences of realisation, the possibility of it arising at all etc. Finally, a list of screened potential sneak paths would be documented in a final report for consideration by the relevant engineers (Miller 1989).

For comparison, Figure 3 shows the MR-1 circuit (originally depicted in Figure 2) in topographic form.

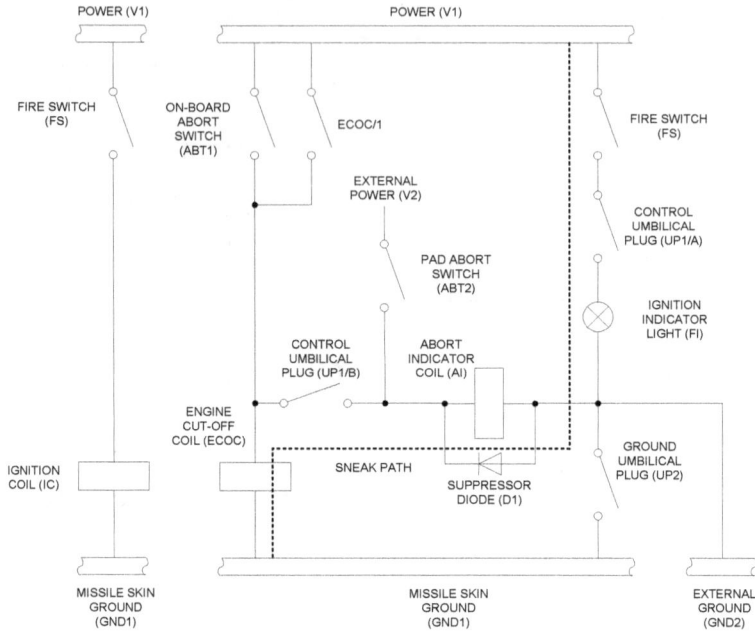

Fig. 3. MR-1 circuit depicted in topographic form

2 The Evolution of Sneak Analysis

During the development of the technique, Boeing came to assign four basic categories of sneak condition. According to Miller, these are "sneak path, sneak timing, sneak indication and sneak label" (Miller 1989). Sneak circuits are specific examples of sneak paths in electrical or electronic systems. More generically, a sneak path can be considered a latent path that can lead to undesired and often unpredicted behaviour. Sneak paths can potentially occur in any system that relies on the correct flow of energy; chemical processes, fluid power and stored energy applications are also susceptible to sneak paths. Rankin and White state that sneak timings "may cause current or energy to flow or inhibit a function at an unexpected time", while sneak indications "may cause an ambiguous or false display of system operating conditions" (Rankin and White 1970). Sneak indications are most often found when incorrect measurements are made, or correct measurements are incorrectly interpreted. Sneak labels, according to Whetton, are "the incorrect or ambiguous labelling of system functions, often resulting in one of the other sneak categories" (Whetton 1993, p170).

According to the Independent Design Analyses, Incorporated (IDA) website, the original Boeing technique was developed throughout the 1970s to include a wider range of circuit types, including digital, analogue and hybrid circuits as well as low-level software assembly language. Following these developments, the technique was subsequently renamed as Sneak Analysis (SA). In the late 1970s, however, Boeing made the technique and all of the supporting information proprietary, meaning it was effectively removed from the public domain.

Boeing continued to develop the SA technique in-house, extending it to cover higher-level software languages; Programmable Array Logic (PAL); Complex Programmable Logic Devices (CPLDs); Application Specific Integrated Circuits (ASICs) and integrated hardware / software systems (Vogas 2016). In 1994, IDA took ownership of the technique, and has retained the intellectual property rights ever since.

This somewhat secretive approach is relatively unique in the field of safety-critical systems engineering; the majority of analysis methodologies are openly publicised to enable the community to understand exactly what is being achieved, to share knowledge and lessons learned, and to assist in developing the techniques over time. The reluctance of Boeing, and latterly IDA, to make their methods publically available appears to be one of the primary reasons SA has not become widely accepted.

The European Space Agency (ESA) Methodology

In October 1997 the ESA introduced a new standard and associated clue list for SA, which steered away from the original topographic approach developed by Boeing and IDA. These documents were subsequently converted to publically-available Technical Memorandums (ECSS Parts 1 and 2, 2010); hereafter collectively referred to as "the ESA methodology".

The ESA methodology is a far more generic approach to SA than that described in the historical literature; it can be applied to a range of system types and at various stages of system development. It suggests that SA can be used as a potential design review aide, helping to make the design review process more algorithmic and reproducible. Two distinct "strands" of SA are identified in the ESA methodology – Sneak Path Analysis (SPA) and Design Concern Analysis (DCA). The remainder of this paper focuses on SPA.

The SPA methodology essentially requires a thorough description of the system design intent in the form of an input / output (I/O) matrix, followed by the identification of vulnerable targets (i.e. the safety-critical outputs of the system / sub-system in question) and the sources of energy that could potentially activate or deactivate the targets. Once the sources and targets have been identified, paths can be systematically traced between them using multiple copies of the system drawings. The paths are then assessed by the application of clues and by comparison with the I/O matrix.

However, there is little evidence to suggest the ESA methodology has been applied outside the aerospace domain. Detailed case studies are also difficult to locate, understandably due to commercial sensitivities.

3 Motivation

In general, SPA is not considered to be a "hot topic" in the system safety engineering community. A large proportion of the publicised research and development work associated with the technique is more than 30 years old and it appears that the technique is either unknown or has been dismissed due to its apparent complexity and unmanageable scope of application.

Since its first development, SPA has been viewed with much scepticism from industry. Miller describes how it "acquired the reputation of a 'black art' which can only be practiced by specialists" (Miller 1989). In fact, much of the more recent work carried out by the Rome Air Development Centre (RADC) (Miller 1990) and the ESA (Bougnal et al 1996) was largely motivated by the need to demystify the technique and to make it more openly available.

In general, the reluctance to accept SPA can be attributed to the following perceptions:

- It is highly labour intensive and time consuming;
- It requires detailed design information and can therefore only be applied at the later stages of the design lifecycle when subsequent changes are more expensive;
- It can only be performed by experienced analysts who will generally be employed by an external contractor, leading to misunderstandings and incorrect interpretations;
- In comparison to other established safety analysis techniques there is little available evidence to suggest it is effective.

Perrow explains how complex interactions "…are those of unfamiliar sequences, or unplanned and unexpected sequences, and either not visible or not immediately comprehensible" (Perrow 1996, p.78). It can, therefore, be inferred that the more complex systems become, the more likely they are to exhibit latent, systematic design flaws such as sneak paths. It is also worth noting that conventional safety assessment techniques, such as Fault Tree Analysis (FTA) and Failure Modes and Effects Analysis (FMEA), are not directly capable of identifying sneak paths as they can occur in the absence of equipment failure.

It could be argued that application of SPA to a fairly large and complex system would be excessively costly (particularly if it required the use of specialist contractors) and its effectiveness could be difficult to quantify, therefore it would not be justifiable under the ALARP principle. However, if the SPA technique were more efficient and cost-effective, this argument may no longer apply.

Market Research Questionnaire

In order to ascertain the current industrial perspective on the technique, a questionnaire was distributed to 52 industrial contacts from a range of safety-critical domains and with varying engineering backgrounds (such as mechanical, electrical, control systems and safety engineering). A total of 31 responses were received. The results showed that around 60% of respondents had previously heard of the technique; fewer knew any of the key concepts and only one had experience of the technique. The overwhelming majority of respondents felt that SPA could indeed be beneficial in the justification and assurance of safety-critical systems.

4 Development Approach

The overall aim of the project was to develop a refined, demonstrable SPA methodology which could be readily applied by engineers. The approach taken was to build upon the ESA methodology and incorporate original ideas and previously-researched concepts from the available literature. Initially, the ESA methodology was applied to a simple case study with a known sneak path in order to highlight the strengths and weaknesses of the process. As improvements were made, the new process was applied to other case studies and further refined. Worked examples were developed to demonstrate how the procedure could be applied to various technologies.

During development the draft process was applied to a simple case study involving a sneak path that resulted in the *loss* of a safety-critical function (as opposed to the unintended activation of a safety-critical function). It soon became clear that this type of "omission" scenario was not suitable for manual analysis. Essentially any component lying on any one of the potential sneak paths is capable of preventing the function from being achieved, meaning every component would need to be further analysed to determine how it is controlled. For all but the most simple of systems, this would result in a combinatorial explosion and an excessive workload; thus undermining the original aim of the project. It was, therefore, decided to restrict the scope of the process to "commission" scenarios where sneak paths may result in the unintended *activation* of a safety-critical function.

5 Revised SPA Procedure

The revised SPA procedure is detailed in Figure 4.

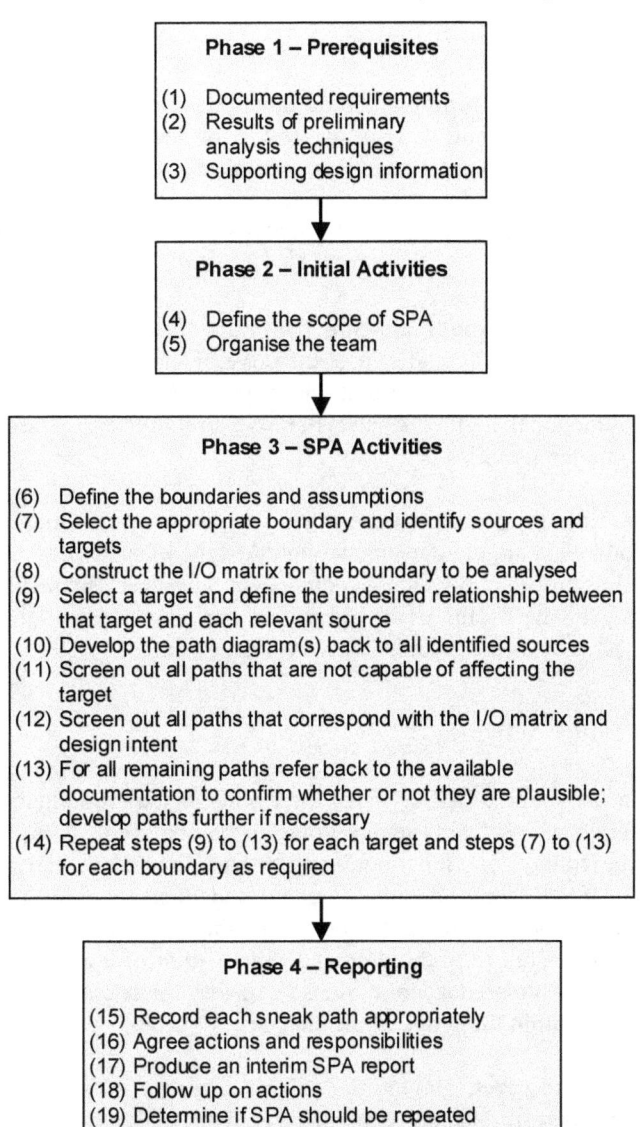

Phase 1 – Prerequisites

(1) Documented requirements
(2) Results of preliminary analysis techniques
(3) Supporting design information

Phase 2 – Initial Activities

(4) Define the scope of SPA
(5) Organise the team

Phase 3 – SPA Activities

(6) Define the boundaries and assumptions
(7) Select the appropriate boundary and identify sources and targets
(8) Construct the I/O matrix for the boundary to be analysed
(9) Select a target and define the undesired relationship between that target and each relevant source
(10) Develop the path diagram(s) back to all identified sources
(11) Screen out all paths that are not capable of affecting the target
(12) Screen out all paths that correspond with the I/O matrix and design intent
(13) For all remaining paths refer back to the available documentation to confirm whether or not they are plausible; develop paths further if necessary
(14) Repeat steps (9) to (13) for each target and steps (7) to (13) for each boundary as required

Phase 4 – Reporting

(15) Record each sneak path appropriately
(16) Agree actions and responsibilities
(17) Produce an interim SPA report
(18) Follow up on actions
(19) Determine if SPA should be repeated following any resulting design modifications
(20) Produce a final SPA report

Fig. 4. Revised SPA procedure

Phase 1 - Prerequisites

(1) Documented requirements: These should be relevant to the level of system to be analysed and should be traceable to the top-level requirements. Requirements should be as comprehensive as possible for the design stage. Poor requirements are likely to lead to either a less rigorous SPA or excessive numbers of "false" paths being identified as the analysts may make conservative assumptions.

(2) Results of preliminary analysis techniques: Applying SPA to an entire system is simply not practical or cost-effective; it should be reserved for safety-critical functions. It is critical that the scope of SPA is determined by either HAZOP, Functional Failure Analysis (FFA), Preliminary Hazard Assessment (PHA) or similar activities.

(3) Supporting design information: This could cover a multitude of different information and is largely dependent upon the type of system, level of maturity, etc. However, information such as component datasheets, manuals and specifications may all be useful when trying to determine whether or not potential paths can actually exist. Detailed drawings are essential.

Phase 2 - Initial Activities

(4) Define the scope of SPA: Using the results of the preliminary analysis activities, define which functions are going to be analysed. SPA should be applied in the context of "functions" rather than directed at individual components. The SPA scope should be agreed and documented. This procedure should only be applied to scenarios where the unintended commission of a function is undesired (e.g. the unintended activation of a relay or pump); scenarios involving unintended omissions (e.g. the unintended deactivation of a valve) are not within the scope of the analysis.

(5) Organise the team: Having the correct expertise is essential. It is likely that the team performing the preliminary analysis activities will be suitable to participate in the SPA. The team must always include the system designer(s) and should be led by a chair with experience in performing SPA.

Phase 3 - SPA Activities

(6) Define the boundaries and assumptions: The system drawings should be evaluated and partitioned accordingly. For extremely large or complex systems, it may be necessary to perform SPA against multiple sections. Partitions are best defined at interface points e.g. where a signal leaves one part of a system before entering another; care should be taken to ensure the interfaces are clearly documented and understood. If multiple functions are to be analysed, it may be beneficial to determine boundaries for each function rather than one large boundary which encompasses all the functionality.

(7) Select the appropriate boundary and identify sources and targets: The sources and targets should be identifiable from the preliminary analysis activities. The targets of SPA are those items or entities which can be considered "vulnerable" to unintended activation. In most cases, these will be outputs or output devices but they could also be vulnerable parts of a system e.g. the local environment could be considered a target, as could a bulk storage tank. The sources are those items or entities that provide a "threat" to one or more vulnerable targets. Sources may be permanent supplies such as electrical power, compressed gas, a continuous supply of a particular chemical, etc. They may also be transient sources such as capacitors, inductors or hydraulic accumulators.

(8) Construct the I/O matrix for the boundary to be analysed: The matrix should be as comprehensive as possible and should include every component within the boundary to be analysed. All components, which during normal operation, could activate or deactivate a path need to be modelled. Components that will only change state in the event of a failure do not need to be modelled (e.g. soldered joints or terminals). This is because SPA is not concerned with random component failure (which is considered using traditional techniques such as Failure Modes and Effects Analysis (FMEA)) but rather latent systematic flaws resulting from insufficiently constrained designs.

List each intended operational mode on the left of the matrix – this should also be as comprehensive as possible (including any maintenance and experimental modes). The components should be listed in individual columns. Populate the matrix with the states of each component during each operational mode (states may be defined according to the application or technology e.g. switches may be defined as "0" or "1" or alternatively "open" or "closed"). If the system incorporates multiple grounds or return paths, these should also be included as separate entities in the matrix.

Time-dependent components should be simplified to discrete states e.g. capacitors could be "discharged", "charging", "charged" or "discharging", depending upon the application. Analogue values such as 4-20mA current loops should, where possible, also be modelled as discrete states e.g. "low / <4mA", "in range", "alarm level", "trip level" and "high / >20mA". The key is to keep things as simple as possible without losing too much definition.

It is important to note that this I/O matrix represents a simplified model of the intended functionality of the equipment and should clearly relate back to the design requirements. If the I/O matrix is difficult to populate, or some items are vague or ill-defined, it is possible that the system design is not fully understood or the requirements are of poor quality.

(9) Select a target and define the undesired relationship between that target and each relevant source: For example, if the intended relationship between a target (X) and a source (Y) exists when two inputs (A, B) are both true (i.e. X should only be activated by Y if both A and B are true), the unintended relationship would be "X = activated by Y when A = false or B = false".

The notation used is very much dependent upon the application. For logic gates, the words "true" and "false" or "high" and "low" are suitable; for hydraulic valves words such as "open" and "closed" or "position 1" and "position 2" may be more suitable. Irrespective of the chosen nomenclature, it is important to be consistent and unambiguous.

(10) Develop the path diagram(s) back to all identified sources: Starting from the target, methodically trace all possible paths back through the system. Where a target is dependent upon connection to a permanent ground or return path (e.g. a path back to the tank for a hydraulic motor), the first "component" should be this ground or return path.

Once a path effectively terminates at a valid point e.g. at ground for an electrical / electronic system or the tank / atmosphere for a hydraulic / pneumatic system, or a "dead end" such as a blank connection, or arrives at any one of the identified sources, it should be considered complete. Consideration should also be given to the physical properties of components within the paths e.g. diodes can only conduct in one direction under normal conditions but, if a source is present that could exceed the breakdown voltage of the diode, reverse flow may well be possible.

If the system incorporates multiple grounds or return paths, a separate path diagram shall be constructed for each ground / return path unless it can be proven that the paths are completely isolated from one another e.g. via an isolation transformer. For example, if a system incorporates two current sources, each with its own respective ground, it is possible that a target that is supposed to be activated by source (Y) could be activated by source (Z) or vice versa.

(11) Screen out all paths that are not capable of affecting the target: The path diagrams are likely to result in a number of paths that terminate at points other than the identified sources. For example, if a path starts and finishes at ground, there is no way in which current can flow through the path (unless, of course, there are multiple grounds). Similarly, if a path terminates at a check valve it would not be possible for flow to occur (remembering that component failures are not considered within the analysis). If a path terminates at a permanent ground and there are additional paths that occur in parallel with this ground, the parallel paths can also be discounted as they cannot function unless the ground path is broken.

(12) Screen out all paths that correspond with the I/O matrix and design intent: The path diagrams will inevitably generate paths which correspond to the operational modes identified in the I/O matrix. These are intended paths and should be screened out appropriately. Equally there may be paths that correspond to the I/O matrix but do not explicitly define the state of every component within the system boundary. In these instances, it may be necessary to manually investigate the state of each remaining component, in order to verify that a particular path does indeed correspond to the matrix.

It may be easy to screen out (reject) a particular path or set of paths. For example, if the undesired relationship is "X = activated by Y when A = false or B = false", any path which requires both A and B to be true can be immediately screened, regardless of the states of any other components.

(13) For all remaining paths refer back to the available documentation to confirm whether or not they are plausible; develop further paths if necessary: All remaining paths at this stage can be considered "candidate sneak paths". In some cases, it may be possible to readily determine if any candidate paths are plausible. In most cases, however, it is likely that further path tracing will be necessary.

For example, a candidate sneak path may incorporate components such as relay contacts. The relays that control these contacts may then require analysis to determine how they might themselves be controlled (they effectively become new targets). New path diagrams may be added to the existing candidate sneak path diagram; this can result in very complex diagrams and should be carefully managed.

At this stage, it may be apparent that the initial definition of the undesired relationship was insufficient e.g. if all paths have been screened in steps (11) and (12) but there are obvious questions as to how one or more components is controlled. In these circumstances the definition of the undesired relationship may require refinement before further analysis can proceed.

(14) Repeat steps (9) to (13) for each target and steps (7) to (13) for each boundary as required: If multiple targets were identified in step (7), it will be necessary to repeat steps (9) to (13) for each of these targets. Once all of the targets have been assessed for a particular boundary, any further boundaries defined in step (6) should be analysed in the same fashion.

Phase 4 – Reporting

(15) Record each sneak path appropriately: If it is determined that a candidate path is genuinely plausible, it should be recorded and documented as a sneak path. The recording method should be suited to the nature of the system in question.

(16) Agree actions and responsibilities: The actions associated with one or more sneak paths should be agreed by the team and documented accordingly. One or more individuals should also be assigned the responsibility for ensuring the actions are closed out within agreed timescales. The actions may require a modification to the design to eliminate the sneak path(s). Other mitigations may, however, be practical depending upon the perceived level of risk associated with the sneak path(s).

(17) Produce an interim SPA report: A report should be produced detailing the findings of the SPA and, if appropriate, a plan for moving forward. If no sneak paths were identified, or if no actions were required, the interim SPA report may be considered a final SPA report and there is no need to continue with steps (18) to (20).

(18) Follow up on actions: This is the responsibility of the individual(s) charged with ensuring actions are closed out within agreed timescales. Once sufficient progress has been made a meeting of the SPA team should be arranged to discuss the overall impact to the design and to move on with step (19).

(19) Determine if SPA should be repeated. If the actions resulted in changes to the design, it is generally recommended that the SPA activity should be revisited. However, much of the background information and artefacts will have already been generated and may only require a review to determine validity (e.g. the I/O matrix for the system may well be identical, unless the design change has significantly altered the component configuration or the operational modes). If the actions did not result in changes to the design but changes to the assumptions made in step 6, it may be necessary to repeat the SPA activity.

(20) Produce a final SPA report: This report should build upon the interim report and should describe the actions that were taken and any subsequent SPA activities that were necessary. The overall aim of this final report is to explain how SPA was carried out, whether any sneak paths were identified, how any identified paths were mitigated and to justify why no further SPA activities are required.

6 Worked Example

This section describes the application of the revised SPA procedure to the Mercury Redstone MR-1 case study. In the interest of brevity, only Phase 3 of the procedure is described and only one target / undesired relationship is considered i.e. the unintended activation of the engine cut-off coil.

Figure 5 depicts the system in the quiescent state. A detailed functional description has also been included to aid understanding and to represent the level of detail which should be available at the start of the SPA.

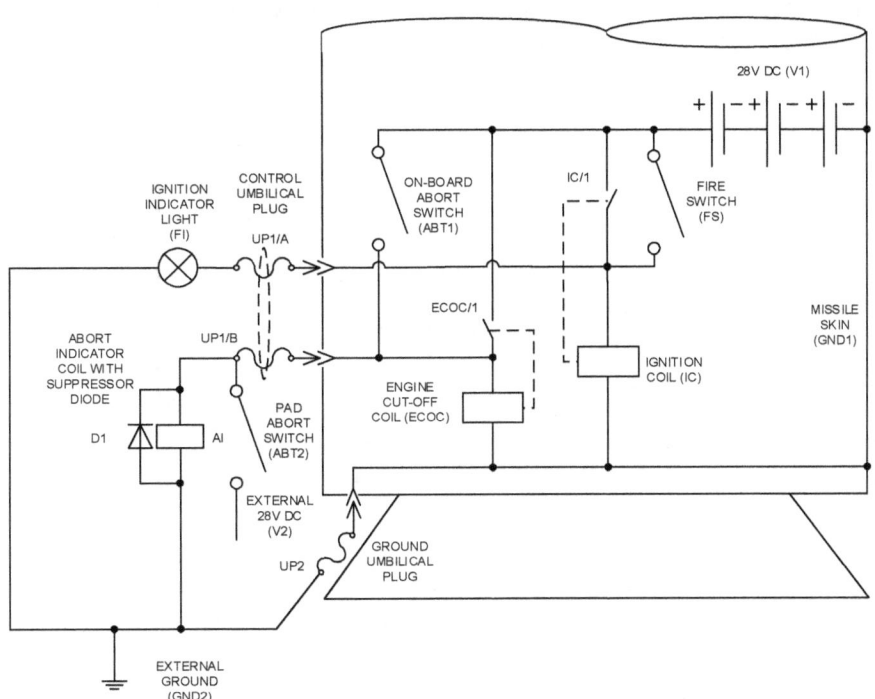

Fig. 5. Mercury Redstone MR-1 system schematic

Functional Description

When the rocket is ready for launch, the system will be connected as shown in Figure 4. The rocket is hooked up to the remote launch pad facility allowing for a remote launch abort in addition to providing the ground crew with information from the rocket. This is achieved by the connection of two umbilical cables – one for the control signals (UP1), which has two pairs of contacts (UP1/A and UP1/B), and one for balancing the on-board and external grounds (UP2). The 28VDC external supply (V2) is continually present whilst the rocket is on the launch pad. The on-board battery pack (V1) is grounded through the missile skin (GND1).

Under normal conditions, once the rocket is ready for launch the pilot will depress the firing switch (FS), energising the ignition coil (IC) from the on-board battery (V1). This coil then self-latches via contact IC/1 and the rocket commences launch. The external ignition indicator light (FI) is illuminated, informing the ground crew of the launch status.

As the rocket lifts away from the launch pad the two umbilical plugs (UP1 and UP2) disconnect, after which time the ground crew are unable to command a launch abort. The post-launch schematic is shown in Figure 6.

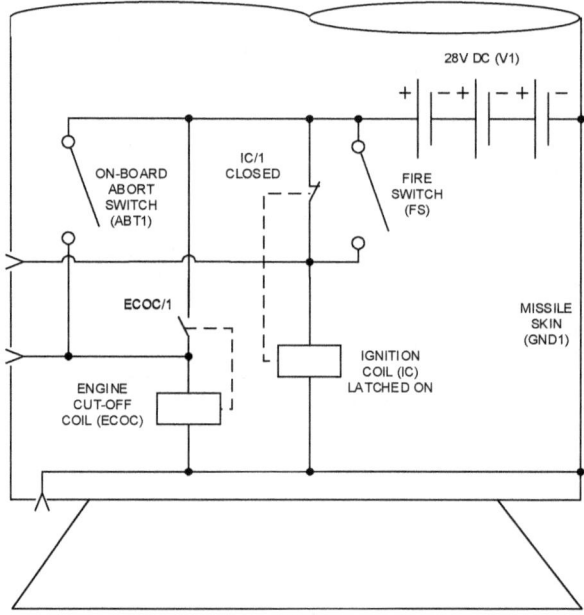

Fig. 6. Post-launch system schematic

If the on-board crew decide to abort the launch, the pilot can press the on-board abort switch (ABT1) which energises the engine cut-off coil (ECOC). This coil also self-latches via contact ECOC/1. This feature is independent of the firing mechanism but uses the same power supply; a loss of power supply would, therefore, cause the launch to abort by de-energising the ignition coil. The on-board abort function can be achieved at any time, irrespective of the launch status.

If the on-board abort is carried out whilst the rocket is still on the launch pad (either prior to launch or between the launch command and physical lift-off), the abort will be indicated to the ground crew via the umbilicals. The abort indicator coil (AI) is energised through UP1/B and grounded via UP2 – this signal is repeated back to the main control room and other locations as necessary.

The remote abort may be carried out whilst the rocket is still on the launch pad (again either prior to launch or between the launch command and physical lift-off) by depressing the pad abort switch (ABT2). This allows the external 28V DC supply (V2) to energise the engine cut-off coil (ECOC) with the ground path being provided by the ground umbilical UP2. Once the coil is energised and latched, the on-board supply (V1) keeps the coil energised, such that if the remote abort is carried out fractionally before lift-off, the abort is maintained. The remote abort operation also energises the abort indicator coil (AI) using the external 28V DC supply (V2) and the external ground (GND2).

Define the boundaries and assumptions: It is not possible to discount any of the components from the analysis scope, so the entire system is considered to be of interest. Figure 7 shows the umbilical connectors modelled as switches.

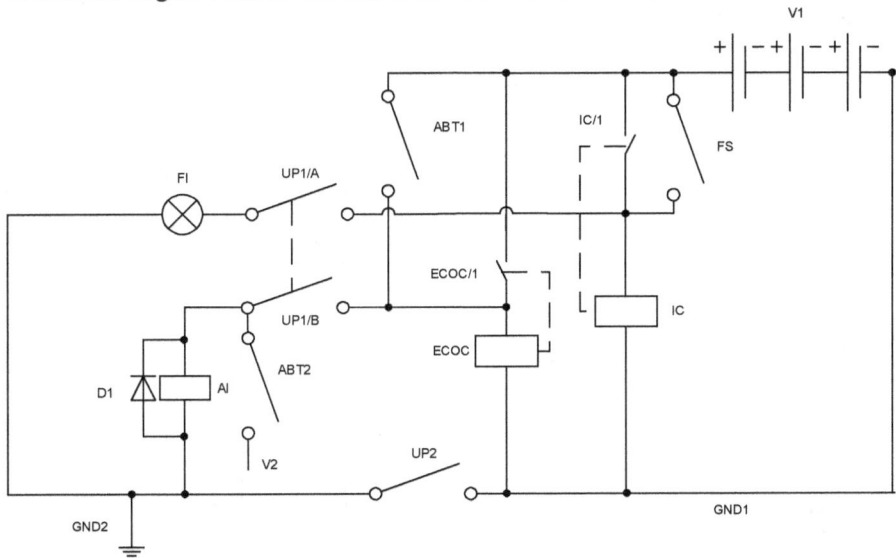

Fig. 7. Boundary for analysis

Select the appropriate boundary and identify sources and targets: There is only one boundary to consider. The target of interest in this case is the engine cut-off coil (ECOC). Should this coil be inadvertently energised, the launch would abort leading to possible risk to life and substantial equipment damage. Two potential sources are present - the on-board battery supply (V1) and the external 28V DC supply (V2).

An interesting aspect of this case study system is the grounding regime. There are two distinct grounds associated with the system – the on-board ground (GND1) and the external ground (GND2). When the rocket is on the launch pad and electrically connected to the remote pad station, UP2 is closed and GND1 = GND2. Therefore, either source (V1 or V2) can sink to either ground, as they are a single entity. However, once the launch commences and UP2 is disconnected (UP2 = 0), the two grounds are separated (GND1 ≠ GND2). When UP2 is disconnected, V1 can only sink to GND1 and V2 can only sink to GND2.

Construct the I/O matrix for the boundary to be analysed: The I/O matrix is populated based on the functional description of the system. For the suppressor diode (D1), "0" indicates no current flow and "1" indicates forward current flow. Reverse current flow was discounted as it is assumed that the diode breakdown voltage is greater than the maximum battery voltage. The diode is shown as being "0" in every mode as, in practice, it would only conduct for a very short period of time.

Table 1. I/O matrix

	V1	V2	FS	ABT1	UP1/A	UP1/B	UP2	ABT2	IC	IC/1	ECOC	ECOC/1	AI	DI	FI
Standby	1	1	0	0	1	1	1	0	0	0	0	0	0	0	0
Remote Abort	1	1	0	0	1	1	1	1	0	0	1	1	1	0	0
Local Abort	1	1	0	1	1	1	1	0	0	0	1	1	1	0	0
Fire	1	1	1	0	1	1	1	0	1	1	0	0	0	0	1
Launch	1	1	0	0	0	0	0	0	1	1	0	0	0	0	0
Local Abort Post-Launch	1	1	0	1	0	0	0	0	1	1	1	1	0	0	0

Select a target and define the undesired relationship between that target and each relevant source: The undesired relationship can be defined as "current flows from V1 or V2 to ground via ECOC when not commanded by either the on-board abort switch ABT1 or the pad abort switch ABT2 (i.e. ECOC = 1 when both ABT1 = 0 and ABT2 = 0)".

Develop the path diagram(s) back to all identified sources: From inspection of the circuit, V2 can only be a source for ECOC when UP1/A, UP1/B and UP2 are all closed. V1 can be a source for the target regardless of the state of these three switches. Therefore, when tracing paths back to V2, only paths where UP1/A, UP1/B and UP2 are all closed (=1) are considered to be valid.

Two path diagrams are required for this exercise – one where ECOC is connected to GND1 and another where it is connected to GND2. Tracing all paths backwards from GND1 via ECOC towards either source, taking account of the physical properties of components (e.g. diodes allow forward only flow), yields the path diagram in Figure 8. This path tracing methodology is based upon concepts described in Whetton and Armstrong's research into batch process plant SPA (Whetton and Armstrong 1994).

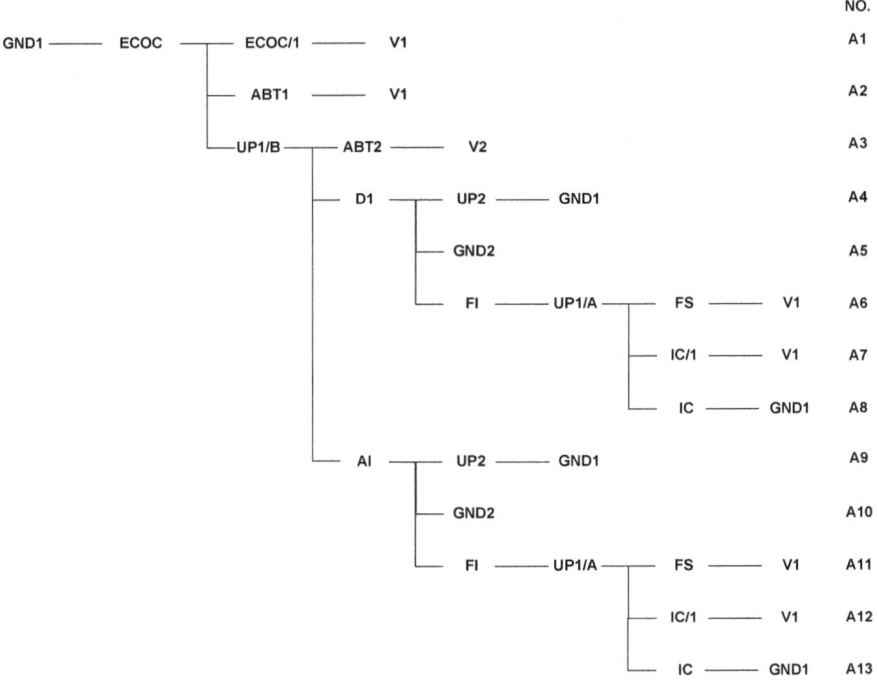

Fig. 8. Path diagram for GND1 via ECOC (path diagram A)

Tracing all paths backwards from GND2 via ECOC towards either source, again taking account of the physical properties of components, yields the path diagram in Figure 9.

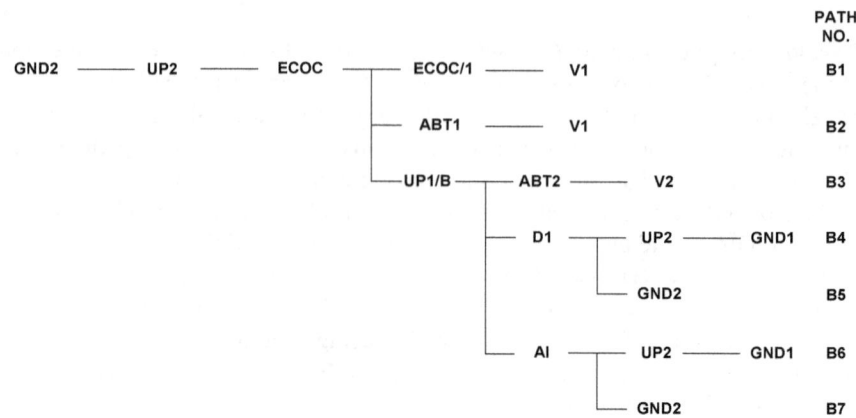

Fig. 9. Path diagram for GND2 via ECOC (path diagram B)

Screen out all paths that are not capable of affecting the target: For path diagram A (Figure 8), those paths not capable of activating the target (as they are not conducive to flow) are shown greyed out in Figure 10.

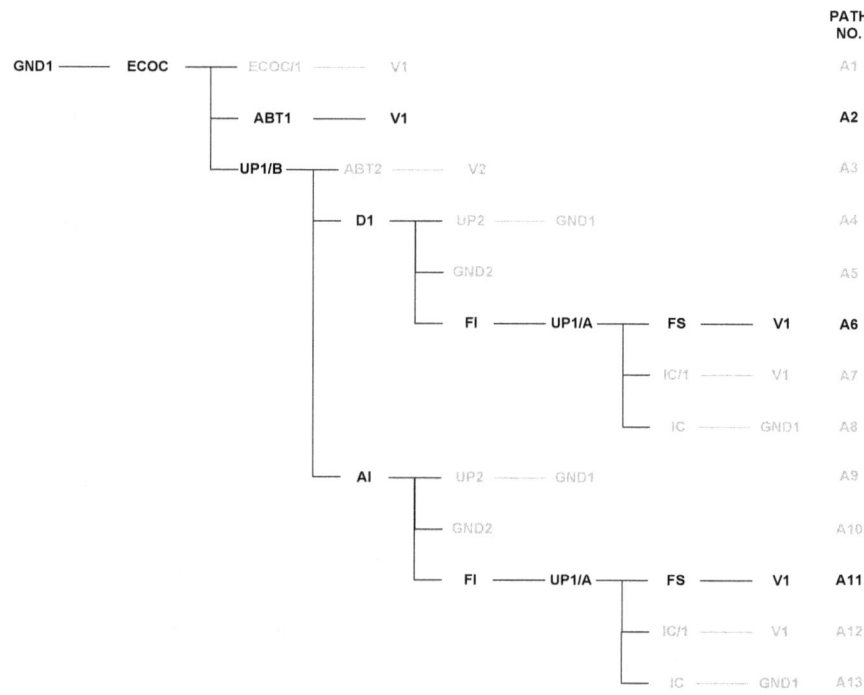

Figure 10. Reduced path diagram for GND1 via ECOC (path diagram A)

Path A1 actually represents the latching functionality of the ECOC coil but it is screened out on the basis that ECOC/1 cannot be closed before ECOC is energised. Path A3 is not conducive to flow as V2 cannot sink to GND1 unless UP2 is closed (the path does not include UP2). Paths A4, A8, A9 and A13 are grounded at both ends. Paths A7 and A12 do not contain the coil IC, so contact IC/1 cannot be closed. Paths A5 and A10 essentially have no final termination point, as they do not pass through UP2 and, therefore, GND1 ≠ GND2.

For path diagram B (Figure 9), those paths not capable of activating the target (as they are not conducive to flow) are shown greyed out in Figure 11. Path B1 is screened out on the basis that ECOC/1 cannot be closed before ECOC is energised. Paths B5 and B7 are grounded at both ends. Paths B4 and B6 are also grounded at both ends as UP2 is closed, connecting the grounds together.

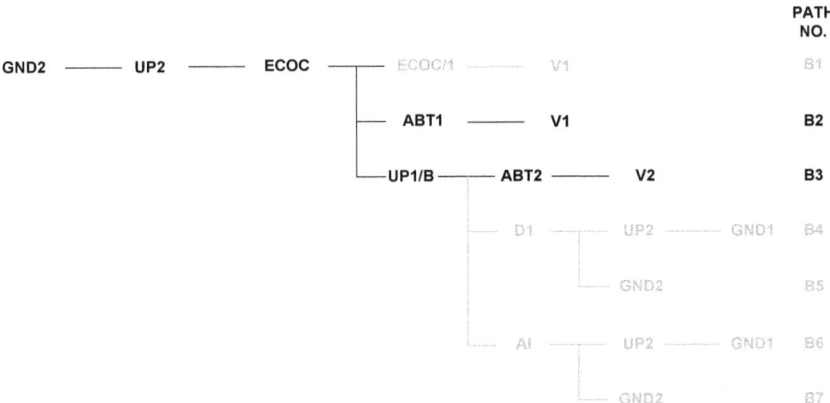

Fig. 11. Reduced path diagram for GND2 via ECOC (path diagram B)

Screen out all paths that correspond with the I/O matrix and design intent:
For path diagram A (Figure 8), additional paths that correspond to the I/O matrix, or fail to satisfy the undesired relationship, are shown greyed out in Figure 12.

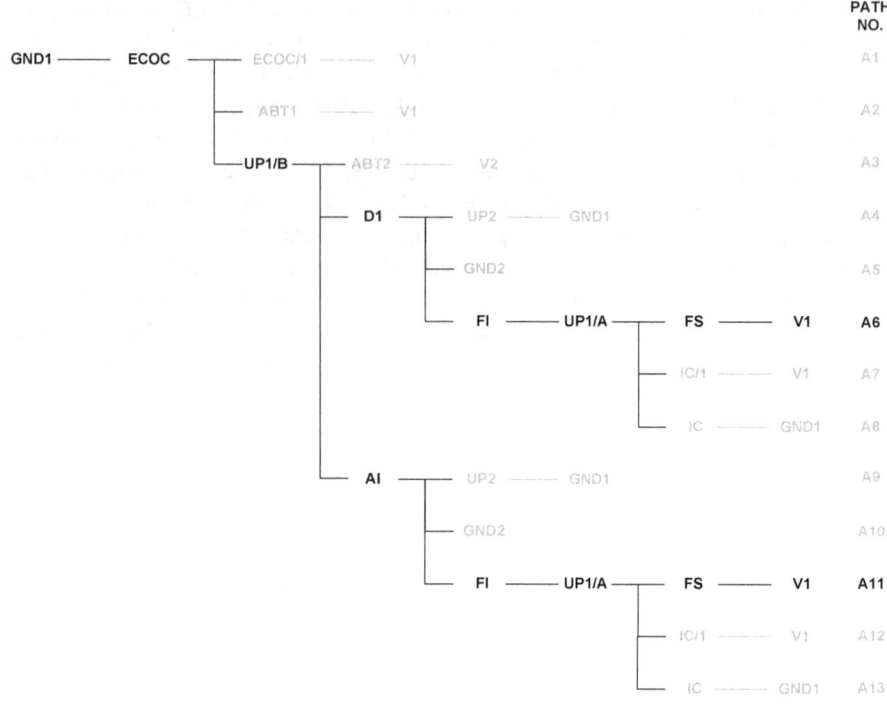

Fig. 12. Further reduced path diagram for GND1 via ECOC (path diagram A)

Path A2 contains the onboard abort switch ABT1, meaning the path could only be realised if a local abort command has been received. This does not correspond to the definition of the undesired relationship and the path is also present in the original I/O matrix (see "Local Abort" mode). For path diagram A (Figure 8), paths A6 and A11 remain as candidate sneak paths.

For path diagram B (Figure 9), additional paths that correspond to the I/O matrix, or fail to satisfy the undesired relationship, are shown greyed out in Figure 13.

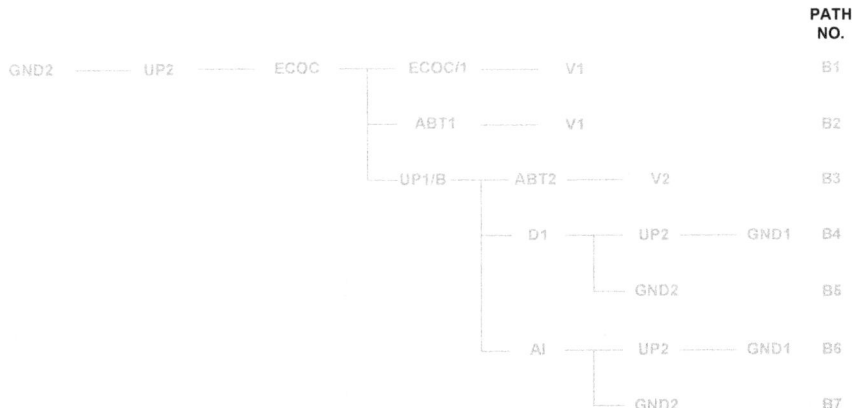

Fig. 13. Further reduced path diagram for GND2 via ECOC (path diagram B)

Paths B2 and B3 each contain at one of the abort switches, meaning the paths could only be realised if a local or remote abort command has been received. This does not correspond to the definition of the undesired relationship and both paths are present in the original I/O matrix (see "Local Abort" and "Remote Abort" modes). For path diagram B (Figure 9), no paths remain as candidate sneak paths.

For all remaining paths refer back to the available documentation to confirm whether or not they are plausible; develop further paths if necessary: A total of two candidate sneak paths remain – A6 and A11. These paths are essentially identical, as the suppressor diode (D1) is in parallel with the abort indicator coil (AI). Neither of the paths are present in the original I/O matrix (Table 1) but are shown in Table 2 for clarity.

The two new paths clearly show that it is possible for ECOC to be energised (=1) whilst ABT1 and ABT2 are both open (=0). In this case path A6 was actually the sneak path that caused the launch to abort – the disconnection of UP2, whilst in the "Fire" mode, led to the incident.

Table 2. I/O matrix including two candidate sneak paths A6 and A11

	V1	V2	FS	ABT1	UP1/A	UP1/B	UP2	ABT2	IC	IC/1	ECOC	ECOC/1	AI	DI	FI
Standby	1	1	0	0	1	1	1	0	0	0	0	0	0	0	0
Remote Abort	1	1	0	0	1	1	1	1	0	0	1	1	1	0	0
Local Abort	1	1	0	1	1	1	1	0	0	0	1	1	1	0	0
Fire	1	1	1	0	1	1	1	0	1	1	0	0	0	0	1
Launch	1	1	0	0	0	0	0	0	1	1	0	0	0	0	0
Local Abort Post-Launch	1	1	0	1	0	0	0	0	1	1	1	1	0	0	0
Path A6	1	1	1	0	1	1	0	0	1	1	1	1	0	1	1
Path A11	1	1	1	0	1	1	0	0	1	1	1	1	1	0	1

7 Integration with HAZOP

The HAZOP study is one of the most extensively used hazard identification techniques and has a number of similarities with SPA (Whetton and Armstrong, 1994). The revised SPA process described in this paper has some distinct parallels with HAZOP. Firstly, both the revised SPA process and HAZOP require a high level of design maturity in order to be effective. Secondly, both processes should be carried out by a team of individuals including the system designer(s), with the team being led by an experienced chair. Thirdly, both processes utilise a "divide and conquer" approach where complex systems are decomposed and analysed separately. Finally, HAZOP considers behavioural deviations along intended paths, whereas the revised SPA process aims to identify unintended paths that could lead to behavioural deviations.

Of all the widely accepted HAZOP guidewords, "OTHER THAN" is the most relevant to SPA. Kletz suggests that "OTHER THAN" can be interpreted as "what else can happen apart from normal operation" (Kletz 1999). During the pilot case study HAZOP, the "OTHER THAN" guideword was interpreted more specifically as *"something other than the design intent"*. It was discovered that application of the standard HAZOP guideword "OTHER THAN" can prompt the participants to consider the possibility of sneak paths and, therefore, derives a requirement to perform SPA.

The sources and targets for SPA can also be identified during HAZOP. Table 3 shows an excerpt from a contrived HAZOP study of the Mercury Redstone system, produced during development of the SPA process. The source (V1) and one of the targets (GND1 via ECOC) are identified by the application of the "OTHER THAN" guideword to the parameter "current path between V1 and GND1 via ECOC".

Table 3. Excerpt from Mercury Redstone HAZOP

Parameter	Guideword	Deviation	Possible Cause(s)	Consequence(s)	Action(s) Required
Current path between V1 and GND1 via ECOC – initial operation of local abort command	OTHER THAN	Current should not flow between V1 (source) and GND1 via ECOC (target) but flow does occur	Possible sneak path(s)	Unintentional abort initiated (potentially catastrophic)	Perform SPA to determine if sneak paths may be present

BS IEC 61882 describes a generic HAZOP methodology split into four stages: Definition; Preparation; Examination and Documentation (BSI 2001). An integrated HAZOP / SPA process, based on this standard, is described in Figure 14. This shows how the unique aspects of SPA might be incorporated as an extension of a widely used technique.

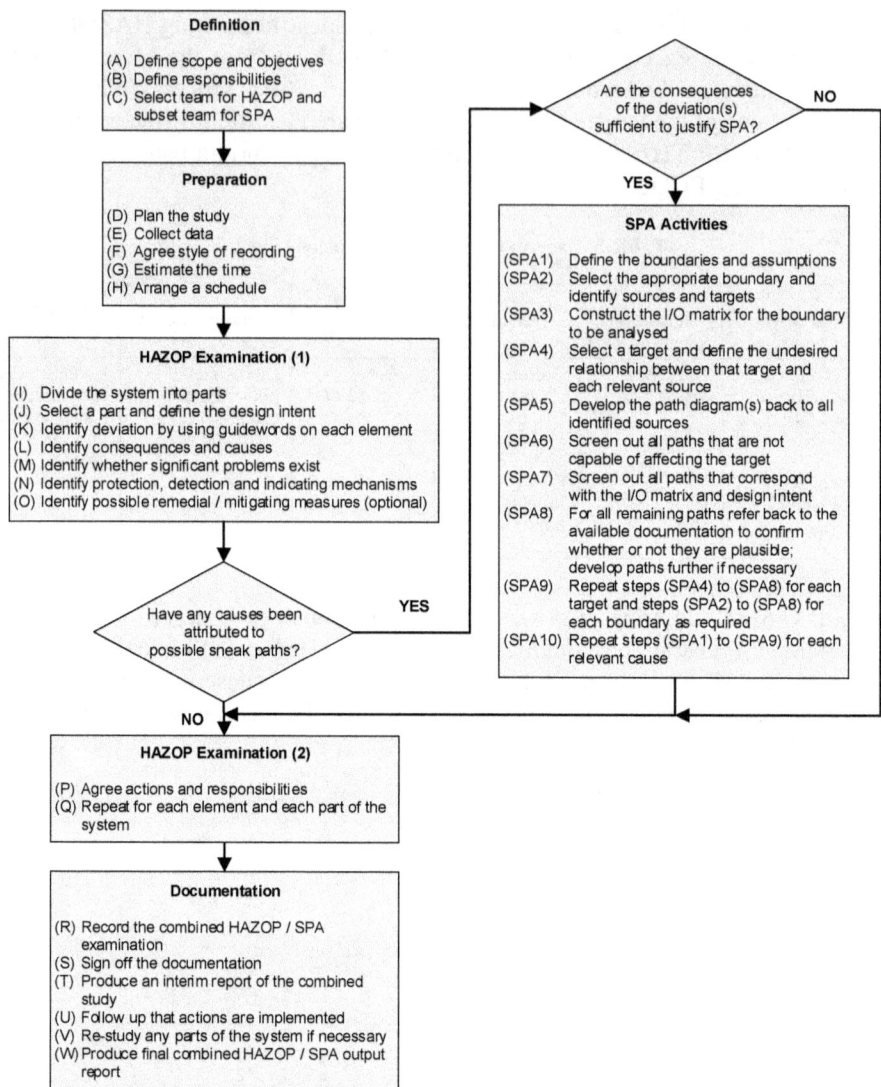

Fig. 14. Integrated HAZOP / SPA process (modified from BS IEC 61882)

The SPA aspects could equally be performed after the main HAZOP has been completed. Due to the time and resource required to perform SPA, it should only be applied to safety-critical functions where the unintended commission of functionality could lead to severe consequences.

8 Evaluation

Training Example

The SPA procedure, MR-1 drawings and functional description (see Section 6) were supplied to a team of two reliability engineers, both of whom had no prior experience of SPA, under supervision of the author (fulfilling the role of SPA chair). The session lasted for approximately three hours, although a third of this time was dedicated to familiarising the participants with the intended functionality of the system. A further 20 minutes was spent at the end of the activity discussing the "model answer" plus follow-up actions, such as reporting and re-designing options. The SPA activity itself lasted for approximately 1 hour 40 minutes.

The participants initially made an unjustified assumption that the two umbilical connectors UP1 and UP2 would always disconnect simultaneously. When asked for the rationale behind this assumption, the participants explained that it was made in order to simplify the analysis. However, when they subsequently revisited the design information, this assumption was removed. This highlighted the importance of oversight by an experienced chair and that the analysts should make their assumptions with due diligence.

The path tracing exercise yielded the most interesting deviations from the model answer. The participants generated a much larger number of paths than was necessary; paths that should have terminated were often continued and superfluous parallel paths were developed. Once again, this was attributed to a lack of familiarity with the procedure. The participants also realised their mistakes during the application of the screening criteria.

Ultimately, the participants successfully identified the same two sneak paths as described in the worked example in Section 6. This activity provided initial evidence of the repeatability of the SPA procedure. Both participants praised the technique for being able to manually identify such a subtle design flaw. The general consensus was that the analysis was not excessively detailed or labour intensive, given the criticality of the system and the intricacy of the sneak path.

Complex System

In order to assess the scalability of the procedure, the author applied it to a complex safety interlock system within AWE's Orion laser facility (see Figure 15). Two distinct elements of the system were subjected to SPA: a dual-leaf laser shutter and a solid state logic solver that is used to safely control the shutter. A single safety function relating to the remote opening of the shutter was analysed as part of the SPA.

The laser shutter circuit mainly consists of basic switches but also includes some more complex, time-dependent components, such as capacitors. The circuit diagram for the logic solver was much more complex, spanning a total of 32 pages. The logic solver consists of solid state AND, OR and inversion gates, as well as input and output modules and hybrid logic modules such as 2-input OR / 3-input AND gates. Many of the pages also contain cross-references to different subsystems on different pages.

Fig. 15. High-power laser beam lines within the Orion laser hall

In total, the SPA took around three hours to complete. Five boundaries were identified, three of which were fully analysed. The importance of defining boundaries and assumptions was reiterated during the analysis. Due to the large number of components and subsystems within the logic solver schematics it was necessary to spend a significant amount of time partitioning the diagrams in order to make the analysis manageable. This "divide and conquer" approach proved to be essential in carrying out the analysis.

It was also necessary to rationalise the amount of equipment within the individual boundaries (had a comprehensive I/O matrix been constructed for one of the boundaries it would have contained 57 columns). Applying the procedure to numerous boundaries in succession was somewhat tedious. There appeared to be a compromise between partitioning the system, in order to minimise the complexity, and analysing the system as a whole, in order to minimise the amount of work.

Following the division of the system into multiple boundaries, it was necessary to identify sources, targets and undesired relationships for each of these boundaries. It soon became apparent that meticulous care was required when defining these parameters, such that they were appropriate for the boundaries but were also traceable back to the system-level sources, targets and undesired relationships.

The system included components that had not previously been analysed during the development case studies. This required some reinterpretation and assumptions to be formed. For example, the capacitors within the shutter circuit were designed to allow high inrush currents when opening the shutter blades. However, the capacitors can also act as sources within their own rights (although this was not the design intent).

Overall this exercise suggested that the procedure is scalable to a complex system, providing the analysis is sufficiently focused. Whilst no sneak paths were identified, the exercise covered all of the major steps in the revised SPA procedure. However, it is acknowledged that this exercise was susceptible to a degree of confirmation bias as the analysis was performed by the author of the procedure.

9 Conclusions

Positive Conclusions

The revised SPA procedure is applicable to a range of different technologies. Providing the SPA is supervised by an experienced chair, the procedure is applicable to electrical, electronic, fluid power, solid-state logic, process plant systems and perhaps many more technologies (effectively any application involving flows of energy).

The revised SPA procedure could also be used to supplement a manual detailed design review. In the author's own experience, design reviews naturally focus on the validity of the presented material (e.g. drawings, control philosophies, calculations, etc.), usually against formal design and safety requirements. The primary aim of SPA is to investigate and analyse unintended (or emergent) behaviours i.e. the behaviours and features that were not foreseen nor covered by requirements. The feedback received from the wider community suggests that the application of the revised SPA process can identify subtle systematic design flaws that may otherwise pass through the design review process undetected.

It is believed that the revised SPA procedure could be applied in a more cost-effective manner in comparison to the original Boeing / IDA approach. The cost of performing the analysis was one of its most prohibitive factors; the original methodology placed little emphasis on rationalising the boundaries for analysis and effectively analysed the entire system. The revised SPA procedure leans heavily towards rationalising the boundaries, resulting in a more focused analysis.

The integration of SPA with the widely utilised HAZOP study has resulted in the potential for a more thorough and powerful analysis than would otherwise be achieved by applying HAZOP alone.

Limitations and Recommendations for Further Work

The revised SPA procedure is only applicable to "commission" scenarios, whereby the unintended activation of a target or provision of a function is considered to be detrimental. Fortunately, in the author's own experience, a high proportion of safety-critical systems are designed to be "fail-to-safe", where the loss of an output or function (i.e. "omission") is generally considered to be safe. Of course, there are also many examples where the omission of an output or function can be hazardous, such as the loss of cooling water to a nuclear reactor or the loss of an active braking system. Further investigations could aim to develop a procedure to deal with "omission" scenarios.

In general, the revised SPA procedure can only be applied once a detailed design has been achieved. Unfortunately, the technique inherently requires detailed design information; sneak paths are, by their very nature, complex and subtle; if they were easy to identify they would not exist. This is a limitation of SPA, although it is by no means a unique limitation. SPA can be considered part of a suite of techniques that can be applied during the lifecycle, with higher-level techniques such as Functional Failure Analysis (FFA) being performed in the earlier stages.

The assumptions and dependencies associated with SPA are recognised to be particularly critical. The misinterpretation of any associated system or contextual information could completely invalidate the analysis e.g. by making an incorrect assumption regarding external conditions. Once again, however, the same limitation applies to any safety analysis technique. Failure to sufficiently describe the top event of a fault tree can lead to the omission of key basic events or, alternatively, the inclusion of spurious events that have no relevance.

The revised SPA procedure is labour intensive, although not to the extent of the original SCA methodology or indeed the ESA methodology. However, it is not considered to be more labour intensive than the widely utilised FMEA. In addition, the amount of effort required can be significantly reduced if the scope is defined and the boundaries are sufficiently rationalised. Performing SPA as an extension of HAZOP, utilising the same team members and environment, can also help to minimise the amount of work required.

SPA is a thorough and comprehensive analysis technique that inevitably generates a large amount of documentation. It could be beneficial to develop a software tool to assist in recording and managing the analysis artefacts. There are a number of software tools available to assist in the conducting of a HAZOP study; a SPA software tool could follow similar principles, covering:

- Single-point collation of system information, drawings, requirements specifications etc.;
- Links to other common software packages such as requirements databases, safety analysis artefacts such as FMEA reports and fault trees;
- Assistance in constructing the I/O matrices and path diagrams;
- Clue-based prompts when considering individual components;
- Systematic application of the path screening criteria, based on the definition of the undesired relationships and the design intent.

Closing Statement

On balance, the revised SPA procedure enables users to perform the analysis in a more efficient and informed manner than previous methodologies, increasing its appeal and prospects for further use in the field. SPA is a viable and beneficial technique, providing it is applied effectively and proportionately.

Acknowledgments   I would like to thank AWE plc for sponsoring my MSc; Steve Bates, Steve Kelly and Helen Emery for proofreading this paper; and Dr David Pumfrey for his guidance during the project.

References

Rankin J P (1984), "Sneak Circuits – A Class of Random, Unrepeatable Glitch". Plant / Operations Progress, volume 3, no. 3, pp. 175-178

NASA (2013), Image S63-00193. Curator: JSC PAO Web Team, Responsible NASA Official: Amiko Kauderer, updated 25/09/2013 (online). Accessed 06 October 2016. http://spaceflight.nasa.gov/gallery/images/mercury/mercury_ov/hires/s63-00193.jpg

Miller J (1989), "Sneak Circuit Analysis for the Common Man". Rome Air Development Centre (RADC) (SoHaR Incorporated), New York

Rankin J P, White C F (1970), "Sneak Circuit Analysis Handbook". The Boeing Company, Houston, Texas, USA

Whetton C P (1993), "Sneak Analysis of Process Systems". IChemE Transactions on Process Safety and Environmental Protection, vol. 71b, pp. 169-179

Independent Design Analyses Inc. (IDA), "What is Sneak Analysis?" (online). Accessed 02 September 2016. http://www.ida-inc.com/whatissneaks.html.

Vogas J L, "Sneak Analysis of Process Control Systems". Independent Design Analyses, Inc. Accessed 25 September 2016. http://www.ida-inc.com/SA%20Papers/ProcContSA.pdf.

European Cooperation for Space Standardisation (ECSS) (2010), "Space Product Assurance - Sneak Analysis - Part 1: Principles and Requirements ECSS-Q-40-04a". ESA Requirements and Standards Division, Noordwijk, Netherlands

European Cooperation for Space Standardisation (ECSS) (2010), "Space Product Assurance - Sneak Analysis - Part 2: Clue List ECSS-Q-40-04a". ESA Requirements and Standards Division, Noordwijk, Netherlands

Miller J (1990), "Integration of Sneak Analysis with Design". Rome Air Development Center (RADC), New York

Bougnol C, Dore B, Taylor J R (1996), "Lessons Learned from Pilot Applications of Sneak Analysis in Space Projects," in Proceedings of the ESA Product Assurance Symposium and Software Product Assurance Workshop, Noordwijk, Netherlands

Perrow C (1999), "Normal Accidents - Living with High-Risk Technologies". Chichester, West Sussex: Princeton University Press

Whetton C, Armstrong W (1994), "Sneak Analysis of Batch Processes". Journal of Hazardous Materials, vol. 38, pp. 257-275

Kletz T (1999), "Hazop and Hazan - Identifying and Assessing Process Industry Hazards", Fourth ed., Rugby, Warwickshire: Institution of Chemical Engineers (IChemE)

British Standards Institute (BSI) (2001), "BS IEC 61882 Hazard and Operability Studies (HAZOP Studies) – Application Guide", BSI

HFACS: Helicopter Operations' Safety

José Corrêa de Sá

Freelance

Portugal

Abstract *Human Factors Analysis and Classification System (HFACS) is an accident investigation methodology. It is based on Reason model, the causality perspective commonly known as "Swiss Cheese" model. HFACS methodology became necessary because the Reason model is primarily descriptive, not analytical. It was designed to facilitate the application of this model to accident investigation and analysis. It was, specifically, developed to define the latent and active failures implicated in Reason model; and, refined through the analysis of hundreds of Aviation accident reports, containing thousands of human causal factors. The present work aims to evaluate the use of HFACS for the improvement of the Brazilian helicopter offshore transport industry safety performance. To do so, the present paper presents the Reason model, the HFACS and the HFACS-HE frameworks, considering evolution, limits, critical reviews, and applications (particularly, in Aviation). A documentary analysis of selected helicopter offshore transport industry accident reports from Brazil was carried out using HFACS-HE. Then, the study was compared with other studies already carried out for similar operations. It concludes that the causality model and culture frameworks have an impact on the application of HFACS, and on the safety performance.*

1 Introduction

To err is human! Since it is an inevitable consequence of being human the fact that humans make errors. Some studies report that up to 70%, or even 90%, of aviation accidents are due to some type of human error. But, of course, the percentage depends on the definition of human error / human factor…

Even so, the greatest potential, at this point, for reducing aviation accidents lies, without doubt, in understanding the human contribution to accidents (Wiegmann and Shappell, 2001), and using that understanding to develop safer systems, considering all the humans involved.

To do so, one needs to be able to analyse the existing aviation accident (and incident) data and information in a consistent, objective, and systematic way. But aviation systems are highly complex, and, consequently, so are aviation accidents (and incidents). For that reason, any analysis framework needs to be able to capture that complexity, without being too complex itself.

In this paper we evaluate the use of one such analysis framework: Human Factors Analysis and Classification System (HFACS); for the improvement of helicopter operations, in particular the safety performance of the helicopter off-shore transport industry. HFACS was chosen since it is the most extensively used human factors accident analysis framework (Harris and Li, 2011).

This paper is divided into eight parts. The first part is the introduction. In the second part the Reason model is briefly presented. The third part, in which HFACS (the classical version) is presented, is divided into its direct antecedents, evolution, framework, procedure for application, and validation and its extensions, adaptations, and application in Aviation. In the fourth part, the HFACS-HE framework is presented, divided into its framework, procedure for application, validation, and application in Aviation. The fifth part contains the brief summaries of the case studies considered. In the sixth part, a comparative analysis is presented, divided into data, procedure, results, and analysis. In the seventh part a discussion of the use of HFACS is given, in particular causality, culture, and Safety performance. Finally, the eighth part is the conclusion.

Next, the Reason model is presented, since it is the theoretical foundation of HFACS.

2 Reason Model

One approach to the genesis of human error is the one proposed by James Reason (1990; 1997), generally referred to as the "Swiss cheese" model of human error. It is a hierarchical, linear, one-directional, sequential, model of causality. Reason describes four levels of human failure, each influencing the next (Figure 1).

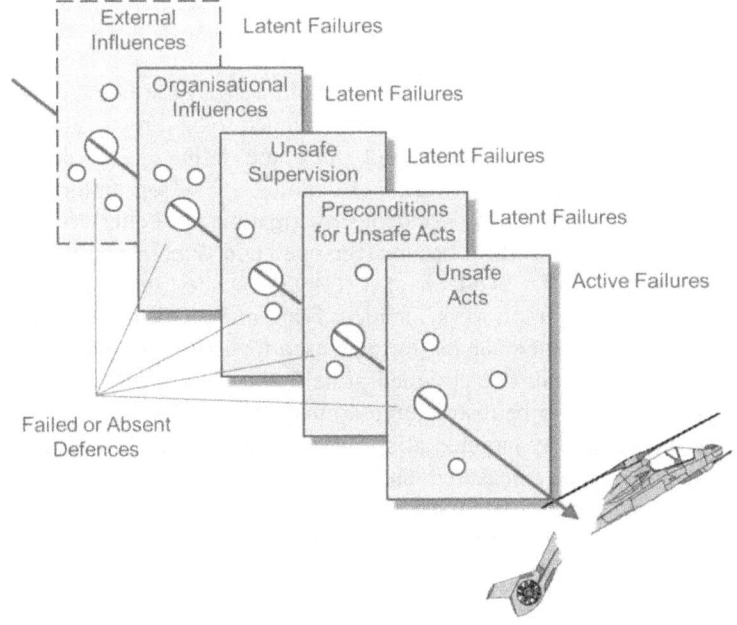

Fig. 1. Reason Model schema (six Levels of Failure).
(adapted from Reason, 1990; Shappell and Wiegmann, 2000, and Omole et al., 2015).

Working backwards in time from the accident, the first level depicts those Unsafe Acts of Operators that ultimately led to the accident, commonly referred to in aviation as aircrew/pilot error. In other words, the actions or inactions of aircrew members, which are directly linked to the accident (or incident). Represented as "holes" in the cheese, these active failures are typically the last unsafe acts committed.

Reason described three more levels of human failure, related to latent failures. As their name suggests, latent failures, unlike their active counterparts, may lie dormant or undetected for long periods of time, until they adversely affect the aircrew performance. The second level, Preconditions for Unsafe Acts, involves the condition of the aircrew as it affects performance. The third level, Unsafe Supervision, involves the management actions which affect performance. The fourth, and last, level, Organisational Influences, involves the organization itself in that it can impact performance at all levels.

It should be noted that the fact that other levels (other than the operators' one) are considered does not lessen the role played by the aircrew, their responsibility and accountability, it only means that intervention and mitigation strategies might lie higher within the system (Shappell and Wiegmann, 2000).

3 HFACS

The Reason model has, in many ways, revolutionized the common way to view accident causation. Even so, it was not intended to be directly applied in a real-world situation. That is, the model does not define the "holes in the cheese". It is necessary to know what these "holes" or system failures are, so that they can be identified during accident investigations or, better yet, detected and corrected be-fore an accident occurs. (Shappell and Wiegmann, 2000)

Douglas Wiegmann and Scott Shappell (1997) detected the necessity for a comprehensive, general framework for identifying and analysing human error. That allows existing accident databases to be used for analysis (since most acci-dent reporting systems are not designed around any theoretical framework of human error, they need to be restructured around the new framework) so inter-ventions can be accurately targeted at specific human causal factors, and their effectiveness objectively measured and assessed. What was needed was a framework around which a needs-based, data-driven safety programme could be developed. With that in mind, they developed HFACS. HFACS identifies the holes, that is, the failures, in the Reason model, and thus provides the means of methodologically categorizing the causes of incidents/accidents. (Shappell and Wiegmann, 2001a; Wiegmann and Shappell, 2003)

The HFACS framework has been used within the military, commercial, and general aviation sectors to systematically examine underlying human causal fac-tors and to improve aviation accident investigations. (Shappell and Wieg-mann, 2000)

3.1 Genealogy / Evolution

The classic version of HFACS (Wiegmann and Shappell, 2003) is the result of more than ten years of work. That evolution is presented next.

3.1.1 Taxonomy of Unsafe Operations

Wiegmann and Shappell (1995) first presented the Taxonomy of Unsafe Opera-tions in 1995 the reliability of which they later evaluated (Shappell and Wieg-mann, 1997). They applied it in the U.S. Naval aviation domain, in particular to Controlled flight into Terrain (CTI) (Shappell and Wiegmann, 1995); Walker (1996) did the same. Shappell and Wiegmann (1996) used it in U.S. naval avia-tion mishaps, considering single- and dual-piloted aircrafts, and Rabbe (1996) used it in U.S. Air Force F-16 mishaps.

3.1.2 Taxonomy of Unsafe Operations II

Shappell and Wiegmann (1997a) further developed the taxonomy, and present-ed the Taxonomy of Unsafe Operations II. Ranger (1997) evaluated its reliabil-ity. Plourde (1997) applied it to fighter-bomber mishaps, and Shappell and Wiegmann (1997b) to CFIT.

3.1.3 Failure Analysis and Classification System (FACS)

The taxonomy further evolved, and Johnson (1997) applied the Failure Analysis and Classification System (Shappell and Wiegmann, 1998) to A-10 class A mishaps. Wiegmann and Shappell (1999) applied it to U. S. Naval aviation mis-haps.

3.1.4 Human Factors Analysis and Classification System - Military Aviation (HFACS-"MA")

Finally, Shappell and Wiegmann (1999a; 2000a; 2001; 2001a) presented the Human Factors Analysis and Classification System, originally a military (Navy) version which Shappell et al. (1999) and Shappell and Wiegmann (2000) ap-plied in the U.S. Navy/Marine Corps to TACAIR and rotary wing mishaps. Shappell and Wiegmann (1999) and Wiegmann et al. (2000) applied it to com-mercial avia-tion, which led to the next evolution.

3.1.5 Human Factors Analysis and Classification System - Civil Aviation (HFACS-"CA")

HFACS was adapted for civil aviation (Wiegmann and Shappell, 2001a; 2001b; 2001c; 2003a), finally, Shappell and Wiegmann (2003) presented the classical version of HFACS.

3.2 Framework

Drawing upon Reason's (1990) model, in particular in the concept of latent and active failures, and the four levels of human failure, HFACS describes four lev-els of failure, with each level correlating with each layer within the model, namely:

1. Unsafe Acts,
2. Preconditions for Unsafe Acts (or unsafe acts, pre-conditional),
3. Unsafe Supervision, and
4. Organizational Influences (or unsafe organisational factors).

The HFACS framework can be seen in Figure 2. Each level is described next.

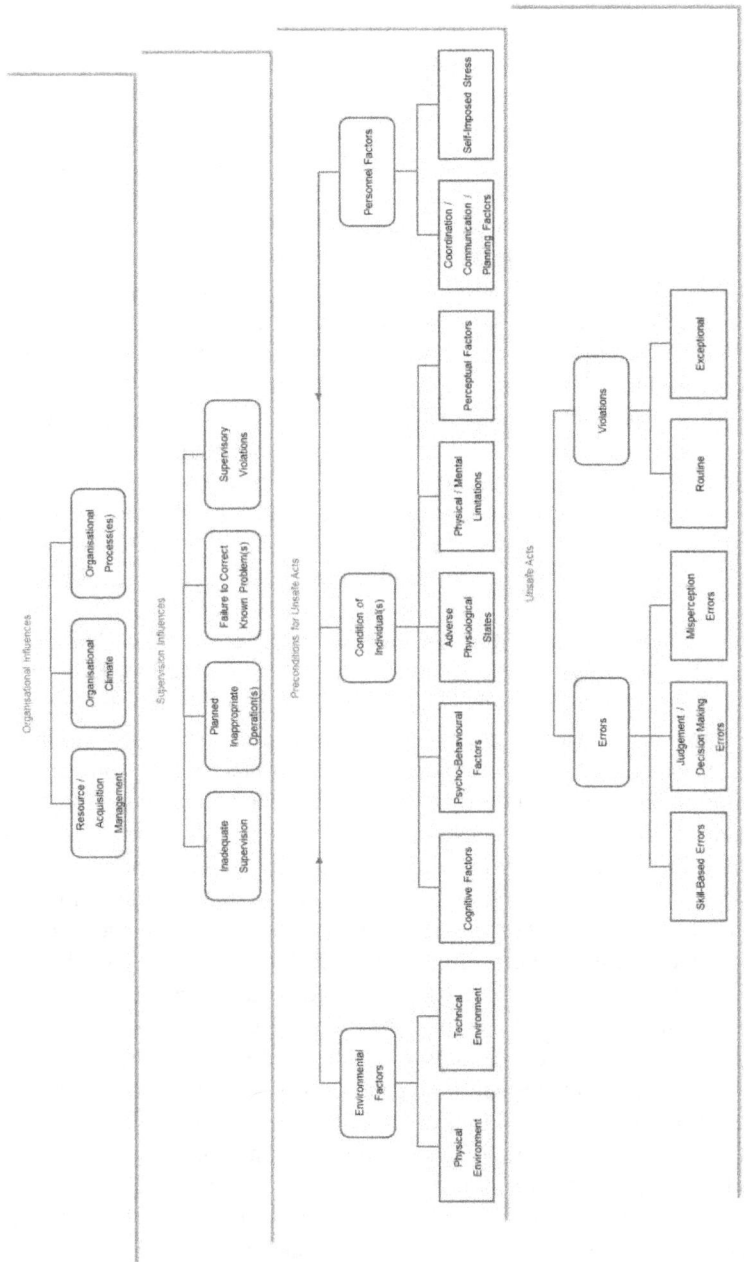

Fig. 2. HFACS Framework schema
(adapted from Reason, 1990; Wiegmann and Shappel, 2003; and, Omole et al., 2015).

The first level, starting in the accident (or incident) and moving backwards, describes the unsafe acts of the operator (aircrew), which directly led to an incident or accident. This level is typically referred to as operator error(s) or pilot errors. The unsafe acts are, usually, easy to identify. They are further, loosely, classified into two categories: errors and violations. Errors are (mental or physical) activities that fail to achieve the desired outcomes. They can be expanded into three parts: skill-based, decision, and perceptual errors. Skill-based errors are actions that occur without significant conscious thought. They, mainly, consist of attention and/or memory failures, and technique errors. Decision errors represent intentional behavior (actions or inactions) that proceeds as intended, yet the plan proves inadequate or inappropriate for the situation. Mainly they consist of procedural, choice and problem solving errors. Perceptual errors occur when one's perception of the world differs from reality and leads to an erroneous response. They, mainly, consist of sensorial errors. Violations can be viewed as conscious activities that disregard established rules and regulations. They can be expanded into two parts: routine and exceptional violations. Routine violations are actions that are habitual by nature and often tolerated, if not sanctioned, by supervisory and/or governing authority. Exceptional violations are isolated departures from authority, not necessarily indicative of individual's typical behavior pattern or, condoned by supervisory and/or governing authority.

The second level, describes the preconditions which lead to unsafe acts of the operator (aircrew). Preconditions are usually latent system failures that lay inactive for long periods of time before ever contribute to an incident or accident. They are further, loosely, classified into three categories: environmental factors, conditions of the operator, and personnel factors. Environmental factors refers to the ambient around the operator. They can be expanded into two parts: physical environment, and technological environment. Physical environment accounts for both operational environment and the ambient environment. Technological environment accounts for all technology and its interaction with the operator. Conditions of the operator refers to the operators capacities. They can be expanded into three parts: adverse mental state, adverse psychological state, and physical/mental limitation. Adverse mental state accounts for those mental conditions that adversely affect performance and contribute to unsafe acts. Adverse psychological state accounts for states that preclude the safe conduct of flight. Physical/mental limitation accounts for those instances when necessary sensory information is either unavailable, or if available, individuals simply do not have the aptitude, skill, or time to safely deal with it; and instances when an individual simply may not possess the necessary aptitude, physical ability, or proficiency to operate safely. Personnel factors accounts for things that personnel (aircrew) do to themselves and which lead to undesirable conditions and unsafe acts. They can be expanded into two parts: crew resource management, and personal readiness. Crew resource management accounts for failures in coordination and communication. Personal readiness accounts for behaviours which, disregarding or not rules and regulations, may reduce the operating capabilities of the individual.

The third level, describes the preconditions, specific to management, which lead to unsafe acts of the operator (aircrew). They are further, loosely, classified into four categories: inadequate leadership, planned inappropriate operations, failure to correct known problems, and leadership violations. Inadequate leadership refers to situations where lack of guidance and/or oversight, by management, are/is present. Planned inappropriate operations refers to situations when performance is adversely affected due to operation planning. Failure to correct known problems refers to situations where deficiencies in related safety areas are "known" to the supervisor, yet are allowed to continue unabated; failure to consistently correct or discipline inappropriate behaviour; or failure to report unsafe tendencies and initiate corrective actions. Leadership violations refers to situations when existing rules and regulations are consciously disregarded; failure to enforce existing rules and regulations; or flaunting authority by management.

The fourth, and last, level describes the preconditions, specific to the organisation (or upper-level management), which lead to unsafe acts of the operator (aircrew). They are further, loosely, classified into three categories: resource management, organizational climate and organizational process(es). Resource management refers to corporate-level decision making regarding the allocation and maintenance of organizational assets such as human resources (personnel), monetary assets, and equipment/facilities. Organizational climate refers to a broad class of organizational variables that influence worker performance, namely: structure, as reflected in the chain-of-command, delegation of authority and responsibility, communication channels, and formal accountability for actions; policies, as reflected in management's decisions about such things as hiring and firing, promotion, retention, raises, sick leave, drugs and alcohol, overtime, accident investigations, and the use of safety equipment; and, culture, as reflected in the unofficial or unspoken rules, values, attitudes, beliefs, and customs of an organization. It can be viewed as the working atmosphere within the organization. Organizational process(es) refers to corporate decisions and rules that govern the everyday activities within an organization, including the establishment and use of standardized operating procedures and formal methods for maintaining checks and balances (oversight) between the workforce and management. (Reason, 1990; Shappell and Wiegmann, 2003; Omole et al., 2015)

3.3 Procedure

As an analytical framework, HFACS can be used to analyse accident reports. Usually, it is best for the analyst to begin as if he/she was an investigator in the field, investigating the accident, and work backward in time, conducting a systematic analysis. First, the analyst needs to determine the 'point of no return', that is, the operator(s) active action(s) that, unequivocally, led to the accident. Then, the classification process initiates, with a continuous sequence of "why" questions. HFACS codes are identified using the sequences of events leading to and after the accident, the immediate and the causal factors, contributory and underlying causes and the investigation recommendations, present in the report. Each HFACS category is counted a maximum of one time per code. This count acts as an indication of the presence or absence of a given category for the accident. (Shappell and Wiegmann, 2003; Omole et al., 2015)

3.4 Validity / Evaluation

The effectiveness of any (human error) framework is based on its validity, which refers to the extent to which a framework is well-grounded and corresponds accu-rately to the real world. Two types of validity are important: external validity, and internal validity. External validity refers to the extent to which a framework can be generalized to other settings (i.e., industries, or areas in an industry). Internal validity refers to the extent to which a framework is valid within a specific setting. (Fleishman et al., 1984) Three types of internal validity are important: content validity, face validity, and construct validity. Content validity refers to whether a framework covers all the major issues within the topic. Face validity refers to whether a framework has a reasonable approach and common sense for users (Wiegmann and Shappell, 2003). Construct validity refers to the extent to which a framework performs in accordance with its theoretical expectations (Carmines and Zeller, 1979).

Wiegmann and Shappell (2001b; 2001d) suggested that four criteria need to be considered when evaluating a human error framework: comprehensiveness, usa-bility, diagnosticity, and reliability. Comprehensiveness is the framework's ability to define and/or identify all significant information relating to an incident/accident. Usability is the framework's ability to be applied for practical use in industry. Diagnosticity is the framework's ability to show the relationships among errors and their trends and causes (Shappell and Wiegmann, 2001). Relia-bility is the framework's ability to yield the same result over repeated trials (Car-mines and Zeller, 1979). Two types of reliability are important: inter-rater, and intra-rater reliability. Inter-rater reliability refers to the framework's ability to ob-tain the same results irrespective of the rater. Intra-rater reliability refers to the consistency of each rater. (Shappell and Wiegmann, 2003; Ergai, 2013)

3.4.1 Evaluations

HFACS was evaluated even during its development (Shappell and Wiegmann, 2001; Wiegmann and Shappell, 2001b; 2003). Li and Harris (2005) performed ROC Air Force reliability analysis and cross-cultural comparison. Further, a cross-cultural comparison of Aviation mishaps, due to technology cultural influences (Li and Harris, 2006b). Olsen and Shorrock (2010) research HFACS's inter-coder consensus and intra-coder consistency. Then, inter-coder consensus of ATC (Ol-sen, 2011). O'Connor and Walker (2011) evaluated the inter-rater reliability of DoD-HFACS, and concluded that the reliability and validity of mishap coding systems needs to be pondered prior to application. Ergai (2013) evaluated intra-rater and inter-rater reliability of HFACS. Al Wardi (2013) evaluated the practa-bility and reliability of HFACS, considering Royal Air Force of Oman (RAFO) accidents. A cultural comparative analysis of helicopter accidents, in Nigeria and UK, was performed by Omole (2015). And, by Al Wardi (2016) in Arabian, Asian, and Western cultures.

3.5 HFACS Extensions / Adaptations

Over time, several adaptations of HFACS to different industries, and specific are-as in Aviation, have been developed. The level of modifications varies from mi-nor terminological refinements to defining a new structure to root cause taxono-my. Some are due to natural limitations of the framework, but others are related to misunderstandings of its contents. The ones mainly related to Aviation will be presented next.

Pounds et al. (2000) and Scarborough and Pounds (2001) adapted HFACS to Air Traffic Control (HFACS-ATC).

Wu and Zhao (2001) considered that hardware and human-machine interaction failure were not considered in HFACS. Based on the HFACS and the energy-based accident causation theory, they developed the Complex Human Factor Analysis and Classification Framework (C-HFACF).

Rothblum et al. (2002) adapted HFACS for offshore and maritime operations.

Luxhøj (2003) reported on advanced risk analytics, which combine the use of HFACS, Bayesian Belief Networks, and case-based scenarios to assess a relative risk intensity metric.

Krulac (2004) adapted HFACS to Aviation Maintenance (HFACS-ME). Like Wood (2000), Nelson (2001), Zolla et al. (2001a), Flanders and Tufts (2001) and Boex (2001) had used before.

In 2003, the Royal Canadian Air Force (RCAF) introduced the Canadian Forces HFACS (CF HFACS) (RFAC, 2013).

Shappell and Wiegmann (2006) developed the Human Factors Intervention Matrix to be used with HFACS as a tool for evaluating current and proposed avi-ation safety programs.

Shappell, et al. (2007) included specific aircrew, environmental, supervisory, and organizational factors associated with two types of commercial aviation (air carrier and commuter/on-demand) accidents, to provide support for the contin-ua-tion, modification, and/or development of interventions aimed at commercial aviation safety.

The Department of Defense (O'Connor, 2008; O'Connor et al., 2010; DoD, 2015), has joined the several HFACS versions existing in DoD (Navy/Marine, Army, and Air Force) to create a common framework: DoD-HFACS, which adds a level of fine grain classification, with (147) nano-codes, which are to be used to identify the mishap causes.

Aas (2008) has merged HFACS and the Man-Technology-Organization (MTO) into a merged framework (HFACS/MTO), for the Oil & Gas industry. It is used for investigating accidents on installations for production or drilling, considering employees with different cultural context and understanding of procedures/documentation and communication.

Hardman and Colombi (2009) used the DoD-HFACS in the Domains of Human Systems Integration (HSI).

Paletz, et al. (2009) incorporated Social Psychological Phenomena into HFACS.

Kavade (2009) used the concept of event severity along with the HFACS framework to help better identify target areas for intervention among unsafe events, in wind turbine maintenance.

Patterson and Shappell (2010) extend HFACS to include latent conditions which exist outside the organization and influence its safety performance, in the mining industry (HFACS-MI). This fifth level, Outside Factors, accounts for external factors such as pressure from environmental groups and legal govern-mental influences.

The Australian Defense Force also has its own version of HFACS: HFACS-ADF. (Olsen and Shorrock, 2010)

Hawkins (2010) has combined HFACS with the Fuzzy Analytical Hierarchy Process (FAHP), to create an analytical HFACS. The HFACS/FAHP combines the judgements of multiple experts on the relative importance of human factors in the accident sequence and converts these judgements into relative importance values that may provide additional insight for accident investigators.

Berry (2010) performed a multi-industry accident analysis which allowed for common high-level human error patterns to emerge and for benchmarking stand-ards to be created, that is, to determine the size of the holes and the interac-tions/error pathways among the holes.

Omole (2012; 2016b) and Omole and Walker (2016) integrated the cultural framework into HFACS (HFACS-CUL).

Hsiao et al. (2013) adapted HFACS for maintenance audit (HFACS-MA).

Bilbro (2013) indicated the Human Factors Analysis and Classification System Maritime (HFACS-M).

Omole et al. (2015), following a systems approach, extended HFACS to include other external factors outside the organisation level causal to unsafe acts and unsafe situation (HFACS-HE).

Omole (2016) indicated a new diagnostic tool, which integrates HFACS-CUL and HFACS-SYS, where the cultural framework explores the contributory factors of each part of the system from a cultural perspective, and the HFACS framework is used as a tool to establish the gaps and ultimately how the cultural components are integrated into the framework.

3.6 Application: Aviation

HFACS and its extensions / adaptations have been applied in Aviation. These applications can be divided into three main categories:

1. (accident or incident) data analysis;
2. (accident or incident) report analysis; and,
3. Safety intervention evaluation.

A fourth, non-academic, category should be considered:
4. (accident or incident) investigation.

Some relevant examples are presented next.

3.6.1 Data Analyses

HFACS accident or incident data analyses use treated data of accident databases to determine trends in human factors.

Wiegmann and Shappell (2001) assessed the use of HFACS as an error analysis and classification tool applied to commercial aviation accident records. They considered the National Transportation Safety Board (NTSB) records between 1990-1996.

Broach and Dollar (2002) analysed the relationship of organizational factors to en route operational error (OE) rates was investigated using data from the Na-tional Airspace Incident Monitoring System (NAIMS), between 1997 and 2000, and HFACS-ATC.

Shappell and Wiegmann (2003) applied HFACS to analyse the remaining non-fatal, general aviation, accidents data, between 1990-1998. These analyses provide unique insight into the genesis of GA accidents. Implications for GA initial and recurrent training are discussed.

Boquet et al. (2004) considered general aviation maintenance accidents in their analysis.

Shappell and Wiegmann (2004) compared military and civilian aviation acci-dents, in the USA.

Gaur (2005) analysed civil aircraft accidents in India.

Scarborough et al. (2005) used HFACS to examine ATC Operational Errors (OEs).

Tvaryanas et al. (2005) used DoD-HFACS to analyse human factors in military Unmanned Air Vehicle (UAV). They concluded that failures in the organizational culture, management, and structure of DoD's acquisition processes for UAVs existed.

Wiegmann et al. (2005) performed a general aviation fine-grained analysis.

Li and Harris (2006a) analysed Republic of China Air Force accidents, between 1978 and 2002.

von Thaden (2006) analysed organizational factors in commercial aviation accidents.

Boquet et al. (2007) compared general aviation accidents in Alaska versus the rest of the United States.

Garr (2007) used HFACS to analyse experimental homebuilt flight test in comparison to professional flight test. He concluded that, ideally, flight test is left to trained professionals.

Lenné et al. (2008) analysed general aviation accidents in Australia.

Li et al. (2008) considered civil aviation in China. Has did Ting and Dai (2011).

Tvaryanas and Thompson (2008) analysed remotely piloted aircraft's Recurrent error pathways.

Jarvis (2009) evaluated the relation between inexperienced glider pilots' acci-dent causes and instructor decisions, in the UK.

Rashid et al. (2010) analysed helicopter maintenance error's organisational in-fluences, using HFACS-ME.

Liu et al. (2013) analysed general aviation accidents with Helicopters.

Taranto (2013) analysed USAF remotely piloted aircraft mishaps.

Dayyabu (2014) analysed aircraft accidents, and effects on aviation industry, in Nigeria.

Cistone (2014) used the Runway Safety database and the National Transportation Safety Board (NTSB) database to determine causal factors for airport surface deviations over a 12-year period (2001-2012).

Luxhøj (2015) used his (2003) advanced risk analytics, which combine the use of HFACS, Bayesian Belief Networks, and case-based scenarios to assess a rela-tive risk intensity metric for rotorcraft-type Unmanned Aircraft Systems.

3.6.2 Report Analyses

HFACS accident or incident report analyses use accident reports to obtain data to determine human factors trends.

Li et al. (2007) used HFACS to compare aircraft accidents in China with that in the United States and India, for cultural difference with regard to the causal factors of the reported accidents. There was a significant difference in the accident outcome of the three countries in seven HFACS categories.

Li et al. (2010) analysed, using HFACS, Aeronautical decision-making (ADM) contribution to accidents, in the Republic of China, showing how actions and de-cisions at higher managerial levels in the operation of commercial aircraft result in decision errors on the flight deck and subsequent accidents.

Wang et al. (2011) performed reliability studies for coding contributing factors of incident reports of military ATC. Results show similarly low consensus for both groups of participants. Several reasons for the results are proposed associated with the HFACS model, the context within which incident reporting occurs in real organizations and the conduct of the studies.

Bryan (2014) analysed helicopter emergence medical service (HEMS) accidents between 2000-2012.

Omole et al. (2015) performed a comparative and documentary analysis of a selected offshore helicopter transport industry accident report from the United Kingdom and Nigeria was carried out using the HFACS-HE model. The outcome of the comparative analysis of the HFACS-HE model via the documentary investigations revealed that causal factors such as public, regulatory and politics have not been included in the original HFACS framework. These factors would be useful during the investigation process.

3.6.3 Safety Intervention Evaluations

HFACS has been used to determine, and to evaluate, Safety interventions pro-grams.

Zolla et al. (2001) evaluated maintenance investigation and intervention cases.

Wiegmann and Rantanen (2002; 2003) evaluated technology intervention strategies.

Inglis et al. (2007; 2010) evaluated the use of HFACS as a predictive accident model.

Cowan (2009) evaluated Naval Aviation mishaps and developing intervention strategies.

Rashid (2010) evaluated proactive monitoring and controlling techniques in Helicopter Maintenance.

Li and Harris (2013) evaluated training deficiencies in military pilots.

Taylor (2014) evaluated the increasing of, general aviation, Safety through improved training, for UK.

3.6.4 Accident Investigations

HFACS is being used as an investigation framework, in Aviation, but, as expected, no research work was found.

Saarelainen and Jäntti (2015) studied the improvement of incident investigation process in IT.

4 HFACS-HE

Omole, Walker and Shappell (2015) extended the HFACS so as to include, potentially relevant, latent conditions which exist outside the organization and influence its safety performance. Based on Rasmussen (1997) system approach, this includes, for example, conditions due to government, regulators, manufacturer, social, environmental, political, economic, and customers influence. These conditions are not usually considered during accident investigations. This is so because these conditions are difficult to find unless a clear understanding of the organization and its (operational) environment exists. The Human Factors Analysis and Classification System - Helicopters, (HFACS-HE) framework allows to diagnose important latent conditions which exist outside the organization and could lead to unsafe acts and unsafe situations. HFACS-HE thus enables organizations to learn from prior experience and become (more) resilient in the future. The HFACS-HE framework can be seen in Figure 3.

A point needs to be made here, the idea to extend HFACS to include latent conditions which exist outside the organization and influence its safety performance had already been presented by Patterson and Shappell (2010).

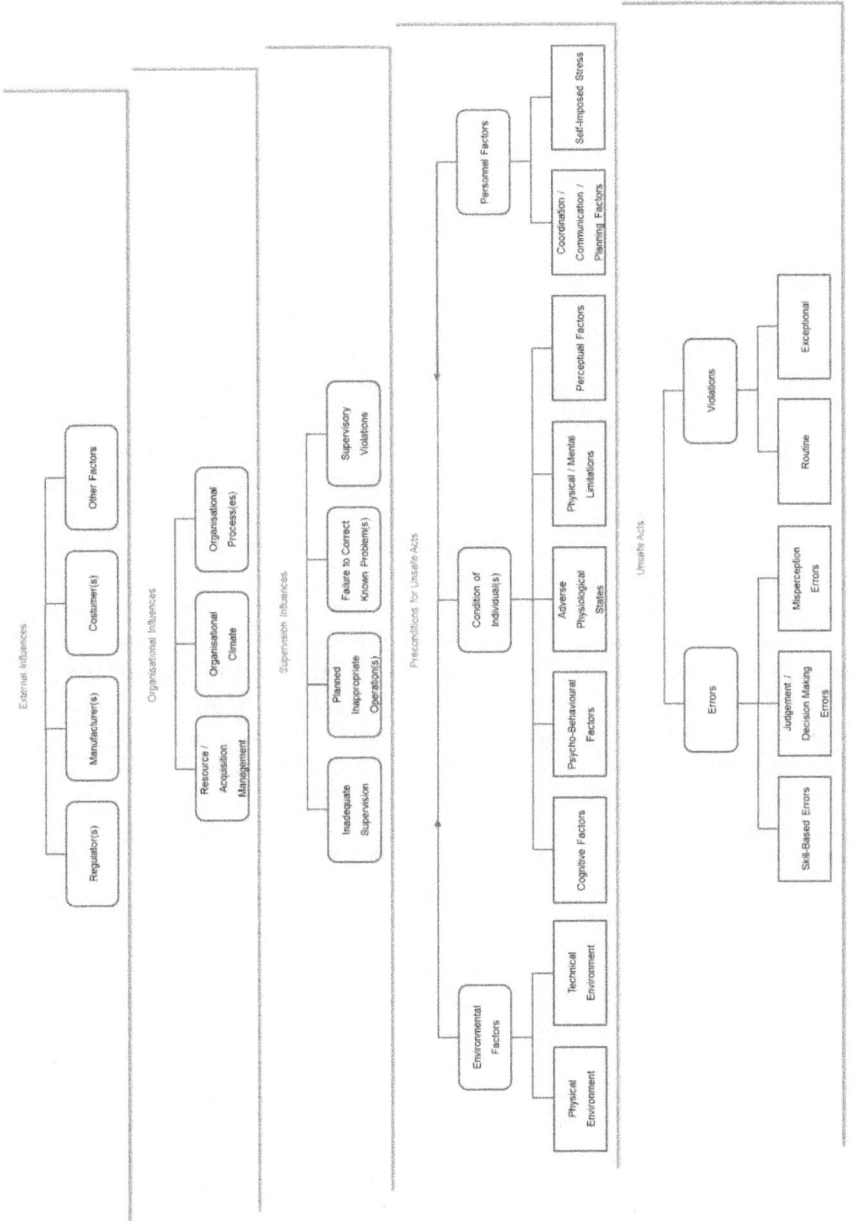

Fig. 3. HFACS-HE framework Schema.
(adapted from Reason, 1990; Wiegmann and Shappel, 2003, and Omole et al., 2015).

4.1 Framework

Only the fifth level is described next, since the other are identical to the ones in the HFACS framework and have been described before (see 4.1).

The fifth, and now the last, level, describes the preconditions, external to the organisation, which lead to unsafe acts of the operator (aircrew). They are further, loosely, classified into five categories: government, regulators, manufacturers, customers, and other factors. Government is responsible for creating the laws governing the industry, based on the safety review of accident analysis; and, enables to regulatory bodies to work independently to oversight the industry, by regulating and investigate accident and incidents, and providing the necessary budget to carry out its activities. They are further, loosely, classified into three categories: government resource management, governing climate, and governing process. Government resource management refers to government-level decision making regarding the allocation and maintenance of governmental assets such as human resources (personnel), monetary assets, and equipment/facilities. Governing climate refers to a broad class of governmental variables that influence worker performance. Governing process(es) refers to government decisions and rules that govern the everyday activities within a country, and its organizations, including the establishment and use of standardized operating procedures and formal methods for maintaining checks and balances (oversight) between the workforce and management. Regulators are responsible for the industry and the workers it oversees. The regulatory bodies split into many groups, but the two main groups are accident investigators and the inspectorate. The investigators investigate accidents (and incidents) and provide analyses and recommendations. The inspectorate regulates the industry, supervises, and provides advice and guidance. They are further, loosely, classified into three categories: regulatory resource management, regulatory climate, and regulatory process. Manufacturers are responsible with providing the industry with the up-to-date aircraft design, technology and maintenance guidance for safe operation. Customers, while exerting commercial and economic pressure on the organization, have (particularly, in the helicopter offshore transport industry), responsibility to regulate, supervise, and cooperate for safe operation. Other factors refers to outside influences, not included before, that have the ability to, adversely or positively, affect the safe performance of a company, for example legal pressure is always a concern, because of blame and liabilities; or environmental issues, because of public opinion and brand value. (Omole et al., 2015)

4.2 Procedure

The HFACS-HE used procedure is identical to the one of HFACS, described be-fore (see 3.3).

4.3 Validity / Evaluation

HFACS-HE framework validity it is equal to HFACS, described before (see 3.4).
The author has not found any work which evaluates the validity of HFACS-HE.

4.4. Application: Aviation

HFACS-HE has been applied by Omole and Walker (2015; 2015a), and Omole et al. (2015), in report analysis.

5 Case Studies

The HFACS-HE was applied to three formal air accident reports, related to (offshore) helicopter operation. These reports were selected because they were availa-ble case studies that would offer HFACS-HE a "good stress test". The HFACS instructions were followed rigidly. The same criterion for each level as in the origi-nal framework, were considered, plus the new HFACS-HE level ones. (Omole et al., 2015; 2015a)

5.1 Brazil

The Brazilian accident took place near the platform P-18 (SS-44), located in Bacia de Campos, Rio de Janeiro coast, at 16:18 hours, on February 26, 2008, during the platform take-off. It involved Eurocopter AS332 L2 Super Puma helicopter (PP-MUM), of BHS Helicópteros S.A., at the service of PETROBRAS. The helicopter took-off platform P-18 (BCP), at 16:16 hours, heading for Macaé Airport (SBME), on the last leg of a, IFR (Instrument Flight Rules), schedule flight, within Bacia de Campos. During the climb, in VMC (Visual Meteorological Conditions), turning left, lost velocity and started an anti-clockwise spiral descent and struck the surface of the sea. The fuselage broke on impact, the tail section separating from the fuselage, and the floats, except the lateral back left, burst. There were 5 fatalities. (CENIPA 2012)

5.2 Nigeria

The Nigerian accident, considered in Omole et al. (2015), took place in a field of the Qua Iboe Terminal heliport (QIT), located in Eket, Akwa Ibom State, Nigeria, at 07:39 hours, on August 3, 2007, during a "local flight". It involved Bell 412 EP (5N-BIQ), of Bristow Helicopters (Nigeria) Limited, at the service of Mobil Producing (Nigeria) Unlimited (MPN). The Bristow line training captain boarded the helicopter at 07:30 hours, without the co-pilot, with no previous flight schedule or request. He started the aircraft engines rapidly, made a radio call at 07:32.16 hours for a "local flight", and lifted rapidly at 07:35 hours. He made two fast fly passes over the airfield and on the third fly pass the aircraft descended steeply over the west of the airfield at a high speed impacting the ground at 07:39 hours. The pilot died of injuries shortly after being rescued from the wreckage. (AIB, 2008)

5.3 United Kingdom

The UK accident, considered in Omole et al. (2015), took place near the North Morecambe gas platform, located in Morecambe Bay, Irish Sea, at 19:06 hours, on December 27, 2006, during the platform approach. It involved Aerospatiale SA365 N Dauphin 2 helicopter (G-BLUN), of CHC Scotia Limited, at the service of Hydrocarbon Resources Limited (HRL). The helicopter departed Blackpool Airport (BLK), at 18:00 hours, on a commercial scheduled flight consisting of eight sectors within the Morecambe Bay gas field. The first two sectors were completed without incident but, when preparing to land on the North Morecambe platform in the dark, the helicopter flew past the platform and struck the surface of the sea. The fuselage disintegrated on impact and the majority of the structure sank. Two fast response craft from a multipurpose standby vessel, which was on position close to the platform, arrived at the scene of the accident 16 minutes later. There were no survivors amongst the five passengers or two crew members. (Tydeman et al., 2008)

6 Comparative Analysis

A comparative and documentary analysis of the HFACS-HE is carried out using content analysis of the selected case studies, focused on depth rather than breadth of analysis. The reports were subjected to an in-depth content analysis. (Omole et al., 2015)

6.1 Data

The data was taken from the selected air accident reports. The reports were up for public access, on CENIPA (http://www.cenipa.aer.mil.br/), AIB (http://www.aib.gov.ng/), and AAIB (http://www.aaib.gov.uk/) websites. For that reason, no de-identification of information was done. Each report provided detailed information about the accident. This included the sequence of events leading to and after the accident, the immediate and the causal factors, contributory and underlying causes, and the investigation recommendations. The reports were selected because of human factor identifiable active and latent factors which lead to the accident. (Omole et al., 2015)

A point should be made. It is not the object of an investigation to determine blame or liability. However, it should be recognised that an investigation report must include factual material of sufficient weight to support the analysis and findings. That material will at times contain information reflecting on the performance of individuals and organisations, and how their actions may have contributed to the outcomes of the matter under investigation, due to the need to properly explain what happened, and why, in a fair and unbiased manner. (Inglis et al., 2007)

6.2 Procedure

The HFACS-HE used procedure is identical to the one of HFACS, described before (see 3.3).

6.3 Results

The results obtained from the analyses are presented in Table 1, and Figures 4 to 9.

The results of these analyses identified causal factors at all (six) levels. The most often identified causal category at the unsafe acts level was skill-based errors for Brazil, and judgement and decision making errors for Nigeria and UK. The most often identified causal category at the preconditions level was technological environment for Brazil, psycho-behavioural factors for Nigeria, and personnel factors for UK. The most often identified causal category at the unsafe supervision level was supervisory violations for Brazil, failure to correct known problems for Nigeria, and inadequate supervision for UK. The most often identified causal category at the organisational influences level was resource management for Brazil, and organisational processes for Nigeria and UK. The most often identified causal category at the external influences level was customer for Brazil and Nigeria, and regulator for UK.

Table 1. Summary of the HFACS-HE factors. (continued over page)

Summary of the HFACS-HE Factors									
HFACS-HE factors		Brazil (PP-MUM)		%		Nigeria (5N-BIQ)		%		UK (G-BLUN)		%	

HFACS-HE factors		Brazil (PP-MUM)		%	Nigeria (5N-BIQ)		%	UK (G-BLUN)		%
Unsafe acts	Skill-based errors	9		11,54	5		4,76	4		2,22
	Judgment and Decision-making errors	6	20	7,69 / 25,64	9	18	8,57 / 17,14	16	24	8,89 / 13,33
	Perception errors	4		5,13	0		0,00	1		0,56
	Violations	1		1,28	4		3,81	3		1,67
Preconditions	Personnel factors	5		6,41	3		2,86	27		15,00
	Perceptual factors	0		0,00	0		0,00	7		3,89
	Physical / Mental limitations	0		0,00	0		0,00	1		0,56
	Adverse psychological states	8	31	10,26 / 39,74	3	20	2,86 / 19,05	0	71	0,00 / 39,44
	Psycho-behavioural factors	2		2,56	11		10,48	0		0,00
	Cognitive factors	0		0,00	1		0,95	4		2,22
	Technological environment	9		11,54	2		1,90	18		10,00
	Physical environment	7		8,97	0		0,00	14		7,78

Table 1. Summary of the HFACS-HE factors. (continued from previous page)

HFACS-HE factors		Brazil (PP-MUM)		%		Nigeria (5N-BIQ)		%		UK (G-BLUN)		%	
Supervision	Inadequate supervision	1		1,28		4		3,81		8		4,44	
	Planned inappropriate operation	0	4	0,00	5,13	3	13	2,86	12,38	3	13	1,67	7,22
	Failure to correct known problems	0		0,00		5		4,76		2		1,11	
	Supervisory violations	3		3,85		1		0,95		0		0,00	
Organisation	Resource management	7		8,97		10		9,52		16		8,89	
	Organisational climate	1	14	1,28	17,95	5	26	4,76	24,76	13	54	7,22	30,00
	Organisational processes	6		7,69		11		10,48		25		13,89	
External	Regulator	2		2,56		4		3,81		15		8,33	
	UK Heli	0		0,00		0		0,00		3		1,67	
	Customer	7	9	8,97	11,54	17	28	16,19	26,67	0	18	0,00	10,00
	Government / Security	0		0,00		7		6,67		0		0,00	
Total		78		100,00		105		100,00		180		100,00	

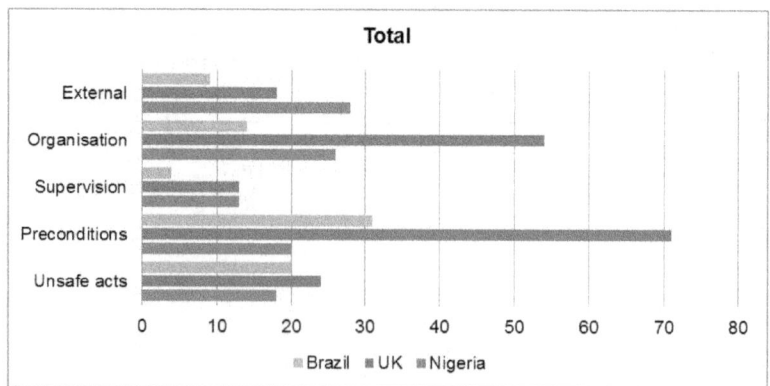

Fig. 4. HFACS-HE factors.

Fig. 5. Unsafe acts.

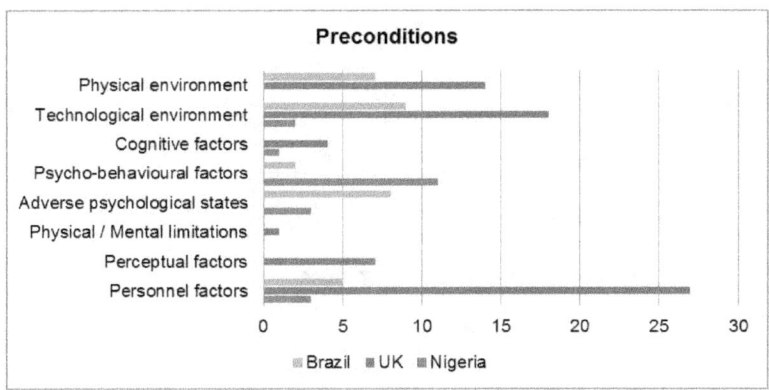

Fig. 6. Preconditions of Unsafe Acts

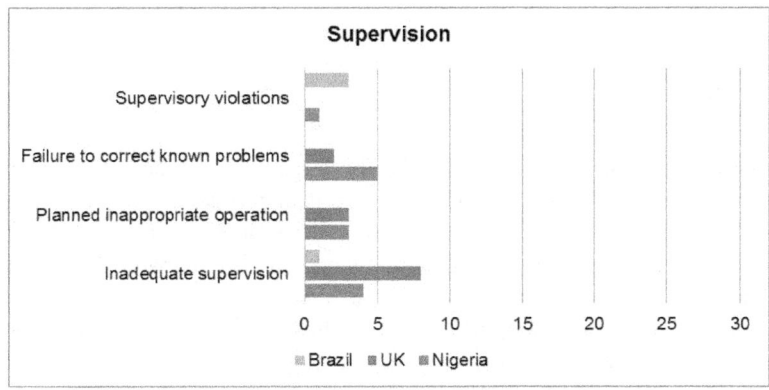

Fig. 7. Supervision Influences

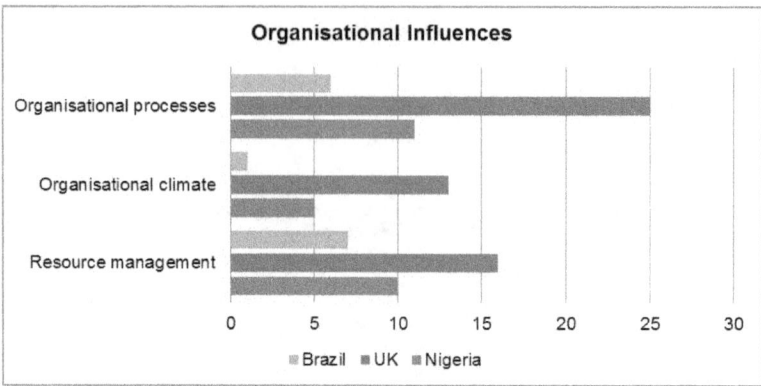

Fig. 8. Organisational Influences

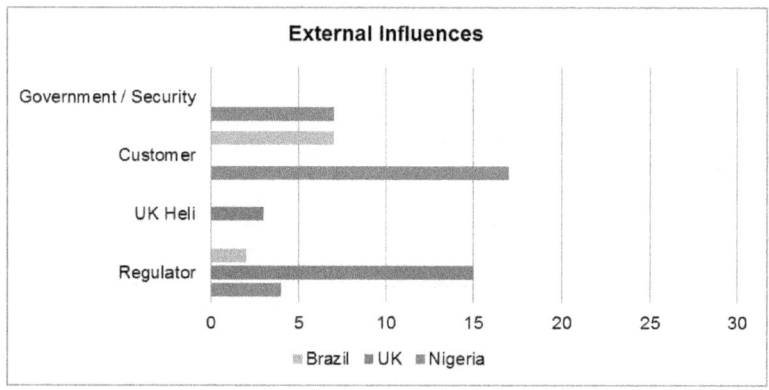

Fig. 9. External Influences

6.4 Analysis

Comparing the results obtained for the three cases is difficult. Several points should be considered before such an analysis is undertaken. First, one is dealing with three different accidents, even if all involve helicopters used in the offshore transport industry. That is, different accidents involving different helicopters, operators, organisations, regulators, etc. Second, one is dealing with data obtained in three different reports, made by three different investigation organisations with different resources, cultures, etc. Third, one is dealing with three different investigation teams using different investigation methodologies, processes, technologies, etc. Fourth, and last, one is dealing with several investigators and experts of three different investigation teams. That is, different people with different knowledge, capabilities, cultures, etc. For those reasons, a quantitative analysis based on the number of codes has no scientific rigour or value. Even, a qualitative analysis of the codes or categories suffers from similar problems.

Probably, based on the results, one could acknowledge that human factors problems exist at all levels (which is obvious since in all levels there are people). Also some attention is necessary, and advantageous for operational safety, in relation to crew knowledge development, training, selection and continuous psychological operational evaluation; management knowledge development; and, regulators personnel knowledge development, training, and selection.

7 Discussion

HFACS is, according to Harris and Li (2011), the most widespread, and used hu-man factors accident analysis framework. In spite of that, several critical reviews of the framework, and the Reason model in which it is based, exist. Next, some of these critical reviews will be considered.

Beaubien and Baker (2002) consider that Reason model could lead to confusion when it comes to categorization and analysis. However, HFACS is the demonstration that this is not so. Even if during its development it has drifted away from Reason's theory (Hendy, 2003). If it has a problem it is oversimplification of reality. This is acknowledged by Shappell and Wiegmann (2000) in their aim to develop a system that would encompass all aspects of human error. Ac-cording to Dekker (2002), HFACS has the problem of analyst hindsight (Dekker, 2001), which could potentially lead to counterfactual interpretation and overgen-eralization of causality. But, that is a problem inherent to all accident analysis frameworks. The same can be said for predictive validity, because there is no ground truth to compare the prediction against, results need to make sense. (Hen-dy, 2003) Omole et al. (2015) point out that some critics of HFACS argue that the framework is effective in identifying the problem, but lacks corresponding solu-tions to solve problems. But, that is not the original intention.

On other hand, Wang et al. (2011) doubt the accuracy and comparability of results in analyses between reports, due to aspects of investigation methodology. That is, it is highly dependent on the quality of the data available (Walker et al., 2011) And, Strauch (2010) points out that cultural differences, with regard to causal factors, cannot be accounted for in HFACS.

Leveson (2004; 2012) goes further, pointing out that Reason model is easy to understand, but it is weak due to the oversimplification of the causality. And, for that reason, HFACS framework can easily miss subtle and complex coupling and interactions among failure events. Therefore, it does not seem to effectively explain accidents involving human failures at all levels as well as technological design in highly adaptive tightly-coupled, complex interactive socio-technical systems such as the aviation industry.

Based on these critical reviews, some comments are made below in relation to: causality model, culture framework, and safety performance; all of which have implications for the credibility of HFACS analyses.

7.1 Causality model

HFACS is based on the Reason model of causality, which is a hierarchical, linear, one-directional, sequential, model of causality. But, causality in aviation accidents (or incidents) is not like that... it is more complex. The causality model needs to capture that complexity, that is, it needs to capture "reality"; that will, surely, imply that the causality model needs to be more complex... which does not mean, more complicated.

7.2 Cultural framework(s)

HFACS does not consider the cultural frameworks or cultures of the human involved in the investigation, or the analysis. Omole et al. (2014; 2016) and Omole (2016a) show that cultural factors, although not currently well represented in ac-cident analysis, were prominent in the sampled accident reports analysed. The accident investigator, like any other human, has several cultural frameworks, cor-responding to the several cultures he belongs to, like national or professional cul-ture. This last includes the knowledge, literacy, and proficiency of (Aviation) Safe-ty. These frameworks are reflected in the investigation (and also in the report), and in the data used for HFACS analysis. Similarly, the same holds for the ana-lyst (Oncu and Yildiz, 2014).

7.3 Safety performance

HFACS has been used to detect (main) trends of causal factors which lead to accidents, and, from trends, safety interventions are developed. These can later can be evaluated using HFACS. The question is: If HFACS is not capable to capture "reality", to capture subtle and complex coupling and interactions among failure events, how can it be used to determine and evaluate Safety interventions?

8 Conclusion

HFACS has been evolving but still needs to consider more active and latent failure conditions, and influences. Its limitations may be due to fact that:

a) a more complex causality model needs to be consider for (Aviation) Safety;
b) an empirical approach is not sufficient for (Aviation) Safety; and/or,
c) it is the knowledge of the (Aviation) system which allows development of an efficient framework, and not the literacy of (air) accident reports.

Even so, HFACS can be used, if its limits are recognised and respected.

In all likelihood, the most one can hope to achieve when comparing groups of reports from the same industry in different places, or from different organisations in the same industry, is to use the analysis as an aid for an analyst's evaluation. That is, HFACS analysis effectiveness depends on the analyst's knowledge, and literacy about Safety, a specific industry (Aviation), a specific technology (heli-copters), a specific operation (offshore transport), and a specific culture (Brazil / Nigeria / UK), etc. In other words, HFACS analysis effectiveness depends on the analyst's knowledge, and literacy about Safety, and a specific system.

Finally, a possible way to test the influence of culture(s) in HFACS would be to perform six parallel accident investigations, of the same accident, by six different investigation teams, from three different countries (one of which from the country where the accident happened), of three different regions, two of each country, one following an investigation based on HFACS, and the other not following it. This way it would be possible to deal with the investigators' culture(s) influence per se. This would work to a certain level, since the individual influence, in terms of knowledge, literacy, capacities, and personality, of each investigator could not be discounted: all people are different.

References

Aas, A. L. (2008) The Human Factors Assessment and Classification System (HFACS) for the Oil & Gas Industry. International Petroleum Technology Conference. 03-05 December 2008. Kuala Lumpur, Malaysia. International Petroleum Technology Conference.

AIB (2008) Aircraft Accident Report 01/2008. 94 pages. Accident Investigation Bureau (AIB), Nigeria Civil Aviation Authority (NCAA): Murtala Muhammed Airport, Ikeja, Lagos, Nigeria.

Al Wardi, Y. (2013) The Utility of Human Factors Analysis and Classification System (HFACS) in the Analysis of Military Aviation Accidents. Indian Journal of Aerospace Medicine 57(2):20-37 2013.

Al Wardi, Y. (2016) Arabian, Asian, Western: A Cross-Cultural Comparison of Aircraft Accidents from Human Factors Perspectives. International Journal of Occupational Safety and Ergonomics ?(?):?-?. [published online]

Beaubien, J. M. and Baker, D. P. (2002) A review of selected aviation human factors taxonomies, accident/incident reporting system and data collection tools. International Journal of Applied Aviation Studies 2(2):11-36.

Berry, K. A. (2010) A meta-analysis of human factors analysis and classification system causal factors: Establishing benchmarking standards and human error latent failure pathway associations in various domains. Doctor of Philosophy in Industrial Engineering. Clemson University: Clemson, SC, USA.

Bilbro, J. (2013) An inter-rater comparison of DoD Human Factors Analysis and Classification System (HFACS) and Human Factors Analysis and Classification System Maritime (HFACS-M). Master of Science in Human Systems Integration. Naval Postgraduate School: Monterey, CA, USA.

Boex, A. R. (2001) Web-Based Information Management System for the Investigation, Reporting, and Analysis of Human Error in Naval Aviation Maintenance. Master of Science in Information Technology Management. Naval Postgraduate School: Monterey, CA, USA.

Boquet, A., Detwiler, C., Hackworth, C., Holomb, K. and Pfleiderer, E. (2007) Beneath the tip of the iceberg: A human factors analysis of general aviation accidents in Alaska versus the rest of the United States. Technical Report (DOT/FAA/AM-06/7). Federal Aviation Administration: Washington DC, USA.

Boquet, A., Detwiler, C., Roberts, C., Jack, D., Shappell, S. A. and Wiegmann, D. A. (2004) General Aviation Maintenance Accidents: An Analysis using HFACS and Focus Groups. Office of the Chief Scientist for Human Factors Aviation Maintenance Human Factors, 4-8.

Broach, D. M. and Dollar, C. S. (2002) Relationship of Employee Attitudes and Supervisor-Controller Ratio to En Route Operational Error Rates. Technical Report (DOT/FAA/AM-02/9). Civil Aeromedical Institute, Federal Aviation Administration: Oklahoma City, OK, USA.

Bryan, C. G. (2014) An Analysis of Helicopter EMS Accidents using HFACS: 2000-2012. Master of Science in Human Factors and Systems. Embry-Riddle Aeronautical University: Daytona Beach, FL, USA.

Carmines, E. G. and Zeller, R. A. (1979) Reliability and Validity Assessment. Sage Publications: Thousands Oaks, CA, USA.

CENIPA (2012) Aircraft Accident Report (A - Nº 039/CENIPA/2012). 61 pages. Centro de Investigação e Prevenção de Acidentes Aeronáuticos (CENIPA), Comando da Aeronáutica. [in Portuguese]

Cistone, J. H. (2014) An Analysis of Airport Surface Deviations using the Human Factors Analysis and Classification System (HFACS). Doctor of Philosophy in Aviation. Embry-Riddle Aeronautical University: Daytona Beach, FL, USA.

Cowan, S. R. (2009) A human systems integration perspective to evaluating Naval Aviation mishaps and developing intervention strategies. Master of Science in Human Systems Integration. Naval Postgraduate School: Monterey, CA, USA.

Dayyabu, D. (2014) Analysis of aircraft accidents and effects on aviation industry in Nigeria. Master of Philosophy in Air Safety Management. City University London: London, England, UK.

Dekker, S. W. (2001) The disembodiment of data in the analysis of human factors accidents. Human Factors and Aerospace Safety 1(1):39-57.

Dekker, S. W. (2002) The field guide to Human Error Investigations. Cranfield University Press: Cranfield, England, UK.

DoD (2015) Department of Defense Human Factors Analysis and Classification System (DoD HFACS): A mishap investigation and data analysis tool. version 7.0. Department of Defense: Washington, DC, USA.

Ergai, A. O. (2013) Assessment of the Human Factors Analysis and Classification System (HFACS): Intra-rater and Inter-rater Reliability. Doctor of Philosophy in Industrial Engineering. Clemson University: Clemson, SC, USA.

Flanders, T. P. and Tufts, S. K. (2001) Software Re-Engineering of the Human Factors Analysis and Classification System b1s (Maintenance extension) using object oriented methods in a Microsoft Environment. Master of Science in Computer Science. Naval Postgraduate School: Monterey, CA, USA.

Fleishman, E. A., Quaintance, M. K. and Broedling, L. A. (1984) Taxonomies of human performance: The description of human tasks. Academic Press: Orlando, FL, USA.

Garr, J. S. (2007) Reduction of Human Factors-Related Accidents During the Flight Test of Homebuilt Aircraft Through the Application of Professional Flight Test Practices. Master of Science in Aviation Systems. University of Tennessee: Knoxville, TN, USA.

Gaur, D. (2005) Human factors analysis and classification system applied to civil aircraft accidents in India. Aviation, Space, and Environmental Medicine 76(5): 501-505.

Hardman, N. and Colombi, J. (2009) A Mapping from the Human Factors Analysis and Classification System (DOD-HFACS) to the Domains of Human Systems Integration (HSI). Technical Report (AFIT/EN/TR-09-04). Air Force Institute of Technology: Wright-Patterson Air Force Base, OH, USA.

Harris, D. and Li, W.-C. (2011) An extension of the Human Factors Analysis and Classification System for use in open systems. Theoretical Issues in Ergonomics Science 12(2):108-128.

Hawkins, J. H. (2010) Analytical HFACS for Evaluating Human Factors in an Aviation Accident. Master of Science in Technology. Purdue University: West Lafayette, IN, USA.

Hendy, K. C. (2003) A tool for Human Factors Accident Investigation, Classification and Risk Management. Technical Report (DRDC Toronto TR 2002-057). Defence R&D Canada: Toronto, ON, Canada.

Hsiao, Y., Drury, C., Wu, C. and Paquet, V. (2013) Predictive models of safety based on audit findings: Part 1: Model development and reliability. Applied Ergonomics 44(?):261-273.

Inglis, M., Smithson, M. J., Cheng, K., Stanton, D. R., Godley, S. T. (2010) Evaluation of the Human Factors Analysis and Classification System as a predictive model. Technical Report (Aviation Research and Analysis Report AR-2008-036). Australian Transport Safety Bureau: Canberra City, Australian Capital Territory, Australia.

Inglis, M., Sutton, J. and McRandle, B. (2007) Human Factors Analysis of Australian Aviation Accidents and Comparison with the United States. Technical Report (Aviation Research and Analysis Report – B2004/0321). Australian Transport Safety Bureau: Canberra ACT, Australia.

Jarvis, S. (2009) A Misjudged Approach to a High Accident Rate: Exploration of Accident Causes and Instructor Decisions Relating to Inexperienced Glider Pilots. Doctor of Philosophy in ?. Cranfield University: Cranfield, Bedfordshire, England, UK.

Johnson, W. (1997) Classifying pilot human factor causes in A-10 class A mishaps. Graduation in ?. Embry-Riddle Aeronautical University: Daytona Beach, FL, USA.

Kavade, H. (2009) A Logistic Regression Model to Predict Incident Severity using the Human Factors Analysis and Classification System. Master of Science in Industrial Engineering. Clemson University: Clemson, SC, USA.

Krulak, D. C. (2004) Human factors in maintenance: Impact on aircraft mishap frequency and severity. Aviation, Space, and Environmental Medicine 75(5):429-432.

Lenné, M. G., Ashby, K. M. and Fitzharris, M. (2008) Analysis of general aviation crashes in Australia using the human factors analysis and classification system. The International Journal of Aviation Psychology 18(4):340-352.

Leveson, N. G. (2004) A New Accident Model for Engineering Safer Systems. Safety Science 42(4):237-270.

388   Corrêa de Sá

Leveson, N. G. (2012) Engineering a Safer World: Systems Thinking Applied to Safety. MIT Press: Cambridge, MA, USA.

Li, W.-C. and Harris, D. (2005) HFACS analysis of ROC air force aviation accidents: Reliability analysis and cross-cultural comparison. International Journal of Applied Aviation Studies 5(1):65-81.

Li, W.-C. and Harris, D. (2006) Pilot error and its relationship with higher organizational levels: HFACS analysis of 523 accidents. Aviation, Space, and Environmental Medicine 77(10):1056-1061.

Li, W.-C. and Harris, D. (2006a) Breaking the Chain: An Empirical Analysis of Accident Causal Factors by Human Factors Analysis and Classification System (HFACS). 37th Annual International Seminar. 11-14 September 2006. Cancun, Mexico. International Society of Air Safety Investigators (ISASI). (proceedings pp:70-76)

Li, W.-C. and Harris, D. (2006b) Eastern Minds in Western Cockpits: A Cross-cultural Comparison of Aviation Mishaps by Applying Human Factors Analysis and Classification System (HFACS). 77th Annual Scientific Meeting. 14-18 May 2006. Orlando, FL, USA. Aerospace Medical Association (AsMA). (proceedings pp:?-?)

Li, W.-C. and Harris, D. (2013) Identifying Training Deficiencies in Military Pilots by Applying the Human Factors Analysis and Classification System. International Journal of Occupational Safety and Ergonomics 19(1):3-18.

Li, W.-C., Harris, D. and Chen, A. (2007) Eastern Mind in Western Cockpits: Meta-analysis of Human Factors Mishaps from Three Nations. Aviation, Space and Environmental Medicine 78(?):420-425.

Li, W.-C., Harris, D., Li, L.-W., Hsu, Y.-L. and Wang, T. (2010) The Investigation of Accidents Related to Aeronautical Decision-making in Flight Operations. 41st Annual International Seminar. 07-09 September 2010. Sapporo, Japan. International Society of Air Safety Investigators (ISASI). (proceedings pp:82-88)

Li, W.-C.n, Harris, D. and Yu, C.-S. (2008) Routes to failure: Analysis of 41 civil aviation accidents from the republic of china using the human factors analysis and classification system. Accident Analysis and Prevention 40(2):426-434.

Liu, S.-Y., Chi, C.-F. and Li, W.-C.n (2013) The Application of Human Factors Analysis and Classification System (HFACS) to Investigate Human Errors in Helicopter Accidents. 10th International Conference on Engineering Psychology and Cognitive Ergonomics (EPCE 2013). 21-26 July 2013. Las Vegas, NV, USA. HCI International. (proceedings pp:85-94)

Luxhøj, J. T. (2003) Probabilistic causal analysis for system safety risk assessments in commercial air transport. 2nd Workshop on Investigating and Reporting of Incidents and Accidents (IRIA 2003). 16-19 September 2003. Williamsburg, VA, USA. ? (NASA / Univ. Virginia).

Luxhøj, J. T. (2015) A socio-technical model for analyzing safety risk of unmanned aircraft systems (UAS): an application to precision agriculture. 6th International Conference on Applied Human Factors and Ergonomics (AHFE 2015). ? 2015. ?. ?.

Nelson, D. B. (2001) Information management system development for the investigation, reporting, and analysis of human error in Naval Aviation Maintenance. Master of Science in Information Technology Management. Naval Postgraduate School: Monterey, CA, USA.

O'Connor, P. (2008) HFACS with an additional layer of granularity: validity and utility in accident analysis. Aviation, Space, and Environmental Medicine 79(6):599–606.

O'Connor, P. and Walker, P. B. (2011) Evaluation of a human factors analysis and classification system as used by simulated mishap boards. Aviation, Space and Environmental Medicine 82(?):44-48.

O'Connor, P., Walliser, J. and Philips, E. (2010) Evaluation of a human factors analysis and classification system used by trained raters. Aviation, Space, and Environmental Medicine 81(10):957-960.

Olsen, N. S. (2011) Coding ATC incident data using HFACS: Inter-coder consensus. Safety Science 49(10):1365-1370.

Olsen, N. S. and Shorrock, S. T. (2010) Evaluation of the HFACS-ADF safety classification system: Inter-coder consensus and intra-coder consistency. Accident Analysis & Prevention 42(2):437-444.

Omole, H. H. (2015) Nigeria and UK helicopter accident: A cultural comparative analysis. Annual Meeting 2015. 14-16 October 2015. Gröningen, The Netherlands. Europe Chapter, Human Factors and Ergonomics Society.

Omole, H. H. (2016) Embedding cultural factors into the HFACS framework: Adding a cultural perspective to the Human Factors Analysis Classification System. GRIN Verlag: Munich, Germany.

Omole, H. H. (2016a) Extracting cultural relationships from helicopter accidents. GRIN Verlag: Munich, Germany.

Omole, H. H. (2016b) Cultural Framework: A new perspective to accident investigation and analysis. GRIN Verlag: Munich, Germany.

Omole, H. H. and Walker, G. H. (2015) Offshore transport accident analysis using HFACS. Procedia Manufacturing 3(?):1264–1272.

Omole, H. H. and Walker, G. H. (2015a) Offshore transport accident analysis using HFACS. 6th International Conference on Applied Human Factors and Ergonomics (AHFE 2015). 26–30 July 2015. Las Vegas, NV, USA. AHFE Conference.

Omole, H. H. and Walker, G. H. (2016) Embedding culture into accident investigation models. 7th International Conference on Applied Human Factors and Ergonomics (2016 AHFE International). 27-31 July 2016. Orlando, FL, USA. ?. (proceedings ? pp:?-?)

Omole, H. H., Walker, G. H. and Netto, G. (2014) Extracting cultural factors from helicopter accident reports using content analysis. 5th International Conference on Applied Human Factors and Ergonomics. 19-23 July 2014. Krakow, Poland. AHFE Conference. (proceedings pp:3-14)

Omole, H. H., Walker, G. H. and Netto, G. (2016) Exploring the Role of Culture in Helicopter Accidents. Human Factors in Transportation: Social and Technological Evolution Across Maritime, Road, Rail, and Aviation Domains. pp:271-295. G. Di Bucchianico, A. Vallicelli, N. A. Stanton, S. J. Landry (Editors). CRC Press: Boca Raton, FL, USA.

Omole, H. H., Walker, G. H. and Shappell, S. A. (2015) Helicopter Accident Analysis using HFACS-HE. 25th Annual European Safety and Reliability Conference (ESREL 2015). 07-10 September 2015. Zürich, Switzerland. European Safety and Reliability Association (ESRA).

Omole, H. H., Walker, G. H. and Shappell, S. A. (2015a) Helicopter Accident Analysis using HFACS-HE. Safety and Reliability of Complex Engineered Systems. pp:3061–3068. Luca Podofillini, Bruno Sudret, Bozidar Stojadinovic, Enrico Zio, and Wolfgang Kröger (editors). CRC Press: Boca Raton, FL, USA.

Oncu, M. and Yildiz, S. (2014) An analysis of human causal factors in Unmanned Aerial Vehicle (UAV) accidents. MBA in ?. Naval Postgraduate School: Monterey, CA, USA.

Paletz, S. B., Bearman, C., Orasanu, J. M. and Holbrook, J. (2009) Socializing the Human Factors Analysis and Classification System: Incorporating Social Psychological Phenomena into a Human Factors Error Classification System. Human Factors 51(4):435-445.

Patterson, J. M. and Shappell, S. A. (2010) Operator error and system deficiencies: Analysis of 508 mining incidents and accidents from Queensland, Australia using HFACS. Accident Analysis & Prevention 42(4):1379-1385.

Plourde, G. (1997) Human factor causes in fighter-bomber mishaps: A validation of the taxonomy of unsafe operations. Graduation in ?. Embry-Riddle Aeronautical University: Daytona Beach, FL, USA.

Pounds, J., Scarborough, A. and Shappell, Scott A. (2000) A human factors analysis of Air Traffic Control operational errors. Aviation Space and Environmental Medicine 71(?):329-332.

Rabbe, L. (1996) Categorizing Air Force F-16 mishaps using the taxonomy of unsafe operations. Graduation in ?. Embry-Riddle Aeronautical University: Daytona Beach, FL, USA.

Ranger, K. (1997) Inter-rater reliability of the taxonomy of unsafe operations. Graduation in ?. Embry-Riddle Aeronautical University: Daytona Beach, FL, USA.

Rashid, H. S. (2010) Human Factors Effects in Helicopter Maintenance: Proactive Monitoring and Controlling Techniques. Doctor of Philosophy in ?. Cranfield University: Cranfield, Bedfordshire, England, UK.

Rashid, H. S., Place, C. and Braithwaite, G. (2010) Helicopter maintenance error analysis: Beyond the third order of the HFACS-ME. International Journal of Industrial Ergonomics 40(6):636-647.

Rasmussen, J. (1997) Risk management in a dynamic society: A modelling problem. Safety Science 27(2/3):183-213.

Reason, J. (1990) Human error. Cambridge University Press: New York, NY, USA.

Reason, J. (1997) Managing the risks of organizational accidents. Ashgate Publishing Limited: Aldershot, England, UK.

RFAC (2013) Airworthiness Investigation Manual (A-GA-135-001/AA-001). Version 7. Royal Canadian Air Force (RCAF). Canada.

Rothblum, A., Wheal, D., Withington, S., Shappell, S. A., Wiegmann, D. A., Boehm, W. and Chaderjian, M. (2002) Improving Incident Investigation through Inclusion of Human Factors. 2nd International Workshop on Human Factors in Offshore Operations (HFW2002). 08-10 April 2002. Houston, TX, USA. ?.

Saarelainen, K. and Jäntti, M. (2015) A Case Study on Improvement of Incident Investigation Process. 22nd European Conference on Systems, Software and Services Process Improvement (EuroSPI 2015). September 30 - October 02 2015. Ankara, Turkey. ?. (proceedings pp:17-28)

Scarborough, A. and Pounds, J. (2001) Retrospective human factors analysis of ATC operational errors. 11th International Symposium on Aviation Psychology. 05-08 March 2001. Columbus, OH, USA. The Ohio State University. (proceedings pp:?-?)

Scarborough, A., Bailey, L. and Pounds, J. (2005) Examining ATC Operational Errors Using the Human Factors Analysis and Classification System. Technical Report (DOT/FAA/AM-05/25). Federal Aviation Administration (FAA): Washington, DC, USA.

Shappell, S. A. and Wiegmann, D. A. (1995) Controlled flight into terrain: The utility of models of information processing and human error in aviation safety. 8th Symposium on Aviation Psychology. 24-27 April 1995. Columbus, OH, USA. The Ohio State University. (proceedings pp:1300-1306).

Shappell, S. A. and Wiegmann, D. A. (1996) U.S. naval aviation mishaps 1977-92: Differences between single- and dual-piloted aircraft. Aviation, Space, and Environmental Medicine 67(?):65-69.

Shappell, S. A. and Wiegmann, D. A. (1997) A human error approach to accident investigation: The taxonomy of unsafe operations. The International Journal of Aviation Psychology 7(4):269-291.

Shappell, S. A. and Wiegmann, D. A. (1997a) A reliability analysis of the Taxonomy of Unsafe Operations. Aviation, Space, and Environmental Medicine 68(?): 620-?.

Shappell, S. A. and Wiegmann, D. A. (1997b) Why would an experienced aviator fly a perfectly good aircraft into the ground? 9th International Symposium on Aviation Psychology. 27 April-01 May 1997. Columbus, OH, USA. The Ohio State University. (proceedings pp:26-32).

Shappell, S. A. and Wiegmann, D. A. (1998) Failure analysis classification system: A human factors approach to accident investigation. Advances in Aviation Safety Conference and Exposition. 06-08 April 1998. Daytona Beach, FL, USA. Society of Automotive Engineers (SAE).

Shappell, S. A. and Wiegmann, D. A. (1999) Human error in commercial and corporate aviation: An analysis of FAR Part 121 and 135 mishaps using HFACS. Aviation, Space, and Environmental Medicine 70(?):407-?.

Shappell, S. A. and Wiegmann, D. A. (1999a) Human factors analysis of aviation accident data: Developing a needs-based, data-driven, safety program. 4th Annual Meeting of the Human Error, Safety, and System Development Conference (HESSD). ? 1999. Liege, Belgium. ?.

Shappell, S. A. and Wiegmann, D. A. (2000) Is Proficiency Eroding among U.S. Naval Aircrews? a Quantitative Analysis Using the Human Factors Analysis and Classification System. 44th Annual Meeting. 29 July-04 August 2000. San Diego, CA, USA. Human Factors and Ergonomics Society(HFES). (proceedings 44(27):345-348)

Shappell, S. A. and Wiegmann, D. A. (2000a) The human factors analysis and classification system (HFACS). Technical Report (DOT/FAA/AM-00/7). Federal Aviation Administration (FAA): Washington DC, USA.

Shappell, S. A. and Wiegmann, D. A. (2001) Applying Reason: The human factors analysis and classification system (HFACS). Human Factors and Aerospace Safety 1(1):59-86.

Shappell, S. A. and Wiegmann, D. A. (2001a) Beyond Reason: Defining the holes in the Swiss Cheese. Human Factors in Aviation Safety 1(1):59-86.

Shappell, S. A. and Wiegmann, D. A. (2003) A human error analysis of general aviation controlled flight into terrain (CFIT) accidents occurring between 1990-1998. Technical Report (DOT/FAA/AM-03/4). Federal Aviation Administration (FAA): Washington DC, USA.

Shappell, S. A. and Wiegmann, D. A. (2003a) Reshaping the way we look at general aviation accidents using the human factors analysis and classification system. 12th International Symposium on Aviation Psychology. 14-17 April 2003. Dayton, OH, USA. The Ohio State University. (proceedings pp:1047-1052).

Shappell, S. A. and Wiegmann, D. A. (2004) HFACS Analysis of Military and Civilian Aviation Accidents: A North American Comparison. 35th Annual International Seminar (ISASI 2004). 30 August-02 September 2004. Gold Coast, Queensland, Australia. International Society of Air Safety Investigators (ISASI). (proceedings pp:135-140)

Shappell, S. A. and Wiegmann, D. A. (2006) Developing a Methodology for Assessing Safety Programs Targeting Human Error in Aviation. Technical Report (DOT/FAA/AM-06/24). Federal Aviation Administration (FAA): Washington DC, USA.

Shappell, S. A., Detwiler, C., Holcomb, K., Hackworth, C., Boquet, A. and Wiegmann, D. A. (2007) Human Error and Commercial Aviation Accidents: An Analysis Using the Human Factors Analysis and Classification System. Human Factors: The Journal of the Human Factors and Ergonomics Society 49(2):227-242.

Shappell, S. A., Wiegmann, D. A., Fraser, J. R., Gregory, G., Kinsey, P. and Squier, H. (1999) Beyond mishap rates: A human factors analysis of U.S. Navy/Marine Corps TACAIR and rotary wing mishaps using HFACS. Aviation, Space, and Environmental Medicine 70(?):416-417.

Strauch, B. (2010) Can Cultural Differences Lead to Accidents? Team Cultural Differences and Sociotechnical System Operations. Human Factors: The Journal of the Human Factors and Ergonomics Society 52(2):246-263.

Taranto, M. T. (2013) A human factors analysis of USAF remotely piloted aircraft mishaps. Master of Science in Human Systems Integration. Naval Postgraduate School: Monterey, CA, USA.

Taylor, A. (2014) UK General Aviation Accidents: Increasing Safety through Improved Training. Doctor of Philosophy in ?. University of Leeds: Leeds, West Yorkshire, England, UK.

Ting, L.-Y. and Dai, D.-M. (2011) The identification of human errors leading to accidents for improving aviation safety. 14th International Conference on Intelligent Transportation Systems (ITSC). 05-07 October 2011. Washington, DC, USA. IEEE. (proceedings pp:38-43)

Tvaryanas, A. P. and Thompson, W. T. (2008) Recurrent error pathways in HFACS data: Analysis of 95 mishaps with remotely piloted aircraft. Aviation, Space, and Environmental Medicine 79(5):525-532.

Tvaryanas, A. P., Thompson, W. T., and Constable, S. H. (2005) U.S. Military Unmanned Aerial Vehicle Mishaps: Assessment of the Role of Human Factors Using HFACS. Technical Report (HSW-PE-BR-TR-2005-0001). 311th Performance Enhancement Directorate, United States Air Force. Brooks City-Base, TX, USA.

Tydeman, R., Cook, M., Conradi, K., Jarvis, M., Moss, S., Wivell, P. and Burrows, A. (2008) Aircraft Accident Report 7/2008. 94 pages. Air Accidents Investigation Branch, Civil Aviation Authority (CAA): London, England, UK.

von Thaden, T. L., Wiegmann, D. A. and Shappell, S. A. (2006) Organizational factors in commercial aviation accidents. The International Journal of Aviation Psychology 16(3):239-261.

Walker, G. H., Jenkins, D. P., Rafferty, L. A., Lenné, M. G., Stanton, N. A. and Salmon, P. M. (2011) Human Factors Methods and Accident Analysis: Practical Guidance and Case Study Applications. Ashgate: Farnham, Surrey, England, UK.

Walker, S. (1996) A human factors examination of U.S. Naval controlled flight into terrain 'CFIT' Accidents. Graduation in ?. Embry-Riddle Aeronautical University. Daytona Beach, FL, USA.

Wang, L., Wang, Y., Yang, X., Cheng, K., Yang, H., Zhu, B., Fan, C. and Ji, X. (2011) Coding ATC Incident Data Using HFACS: Intercoder Consensus. International Journal of Quality, Statistics, and Reliability 2011(?):1687-7144.

Wiegmann, D. A. and Rantanen, E. M. (2002) Defining the Relationship Between Human Error Classes and Technology Intervention Strategies. Technical Report (ARL-02-1/NASA-02-1). National Aeronautics and Space Administration (NASA): Hampton, VA, USA.

Wiegmann, D. A. and Rantanen, E. M. (2003) Defining the Relationship Between Human Error Classes and Technology Intervention Strategies. Technical Report (AHFD-03-15/NASA-02-1). National Aeronautics and Space Administration (NASA): Hampton, VA, USA.

Wiegmann, D. A. and Shappell, S. A. (1995) Human factors in U.S. Naval aviation mishaps: An information processing approach. 8th Symposium on Aviation Psychology. 24-27 April 1995. Columbus, OH, USA. Ohio State University.

Wiegmann, D. A. and Shappell, S. A. (1997) Human Factors Analysis of Post-accident Data: Applying Theoretical Taxonomies of Human Error. International Journal of Aviation Psychology 7(1):67-81.

Wiegmann, D. A. and Shappell, S. A. (1999) Human error and crew resource management failures in Naval aviation mishaps: A review of U.S. Naval Safety Center data, 1990-96. Aviation, Space, and Environmental Medicine 70(?):1147-1151.

Wiegmann, D. A. and Shappell, S. A. (2001) A human error analysis of commercial aviation accidents using the human factors analysis and classification system (HFACS). Technical Report (DOT/FAA/AM-01/3). Federal Aviation Administration: Washington DC, USA.

Wiegmann, D. A. and Shappell, S. A. (2001a) Applying the Human Factors Analysis and Classification System to the Analysis of Commercial Aviation Accident Data. 11th International Symposium on Aviation Psychology. 05-08 March 2001. Columbus, OH, USA. The Ohio State University.

Wiegmann, D. A. and Shappell, S. A. (2001b) Assessing the reliability of the human factors analysis and classification system (HFACS) within the context of general aviation. Aviation, Space, and Environmental Medicine 72(3):266-?.

Wiegmann, D. A. and Shappell, S. A. (2001c) Human error analysis of commercial aviation accidents: Application of the human factors analysis and classification system (HFACS). Aviation, Space, and Environmental Medicine 72(11):1006-1016.

Wiegmann, D. A. and Shappell, S. A. (2001d) Human error perspectives in aviation. The International of Aviation Psychology 11(4):341-357.

Wiegmann, D. A. and Shappell, S. A. (2003) A Human Error Approach to Aviation Accident Analysis: The Human Factors Analysis and Classification System. Ashgate: Burlington, VT, USA.

Wiegmann, D. A., Boquet, A., Detwiler, C., Holcomb, K. and Faaborg, T. (2005) Human error and general aviation accidents: A comprehensive, fine-grained analysis using HFACS. Technical Report (DOT/FAA/AM-05/24). Federal Aviation Administration (FAA): Washington DC, USA.

Wiegmann, D. A., Shappell, S. A., Cristina, F. and Pape, A. (2000) A human factors analysis of aviation accident data: An empirical evaluation of the HFACS framework. Aviation Space and Environmental Medicine 71(?):328-339.

Wood, B. P. (2000) Information management system development for the characterization and analysis of human error in Naval Aviation maintenance related mishaps. Master of Science in Information Technology Management. Naval Postgraduate School: Monterey, CA, USA.

Wu, J. and Zhao, T. (2011) C-HFAMF: A New Way to Accident Analysis Considering Human Factor. 9th International Conference on Reliability, Maintainability and Safety (ICRMS 2011). 12-15 June 2011. Guiyang, China. IEEE Reliability Society. (proceedings pp:319-322).

Zolla, G., Boex, T., Flanders, P., Nelson, D., Tufts, S. and Schmidt, J. K. (2001) Distributed Maintenance Error Information, Investigation and Intervention. Aerospace Congress and Exhibition. 10-14 September 2001. Seattle, Washington, USA. Society of Automotive Engineers (SAE): Warrendale, PA, USA.

Zolla, G., Flanders, .P and Boex, T. (2001a) Web-Based Information Management of Maintenance Errors in Aviation Mishaps. 4th International Conference on Electronic Commerce Research (ICECR-4). 08-11 November 2001. Dallas, TX, USA. Institute for Operations Research and the Management Sciences (INFORMS).

Integrating Data into the Safety Assessment Methodology for Defence

Louise Harney

Raytheon UK[1]

Harlow, United Kingdom

Abstract *The work of the Data Safety Initiative Working Group (DSIWG) has been progressing since January 2013 with the aim of integrating the assessment of data safety into the system engineering development process, which incorporates system safety assessment. While the Data Safety Guidance is being continually improved and public emphasis on data in our lives is growing, Raytheon UK is integrating the data safety assessment into the system safety assessment process. Raytheon UK's previous case study focussed on Air Traffic Management Systems (ATMS) but, since there are a multitude of safety assessment requirements which are customer-dependent within ATMS, this methodology paper focusses on defence and aims to develop sector-specific data safety guidance for defence engineering programmes in the UK.*

1 Introduction

1.1 What is Data Safety?

In an increasingly data-driven world, the systems we build and operate are dependent not only on hardware, firmware and software, but on the data they use and process. There are safety risks associated with all these system elements,

---

[1] This work was written while at Raytheon UK, but published while at PA Consulting. Raytheon UK have supported the development and review of this paper, and accepted its publication.

which means that hardware, firmware, software and data may all contribute to a system-level hazard. We need to be able to identify all these causes, and address them, to convince ourselves that we have adequately mitigated all the hazards. Within a standard safety process, data is not currently a focus and current safety management practices and standards do not adequately address the safety risks associated with data. By not considering data explicitly, we give a misleading view on the acceptability of each hazard we have identified because we have not accurately estimated the likelihood of it occurring due to all possible causes. We could be missing chances to reduce the likelihood of hazards occurring.

The Data Safety Initiative Working Group (DSIWG) was formed in January 2013 by the Safety Critical Systems Club (SCSC) to provide an industry-neutral body focussed on giving data appropriate emphasis and priority in system safety assessments. In the working group, data is often referred to as the elephant in the room because current safety assessments, even those which rigorously follow the tried and tested standards and are scrutinised by regulators worldwide, often (if not always) completely neglect consideration of the data used or processed by a system.

In general, there is growing public awareness of data in our lives, as evidenced by the recent Science Museum exhibition on *Our Lives in Data* (Science Museum 2016). More and more information is being collected and analysed which is used in all aspects of our lives. The general public is developing an awareness of data and its importance, although typically with a focus on data security. Alongside these general developments, standards such as Def Stan 00-055 (Ministry of Defence 2016) are gradually evolving to recognise the need for consideration of data in the safety management of our systems, closely linked to software assurance assessments which are already reasonably well-understood. Experience shows us that data errors have the potential to contribute to many accidents and incidents, with examples across all industry sectors, some of which can be found in the Data Safety Guidance (Data Safety Initiative Working Group 2016).

As part of the working group activities, a case study was performed on the Raytheon UK Mk3 Monopulse Secondary Surveillance Radar (MSSR) (The IET 2016) with the aim of understanding how the Data Safety Guidance may be used as part of a real developmental project. From this experience, Raytheon UK have been able to expand our working knowledge of the guidance and propose ways that the guidance can be used as an integral part of the Safety Management System (SMS) for defence projects.

Since this paper is written from the perspective of an equipment manufacturer, it focusses on the equipment system safety assessment process and does not consider the further work needed to integrate this into the operational safety assessments. That is an activity which should be considered by the working group in developing sector-specific guidance for defence.

1.2 Evolution of the Data Safety Guidance

For the purposes of this paper and the initial integrated system safety assessment methodology, version 1.3 of the Data Safety Guidance (Data Safety Initiative Working Group 2016) is referenced. However, the guidance is being continually improved as the DSIWG gains a greater understanding of how it will be used and how it could be more useful.[1]

Additional improvements will be needed to make the guidance easier to use within the time, resource, and budget constraints of each organisation. For example, an online tool or spreadsheet to enable tailoring of DSAL techniques to a system type and assurance requirement has been suggested and requires further investigation.

1.3 Step-by-step: Understanding how data fits into safety assessments

The following steps are being followed by Raytheon UK to understand the data safety 'problem' and ensure that data safety is appropriately considered in its safety assessments:

1. **Recognise the problem**

    Through analysis of recent accidents and theoretical working group discussions, it is recognised that there are specific risks associated with data in the systems that Raytheon UK build. We have both a duty of care and a professional responsibility to assess all sources of safety risk, including specific risks associated with data.

2. **Case study (The IET 2016)**

    Since the Data Safety Guidance is a new concept and requires additional development activity, conducting a case study tests the guidance to identify which parts are most useful. If separate integration is possible, the most useful aspects can then be prioritised for integration into the safety assessment methodology.

---

[1] The following ongoing updates have been considered as part of this methodology:
- Renaming of Data Integrity Levels (DILs) to Data Safety Assurance Levels (DSALs)
- Linking of the DSAL tables with Data Properties.

These updates to the guidance have been integrated in version 1.4 of the Data Safety Guidance, which was yet to be published at the time of writing this paper. The updated Data Safety Guidance will be available to download from the SCSC website: http://www.scsc.uk/p133.

3. **Make the guidance work (see section 1.2)**

The case study identified which aspects of the guidance may not work and proposed improvements to the guidance. There is an ongoing activity in the DSIWG to action these improvements and reissue the Data Safety Guidance, such that all the guidance is usable within existing mature system safety assessment frameworks.

4. **Write the process into safety assessment guidance**

Once the Data Safety Guidance is usable in its entirety, or the useful parts of the guidance have been identified, the data safety assessment process can be written into the Raytheon UK Safety Management System (SMS) to ensure that all system safety assessments undertaken by Raytheon UK will consider data. This paper is a step towards the goal of fully integrating data safety assessments into the Raytheon UK SMS. It may be necessary to have sector-specific aspects of the SMS in the context of data safety.

2 Approach: Integrating Data into the Safety Assessment Methodology for Defence

2.1 Scoping the Initial Activities

It is clear that all system safety assessments take place within organisational, budgetary, resourcing, and timeline constraints. This is why a key principle of system safety management is performing assessments which are commensurate with the perceived or allocated risk associated with the aspects being assessed. It is not helpful to have overly stringent safety assessment requirements which restrict our ability to improve the design of systems in practical ways, while recognising their ability to impact on operational safety.

The approach taken to establish an initial integration of data safety into the safety assessment methodology for defence systems recognises the clear need for prioritisation of the types and uses of data which are assessed. The previous case study on the Mk3 MSSR system (The IET 2016) confirmed that the types and uses of data in a system, even excluding the other interfaced systems, are so numerous that the assessment becomes disproportionately onerous compared with the potential safety risk associated with the system.

In scoping the initial activities, it is essential to consider how the data safety assessment links to both the current system safety assessments undertaken by Raytheon UK and to the system engineering development processes. Raytheon UK uses flexible processes for all projects tailored to ensure that customer needs can be met and to recognise the differences between each type of system. For example, a missile requires a very different type of safety assessment compared with a civil radar system. This paper would become extremely long if we attempted to discuss the links to each detailed safety assessment and provide justification for each safety activity undertaken. Therefore, this paper focusses on a generic application of Def Stan 00-56 (Ministry of Defence 2015), rather than a specific system development. It provides a starting point to develop more application-specific guidance, if that is required.

The data safety assessment will become part of the mature Safety Management System which Raytheon UK has tried and tested over many years, with many equipment deliveries and evolution of technology. The integration of data safety into this methodology should not remove any of the useful assessment processes we currently have in place, but should add to and improve upon them.

2.2 The current approach: Defence Standard 00-56

In providing an overview of the current approach to system safety assessments from the perspective of an equipment manufacturer, many details must be left out in the interests of brevity. It is important to remember that all system safety assessments have a complex range of stakeholders, each of whom holds a specific system safety responsibility. System safety is not limited to the assessment described in this paper, i.e. this paper assumes that the equipment manufacturer already has in place a mature SMS with the provisions to allow for continual improvement of the safety assessment processes (as is the case for Raytheon UK). For those readers not familiar with the concept of defence system safety assessments, the Ministry of Defence has prepared an introductory briefing paper (Ministry of Defence 2011) that represents, in the Author's opinion, a best practice approach to safety which can in many ways be read-across to other industry assessments.

For simplicity sake, this paper assumes that a Waterfall system development is being undertaken. It is recognised that many system developments, particularly in defence when responding to an Urgent Operational Requirement (UOR), require an iterative system development process. The system safety assessment is also iterative and, to place maximum focus on the inclusion of data within the system safety assessment process, this paper avoids the inclusion of multiple iterative processes which become extremely complex once combined. The approach can, however, be tailored to take into account different system development processes.

Figures 1 to 3 provide an overview of the existing safety assessment approach which has been tailored to an equipment manufacturer's perspective from Def Stan 00-56 (Ministry of Defence 2015).

Def Stan 00-56 is split into three main sections:

1. Safety Management
2. Safety Engineering
3. Safety In-Service

Separate diagrams are provided for each of these to avoid a single over-complicated diagram.

Key Table 1 provides a key to understanding the colour-coding used in each diagram in this paper. Note that all activities taken from Def Stan 00-56 have been paraphrased to reduce complexity of the figures.

Table 1. Key

Symbol	Corresponds to
Monitoring and reporting on operational performance and incidents	Activity out of scope of or higher level than equipment manufacturer role
Analyse incident reports provided by Operator / Duty Holder	Activity conducted by equipment manufacturer

Figure 1 provides an overview of the Def Stan 00-56 safety management process.

Fig. 1. Def Stan 00-56 Safety Management Process

Safety management is an iterative process throughout the lifecycle of an equipment development and deployment. We must constantly update our safety planning to recognise the identified safety issues and the evolving project requirements. Figure 1 does not attempt to show these iterations in detail. It remains engineering judgement to ensure that the safety planning is updated regularly enough to remain current. At least annual reviews are recommended.

Product line safety planning may include preparation for compliance with increasingly stringent safety standards which are introduced following improved understanding of the potential safety issues present in our systems. One example of this is the inclusion of data safety.

While safety culture is not a measurable achievement, it is included in Figure 1 due to its importance in the safety management process. Inadequate buy-in from the senior management is often a causal factor leading to accidents. For example, the accident on the Smiler ride at Alton Towers which resulted in 16 people being injured[1], including two amputations, was in part blamed on a lack of safety culture in which management failed to provide the required training in safe operation of the ride. Not only has this had a devastating effect on the victims, it has cost Alton Towers' owners £5 million in fines and has caused severe reputational damage following extensive press coverage. To protect reputation and financial investment, safety culture needs to include a willingness to evolve with growing understanding.

Figure 2 provides an overview of the Def Stan 00-56 safety engineering process.

---

[1]https://www.theguardian.com/business/2016/sep/27/alton-towers-owner-fined-smiler-rollercaster-crash

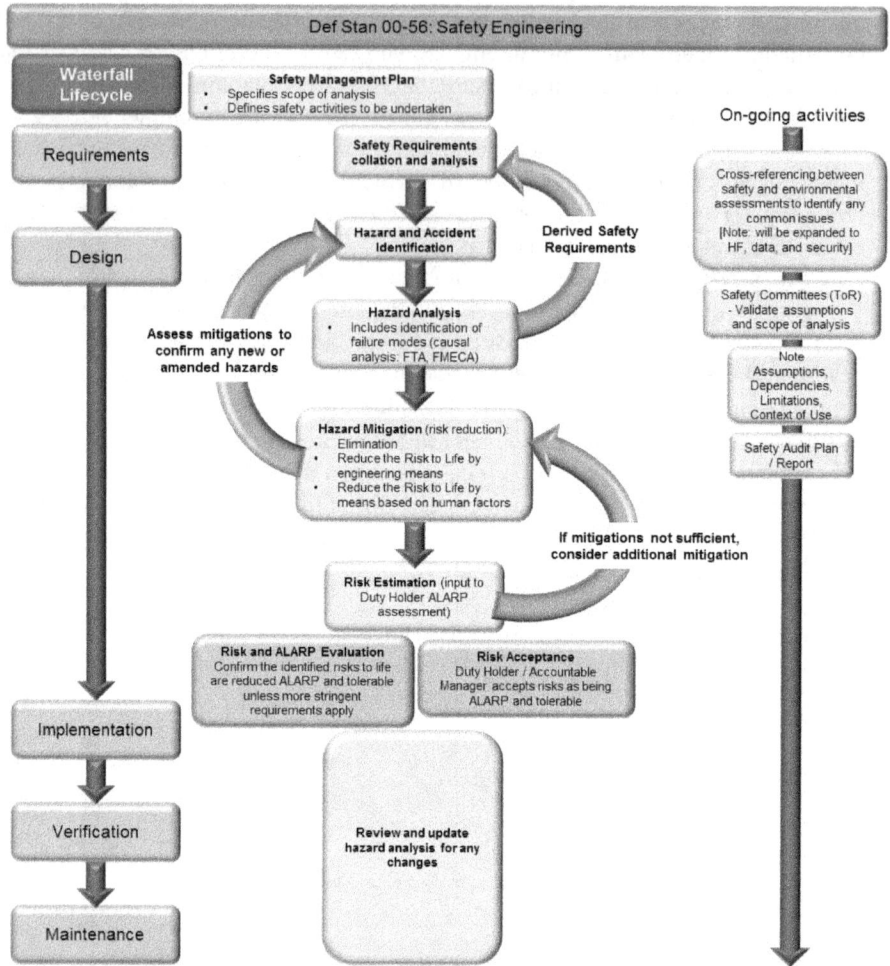

Fig. 2. Def Stan 00-56 Safety Engineering Process

Figure 2 above focusses on the design phase of the Waterfall lifecycle since the tasks required during design are typically updated during the later phases of implementation, verification, and maintenance. Operational use is analysed during the design phase to ensure the equipment is designed for its intended use.

The following system safety assessment steps are required:

1. Safety Planning
    Safety planning identifies the system to be assessed, the scope of the assessment and the detail to which the system needs to be considered to provide an adequate understanding of the potential system safety risks.

2. Requirements Analysis
    Safety requirements are collated from numerous sources, including relevant
    Ministry Of Defence (MOD) Policy; requirements set by MOD (e.g. in the
    User Requirements Document or System Requirements Document); re-
    quirements from legislation, regulations, standards and guidance; high-
    level safety requirements from operational hazard analysis (initial assess-
    ment); and derived safety requirements from the equipment hazard analysis
    (see the iteration arrow on Figure 2).

3. Hazard Identification and Analysis
    All plausible hazards caused by the equipment or its use, storage, or
    transport are identified in a number of complementary ways (e.g. top-down
    functional analysis of the system's intended use; bottom-up analysis of the
    system components; comparison with suitably similar existing systems in-
    service (carefully considering the context and environment in which these
    systems are used); and brainstorming during hazard review meetings).
    Hazards are analysed in accordance with agreed safety criteria in order to
    prioritise which hazards most require mitigation.

4. Hazard Mitigation
    Potential methods of mitigating the identified hazards are considered, based
    on a prioritisation of the following high-level methods (Ministry of De-
    fence 2015):
    *   Elimination
    *   Reduce the Risk to Life by engineering means
    *   Reduce the Risk to Life by means based on human factors
    Any mitigations implemented must be reassessed to confirm whether they
    introduce any new hazards or alter existing hazards.

5. Risk Estimation
    The safety committee reviews the hazards identified and all mitigations
    implemented to assess whether the mitigations are believed to sufficiently
    reduce the risk associated with the hazard. If the mitigations are not con-
    sidered to be sufficient, additional risk mitigation will be required prior to
    placing the equipment on the market. From the equipment manufacturer
    perspective, input is provided to the Duty Holder who is responsible for ac-
    cepting the risk associated with the system, and agreeing that the identified
    risks have been reduced to a level which is at least tolerable and As Low
    As Reasonably Practicable (ALARP).

Results of the system safety assessment are recorded throughout the assessment in working drafts of the Hazard and Accident Log and the Safety Case.
These repositories are updated each time the assessment progresses and will

typically be reported at significant project milestones (e.g. System Requirements Review (SRR), Critical Design Review (CDR)).

Raytheon UK typically provides a single repository of safety information (the Safety Case, summarised in a Safety Case Report at the time of issue) in support of defence programmes, reducing potential duplication while providing the necessary input to the Duty Holder's assessment of Risk to Life[1].

Since Raytheon UK is an equipment manufacturer, most of the Safety In-Service process described in Def Stan 00-56 is out of the managerial control of Raytheon UK. The provider of the service and/or the Duty Holder are responsible for safety in-service. Figure 3, below, shows the support provided by an equipment manufacturer on a typical defence programme

.

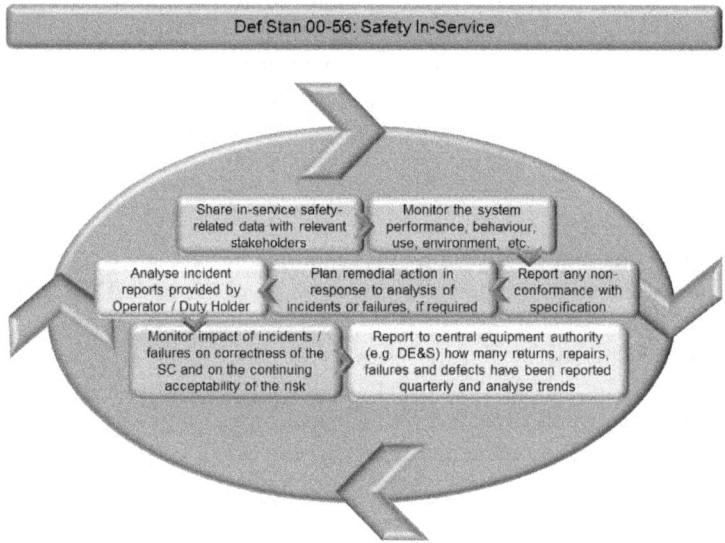

Fig. 3. Def Stan 00-56 Safety In-Service Process

The purpose of safety in-service is to monitor the system once it has been introduced into operations to validate the safety case, and to ensure that no changes are introduced to the system without first being assessed in terms of their potential impact on safety.

When an excessive number of failures are detected, or an incident is reported, it is essential to review the safety case to confirm the following:

---

[1] Note that the Def Stan 00-56 requirement for three different types of documents (Command Summary, Information Set and Information Set Safety Summary (ISSS), and Safety Case and Safety Case Report (SCR)) can be met by providing some or all of the Safety Case information, exported in the required format. Note also the requirement to refer to equipment-level safety information as 'safety assessment' for air platforms.

a)   Was the cause of the failure/incident equipment related?
b)   Was the cause of the failure/incident already identified in the hazard analysis?
c)   Should an equipment detection measure have alerted the user before the failure/incident occurred?
d)   Are the numbers of recorded failures/incidents much higher than predicted/tolerable?

In answering these questions, the cause of an incident can be investigated from the equipment manufacturer perspective. The investigation will assess whether any additional mitigations are required and agree a path for their implementation. Interim measures may need to be provided to the operator to ensure the safety of the wider operations. The safety case should be reviewed in light of reported incidents and updated as required.

3 Integrated Safety Assessment Methodology for Defence

Having described the high level system safety assessment methodology for defence, based on Def Stan 00-56, the next step is to show the integration of the Data Safety Guidance with this methodology such that a single integrated system safety assessment can be undertaken. There are many risks associated with conducting assessments in isolation, most notably the potential for misalignment between the assessments (e.g. a mitigation implemented in one assessment causes a new hazard in the system or impacts an existing hazard). There are also many benefits of conducting an integrated assessment far beyond simply mitigating these programmatic risks. An integrated assessment is quicker, cheaper, easier to understand, and importantly enables reuse of much of the read-across material (e.g. the system description).

This paper is a simple representation of a relatively generic process. It shows how simply data safety can be integrated into the safety assessment methodology for defence, since the activities required to perform a data safety assessment are already done for hardware and software/firmware, just with a different focus.

Figure 5 below provides an overview of the high-level integration of data safety into the safety assessment methodology for defence, providing some examples of the system safety assessment activities required without expanding all the detail of the safety engineering in the design phase. Figure 4 shows a basic safety framework set of activities which cover all system aspects, and a set of specific focus areas (within brackets) upon which the system safety assessment is built. It is essential that the interface between each of these focus areas is considered, since they can influence each other. For example, software with a low integrity processing data of a high integrity may result in the output of low integrity data (i.e. the data has been changed or corrupted in some way, such that it would no longer be suitable to be used for a safety-critical application).

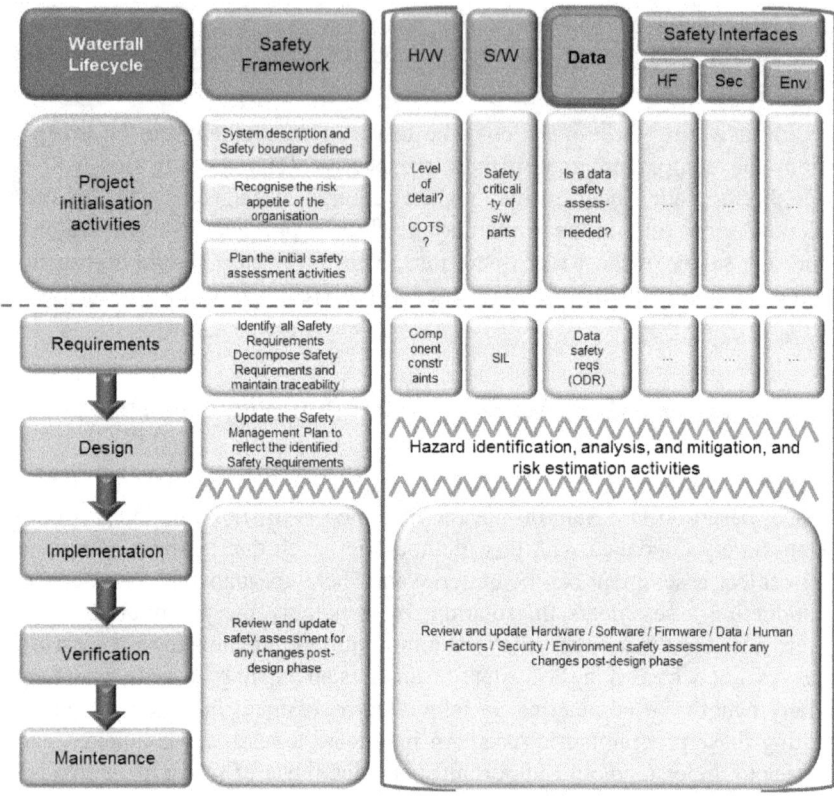

Fig. 4. Data safety assessment approach mapped to Def Stan 00-56

This Section sets out the introduction of an assessment of data into the overall safety assessment. Figures 5 to 7 below update Figures 1 to 3 respectively in Section 2.2 to illustrate the interface of data safety into the existing safety assessment approach.

The approach diagram from the Raytheon UK ATMS Case Study (The IET 2016) provides a starting point for the integration of data safety within the existing system safety assessment process, which has been expanded to ensure all aspects, such as safety planning, are included in this methodology.

Figure 5 below shows the safety management process with simple links to data safety. The same safety management activities are required but augmented with an awareness of data safety, with the recognition that data can have a severe safety impact. Attitudes towards assessing data safety need to fit within a *just data safety culture*, in exactly the same way as we expect organisations to have a *just safety culture*.

Fig. 5. Def Stan 00-56 Safety Management Process with data focus

The Organisational Data Risk assessment, detailed in the Data Safety Guidance (DSIWG, 2016), could be used at product line level to provide an initial understanding of the approximate level of data safety risk associated with that product line, and therefore how much time, resource, and budget should be made available to the product development teams to ensure that the data safety risks are adequately controlled, commensurate to the estimated level of risk.

The Safety Management System (SMS) requires augmentation to explicitly refer to data safety but, as described in the sections that follow, no separate data safety assessment activities are required. In that sense, it will be an organisational decision whether or not it is appropriate to explicitly refer to data safety in the SMS. This is because many organisations will state only the process-based requirements in the SMS and the Suitably Qualified and Experienced Personnel (SQEP) system safety team are expected to analyse all relevant aspects of the system, such as hardware and software/firmware.

Figure 6 below shows the proposed integration of data safety within the Def Stan 00-56 safety engineering process.

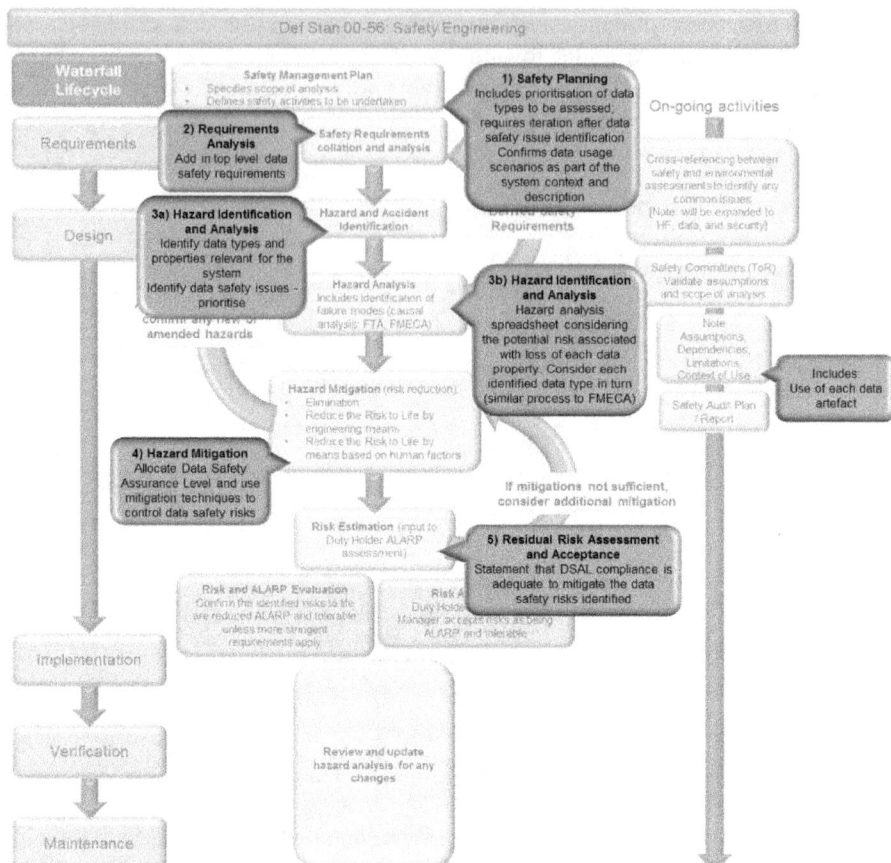

Fig. 6. Def Stan 00-56 Safety Engineering Process with data focus

The data safety assessment fits within the current safety assessment process and the existing safety assessment documentation, in the following ways:

1. Safety planning
    The system and safety boundaries identified as part of the system-level safety planning are the same for the data safety assessment. The safety plan retains the existing structure, with a few additional considerations. The safety activities planned need to be augmented to include the extension of hazard analysis to identification and analysis of data types, associated properties and the potential data safety risks. Without already having identified the data types present in the system, it is very difficult to estimate the required time, resource, or budget. Therefore, the safety plan including data requires more iteration than planning for 'traditional' (hardware and software/firmware) assessments. After identifying the data types, the plan

should be updated to reflect the prioritisation of data safety. It will now be known approximately how many data safety risks require detailed assessment, and thus approximately how long that assessment will take. We typically do not need to do this for hardware and software/firmware because it is easier to identify and structure (e.g. through Hardware Configuration Items (HWCIs) and Computer Software Configuration Items (CSCIs)), whereas data can be hidden in the system.

2. Requirements Analysis
    The top-level safety process requirements need to mandate that all causes of hazards are analysed and the associated risk mitigated to at least a tolerable and ALARP level. These include the data safety issues. More detailed data safety requirements will be derived through the hazard analysis and mitigation (i.e. mandating certain data error mitigation from the DSAL techniques tables), so iteration of the requirements analysis is needed, as is already done for hardware and software/firmware.

3. Hazard Identification and Analysis
    The initial identification activity needs to list all data types and data artefacts in the system. Identification of data types can follow a dual approach: bottom-up identification of all data types present in the system by design, and top-down identification of all data types required to perform system functions and sub-functions. The list of data types from the Data Safety Guidance should be reviewed to ensure this identification activity has been complete.
    Following the initial identification, the list of data properties from the Data Safety Guidance should be reviewed and applicable data properties associated with each data artefact should be listed (and linked to the data artefact). The guidance suggests a set of guidewords linked to data properties which can be used for identifying data safety issues. The set of guidewords should be tailored to the identified applicable data properties and recorded in the hazard analysis documentation.
    The guidewords should be used in the same way as existing hazard analysis, i.e. statements should be made for each data artefact to lose each data property. For example, processed altitude data is not correct. Processed altitude data is the data artefact, and the guideword is correct, corresponding to the data property integrity. An example of this assessment activity was included in the previous Raytheon UK case study paper (The IET 2016). Table 3 below provides a basic structure for recording the data safety issue identification.

Table 3. Data safety issue identification

Data Type	Data Artefact	Data Property	Guideword	Data Safety Issue	Severity
Output Data					
Processed Aircraft Information		*Data Type refined to make the data safety issue identification more detailed*			
Altitude Data	Processed Altitude Data	Integrity	Correct	Processed altitude data is not correct	Catastrophic
Altitude Data	Processed Altitude Data	Integrity	Coherent	Processed altitude data is not coherent	Major
			ETC		

Additional columns could be added to estimate the likelihood of the data safety issue and state the corresponding DSAL to be allocated to that data safety issue. The table would then also be useful for the hazard mitigation activity. The table will become very large when it includes all data types, data artefacts, and data properties.

The assessment can be a time consuming activity, so it could be automated using a tool and then reviewed by the system safety expert, or a prioritisation decision could be made early in the process (e.g. to disregard certain types of data which are already known not to be safety-critical). Early prioritisation decisions are not typically recommended since the use of data can change its safety-criticality and, without an initial assessment, it is difficult to have confidence that the prioritisation decision is valid.

For completeness, the list of generic data safety issues contained in the Data Safety Guidance should be reviewed to ensure all relevant issues have been covered through the hazard identification phase.

Data safety issues are then categorised based on the data safety criteria (the Data Safety Guidance contains some criteria, which should have been tailored to the system during the safety planning phase). The categorisation enables prioritisation of the data safety issues, which enables us to update the safety plan.

Further detailed analysis of the data causes, linked to each priority data safety issue, is conducted by first linking the data safety issue to a system-level hazard, proposing the data-related causes, and taking this discussion to a Hazard Review Meeting (already held as part of the 'traditional' system safety assessment).

4. Hazard Mitigation

Similarly to software, data errors do not occur in the same way as hardware failures, meaning that a failure rate allocated to data would not make sense. Instead, we use Data Safety Assurance Levels (DSALs) which can be thought of as equivalent to Safety Integrity Levels (SILs) for software. DSALs enable us to assure ourselves that we have developed our data and methods of storing, processing and using that data to a level which is appropriate for the risk associated with that data.

DSALs are allocated to each data safety issue (i.e. they are linked to the loss of a data property from one data artefact). This means that we may have many different DSAL allocations within our system. Once the DSALs have been allocated to the priority data safety issues, we can group the DSAL allocations together where they relate to the same data property. The potential mitigation techniques in each data property DSAL table (see the Data Safety Guidance) should be reviewed to decide whether these should be implemented. The mitigation techniques identified should be required by iterating the requirements analysis (i.e. documenting derived requirements of the data-related hazard analysis). The combination of all mitigation techniques proposed with the 'traditional' assessment for each system-level hazard needs to be discussed at a Hazard Review Meeting to determine if it is sufficient to reduce the risk associated with that hazard.

5. Risk Estimation

Risk estimation provides input to the Duty Holder's process to accept (or otherwise) the risks associated with bringing a new or updated system into service. This includes the risks associated with data. In order to claim that sufficient mitigation of the data safety issues has been implemented, compliance with the allocated DSALs, and an assessment that the DSAL compliance is sufficient to mitigate the specific data safety issues identified needs to be made. If the mitigation techniques identified are insufficient, additional mitigation analysis is needed.

To record the data safety assessment, existing safety assessment documentation would be augmented with the additional focus on data. This ensures that there is an obvious need to assess the interfaces between data, hardware and software/firmware. For example, data safety fits into the existing safety argument structures documenting the safety activities described above.

In the hazard analysis phase, it may be necessary to record the results of the data safety assessment in a separate tool (e.g. an MS Excel Spreadsheet) since the analysis may be lengthy and detract from the top-down functional analysis traditionally conducted as part of the hazard analysis. The outcomes of the data safety assessment, including identified data safety issues (typically causes of system level hazards) and mitigations, should be integrated into the Hazard and Accident Log for the system.

Figure 7 below shows the integration of data safety within the Def Stan 00-56 safety in-service process. Similarly to the safety management process links shown in Figure 5, no new activities are required to manage data safety in-service, but a shift of focus is required to ensure that all possible causes of reported failures or incidents are assessed, including data errors.

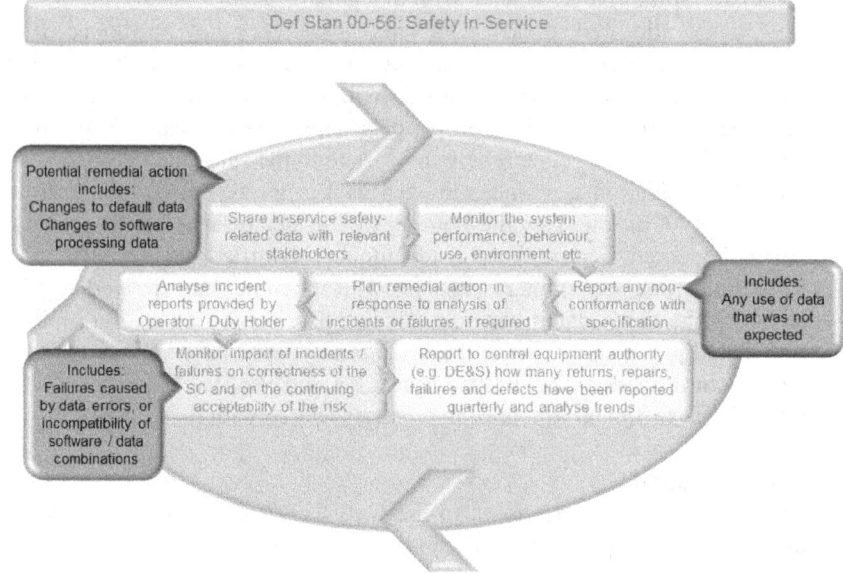

Fig. 7. Def Stan 00-56 In-Service Process with data focus

Data safety can be monitored using the same methods as all other analysis of field data, in particular through the reporting and analysis of incidents. The potential data-related causes of incidents should be reviewed each time an incident or failure report is received. It should then be assessed whether this risk was already included in the hazard analysis and whether the occurrence suggests the hazard analysis was not correct. Potential remedial actions also need to consider the potential to implement data safety mitigation techniques.

4 Next Steps

Def Stan 00-56 states the goal that *civil or open standards should be used as the basis for complying with its requirements* (Ministry of Defence 2015). The clear next step based on this goal is to map the data safety assessment approach to each industry-specific civil or open standard which is typically used and consider any appropriate *military delta* (the additional considerations necessary for a military use of the system) to provide guidance for each type of system provided for defence applications, such as aircraft; airborne systems such as Identification Friend or Foe (IFF); land systems such as Improvised Explosive Device (IED) detection devices; etc. There is a significant amount of work involved here. A potential starting point would be to assess whether the process proposed in this paper meets the recommendations of MIL-STD-882 (Department of Defense 2012), which is specifically referenced as an open standard in Def Stan 00-56.

The process presented in this paper is by no means exclusively relevant to Def Stan 00-56, or indeed to defence systems. It is a relatively generic process which could be utilised in much wider applications, at least in concept. While the terms may change, the general approach to safety assessment does not. The Author and Raytheon UK would greatly appreciate any feedback on the process presented here, which is being integrated into the Raytheon UK SMS as part of continual improvements to the SMS.

In parallel to the development of sector-specific guidance for defence, guidance is being developed as part of the DSIWG activities specific to other industries, such as healthcare. The upcoming working group meetings will be used to establish the lessons learned from these activities, gaining insights from all industries and collaboratively updating the Data Safety Guidance.

Whilst the sector-specific guidance will be continually improved over the coming months, it is highly recommended that organisations across the defence industry use the Data Safety Guidance and this proposed integrated safety assessment methodology and test its implementation. The more case studies that can be gathered, the better the Data Safety Guidance will become, making it more useful and providing a more realistic understanding of the safety of the systems on which our armed forces rely.

Acknowledgments   With thanks to Eric Bridgstock, Head of Engineering Safety, Raytheon UK, and the DSIWG for their support and encouragement in developing this methodology.

References

Department of Defense (2012) Standard Practice System Safety, MIL-STD-882E

L Harney (2016) Implementing the Data Safety Guidance, proceedings of the 11th International Conference on System Safety and Cyber Security, The IET[1]

Ministry of Defence (2016) Defence Standard 00-055 Part 1, Requirements for Safety of Programmable Elements (PE) in Defence Systems, Issue 4

Ministry of Defence (2015) Defence Standard 00-56 Part 1, Safety Management Requirements for Defence Systems, Issue 6

Ministry of Defence (2011) An Introduction to System Safety Management in the MOD, Issue 3

---

[1] Yet to be published. An oral presentation of this paper will be made in October 2016.

Science   Museum   (2016)   Our   Lives   in   Data
http://sciencemuseum.org.uk/visitmuseum/plan_your_visit/exhibitions/our-lives-in-data.
Accessed 12 September 2016

The Data Safety Initiative Working Group (DSIWG) (2016) Data Safety Guidance, the UK Safety Critical Systems Club (SCSC)

Cybersecurity problems in a typical hospital (and probably in all of them)

Harold Thimbleby

Swansea University

Wales

Abstract   *A criminal case balancing on the corruption of patient data in a UK hospital resulted in some nurses being acquitted and some given community service and custodial sentences. This paper explains the background, demonstrates the inability of hospital IT systems to provide reliable evidence, and highlights broader problems with IT culture affecting manufacturers, hospitals, police, legal advisors — and ultimately misleading clinicians and compromising delivery of care.*

The NHS (and healthcare more generally) urgently needs to improve its IT awareness, management and policies. The police and the legal system need a more mature approach to IT. Manufacturers need to provide dependable systems that are fit for purpose for complex hospital environments. Regulators should ensure that systems meet better standards of quality and dependability.

This paper includes recommendations; the most fundamental being that hospitals acknowledge that IT is unreliable and they should procure and manage equipment with this in mind. In particular, mature and effective data protection and cybersecurity policies must be in place and used proactively. When problems occur, evidence derived from IT (whether systems or devices) must not be used in legal or disciplinary investigations without extreme care and independent proof of provenance.

1 Introduction

This paper summarizes my insights from being an expert witness in a criminal case involving alleged fabrication of patient data by nurses.

The outcome and details of the court case are in the public domain, but the aim of this paper is not to tell a story about a hospital or its nurses, but rather to tell a more worrying story: *this could happen anywhere — and probably is happening everywhere.*

The court case collapsed because prosecution evidence was derived from flawed data, flawed IT and flawed management of IT. Nurses were blamed, but the underlying causes must be understood as basic cybersecurity issues that should have been taken seriously as such when they happened. The allegedly incriminating data and the later corruption of data (the final understanding of which led to the trial collapsing) should have been detected as and when they happened.

One wonders how many other cases inappropriately pursue clinicians caught up in fallout from IT chaos, with nobody recognizing or wanting to admit or check that IT can cause such problems. It is worrying that the case here very nearly did not end at all happily, and would not have ended as well as it did without a lot of work correcting widespread misunderstandings of IT — and, harder!, correcting *widespread unconscious and unintentional misunderstandings* of IT.

Although we are not criticizing hospitals or their staff, in view of understandable sensitivities this paper does not provide any citations to the court case or to related evidence. We have avoided using identifiable names in this paper, though we have not changed technical details or standard procedures. However, numbers have been rounded and ward names changed, etc. Although the blood glucometers and databases are made by a company we will call TechCo, we do not think this company is egregious: we think their products are of typical quality and design for the industry. The story is therefore representative of the industry and its regulation, not about any one company; similarly, the story is not about one hospital nor about any nurses in it; it is about *all* hospitals and their staff, and what can unwittingly happen.

At a higher level the story is about the widespread misunderstanding of IT in healthcare, and in particular, about mismanagement of IT by hospitals and by the police. Note that TechCo's systems are used across the NHS and worldwide.

The one sentence take home is that there must be effective, mature procedures and understanding in place to detect and manage cybersecurity problems before they trigger catastrophes. The story here fortunately involved no patient harm or malicious hacking, but that was only by luck. Healthcare cyberattacks are "growing exponentially" (Davidson, 2016) — with 113 million US electronic health records breached in 2015. Pure luck protecting staff and patients cannot hold out for long.

2 A public perspective

Concerns about the quality of patient care in a hospital ward led to a police investigation. For the criminal investigation, the police focused on the treatment of vulnerable adults, which may be criminal even though there is no patient harm. Indeed in this case there was no patient harm caused by poor care.

There was considerable political and public interest in this case, particularly since the powerful and high-profile criticism of Robert Francis's 2013 *Report of the Mid Staffordshire NHS Foundation Trust Public Inquiry* ("The Francis Report"); nobody wanted another Mid Staffs.

The news reported stories of many — over 50 — nurses being investigated and soon of an imminent court trial. (I did not follow the internal investigations.) The nurses were alleged to have fabricated blood glucose readings (that is, not having actually taken any readings from patients) and then written them up in paper patient notes. For vulnerable adult patients this would be criminal. The implication was that the nurses were lazy and dishonest and had put patients at risk. Publically, it was known that three nurses pleaded guilty, but two pleaded not guilty and their case proceeded to a jury trial.

TechCo's blood glucometers are used in the hospital, and they automatically upload glucose readings to a central patient record system. The police established that the central record system had no records of many tests the nurses had written on patient paper notes. Therefore the police concluded that the nurses had written down fictitious readings and not bothered to do their job properly.

The police were thorough in their investigations and considered various ways the nurses may have made accidental errors. The police compared paper records with a computer database, involving around 150,000 test records — a great deal of combined manual and computer work!

In addition to identifying alleged fabrication (that is, paper records with no corresponding computer records) the police also found evidence of many cases of poor operating practice. For example, a nurse is supposed to enter the patient's ID, but sometimes a nurse will scan their own staff card instead. This is easier and enables the glucometer to work, so the nurse can quickly obtain a test reading. From the computer records it is clear there are many cases of this practice. (The next section, below, describes in more detail what nurses are supposed to do.)

The accused nurses had followed such bad practice repeatedly, and under the UK Criminal Justice Act this was considered evidence of "bad character." Put briefly this means that if proven guilty your sentence may be harsher: not only are you guilty of the crime, but you are a bad person. The bad character concept makes sense when a crook is arrested for one, perhaps relatively petty, crime but nevertheless is known to be a hardened criminal.

The prosecution argued that there were no problems with the equipment. There are national databases in the UK and USA for reporting problems, and no related problems had ever been reported with TechCo's systems. Therefore this was a nursing problem, not a system problem, they argued.

The case thus went to trial ... but weeks later the trial collapsed. The two nurses were released.

TV crews were there and filmed a patient victim group protesting outside the court. The media presented the collapse of the trial as a failure; as if the nurses were still guilty because the trial only collapsed on legal technicalities.

The other nurses who had previously pleaded guilty did not change their plea and were later sentenced, some to community service and some to prison.

3 What do nurses do on the ward?

To help patients manage blood glucose levels (particularly if the patients have limited capacity to look after themselves) it is important to take and record blood glucose test readings. Using the TechCo blood glucometer, an outline of the operating procedure is follows:

1. Find a glucometer;
2. The nurse then identifies themselves to the device (by scanning the barcode on their staff card or by typing their ID);
3. The patient ID is scanned from their barcode or typed;
4. Scan a glucose test strip, clean the patient's finger; the patient is pricked and a drop of blood placed on the test strip;
5. The test strip is inserted in the glucometer;
6. The glucometer displays the blood glucose level (or possibly an error);
7. The nurse may then take immediate action to address any clinical issues;
8. The nurse then "contemporaneously" writes down on the paper patient notes the time and reading;
9. One further step, that has no immediate clinical significance, is that the glucometer must be placed in a dock, and then its data will be automatically uploaded to TechCo's central systems.

The TechCo glucometer itself can record over 2,000 readings before it needs to be docked, and it will warn if its memory is full. This memory feature means that TechCo's glucometer can be used for batching tests: a nurse can test a whole ward, respond to patients' needs, then later write up the results using the glucometer review screen to recollect individual test details.

Once successfully uploaded to TechCo's systems, the test readings will later appear in the main patient records available on ward PCs — however, this might take days or longer (see below).

3.1 Patient ID workarounds

Barcode workarounds are a well-known problem (Koppel, et al, 2008), and indeed sometimes it is hard to scan or read the patient ID, so one workaround is to type 000 etc on the glucometer keyboard, or more easily just scan the staff ID barcode again just to get a number the glucometer accepts as a valid "patient ID." Using the staff ID instead of the patient ID is called "double tapping" and (perhaps) coincidentally the number of nurses in the database who double tapped at least twice in the period for which I have data is approximately equal to the number of nurses originally investigated.

The glucometer accepts both of these workarounds (arbitrary IDs and double tapping) and will still give a correct blood glucose reading. Behind the scenes, however, the hospital had configured its systems to reject this data, which then requires manual intervention (which may never happen, or may introduce further errors) to fix it. It should be noted that configuring the system in this way makes sense: the blood glucose reading in the database cannot be reliably associated with any particular patient in the database until the issue is manually resolved.

I would argue that double tapping and other workarounds, since they are trivial to detect, should be sorted out immediately rather than ignored — TechCo's systems can be configured to detect these problems immediately they occur. Since double tapping was so prevalent, I believe nurses were not aware it was problematic nor that it was being monitored. Since double-tapped data at this hospital got lost, this fact would further confirm to nurses the irrelevance of docking the glucometer.

4 An expert witness perspective

I was invited by the defence team to be an expert witness. In fact, an expert witness in the UK works for the court, but in this case the defence thought the truth would help their case.

My first comment was that if so many nurses are all alleged to have made the same mistakes, it is more likely there may be a common explanation, such as an IT failure — which would affect everybody. The legal view, however, was that if many nurses are making the same mistakes "they are all in it together" and the bad character concept applies.

My first task was to analyze the prosecution evidence (presented as a CD of Excel spreadsheets and, later, data logged on blood glucometers and XML files) to see if the police had made any mistakes claiming that the test data was not there.

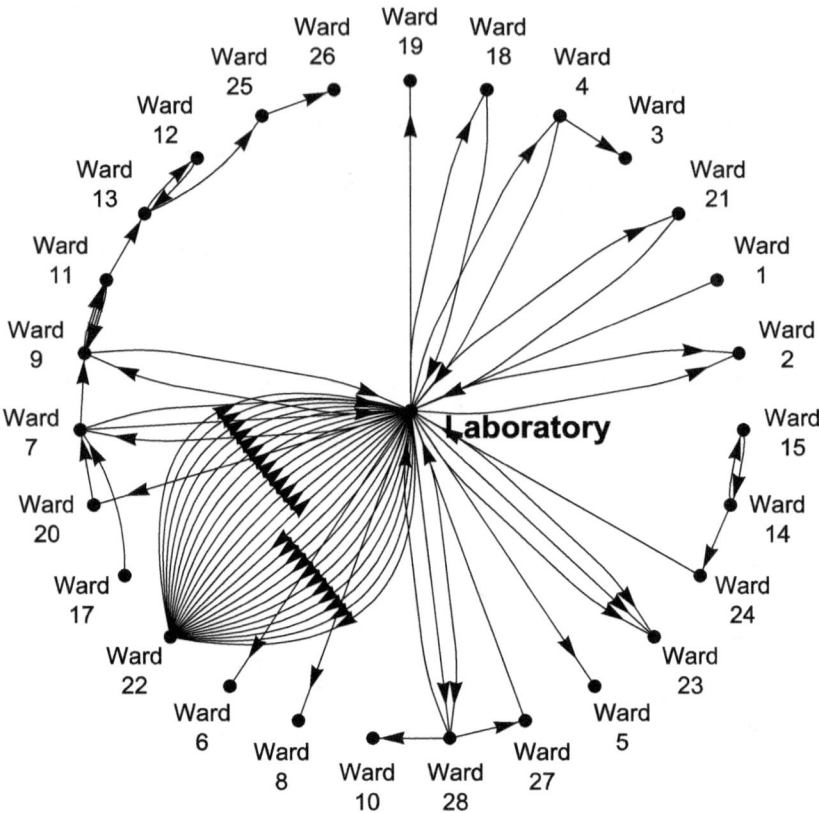

Fig. 1. Recorded movement of glucometer dockings around the hospital over a period of one year. Note the centrality of the laboratory as a hub of movement, and that Ward 22 seems to have a lot of activity — 25 movements. Since wards presumably try to maintain a constant stock of glucometers, there must be other movement that is not being recorded. (The diagram layout is arbitrary and unrelated to the real numbering and locations of wards, and to further help preserve anonymity, some of the "wards" aren't strictly wards at all.)

It was easy to show that the data the police claimed was not there was indeed not there. Nor was closely related data, as might be expected if glucometer clocks were not synchronized with nurse watches, or if other minor transcription errors had been made, and so on.

If data was not present, it implied, so the prosecution claimed, that the nurses had fabricated doing actual tests — for if they had actually done the tests, the data would be present in the spreadsheets they argued. That is possible, but I thought it far more likely that IT problems or even a "technician with a grudge" would be a simpler explanation — indeed, normal operation of TechCo's system *requires* administrators to make changes to data, for instance to sort out double tapping.

That over 20% of the database had an error flag set raised my suspicions; this was becoming a much more complex story than the prosecution painted. Another worry was that a comment field on each test said "Wrong patient" for just 2 entries and *nothing at all* for the remaining hundreds of thousands of entries — suggesting to me that nobody was really paying much attention to the management of the database; indeed, the "reviewed" flag was false for almost all data entries.

I noted that staff names occurred with many implausibly close variant spellings (e.g., differing only in capitalization or spacing, or variants like Jon and John but with the same surname and ID). Many identical staff names occurred with different numeric IDs. All this, and more, suggested the database was not well-managed and might not be reliable for the purposes the prosecution wanted it for. Moreover, the poor staff data suggests that the TechCo features for only permitting authorized staff to use the glucometer might easily have been compromised (how can it reliably lock out unauthorized staff when the staff data is so poor?).

Unfortunately for the police, proving absent data is absent *for a specific reason* is hard when the provenance and quality of the data is in question. Moreover, Excel spreadsheets have no way to audit: it is impossible to tell whether rows or columns have been deleted, been edited, or even had never existed.

The police claimed they had used forensic methods to make the copies of the database. In fact, some being CSV (an Excel data format) files proved there had been manual intervention: the hospital database was SQL so a manual process had converted it. Some of the Excel worksheets had differences further strongly suggesting an unreliable process had been used to create or edit them.

The police had copied Excel spreadsheets at the hospital to a USB stick and only then digitally signed the data and held it securely. Unfortunately, this "forensic rigor" came too late: the police should have made a signed copy of the original database, not a manually created copy on a USB stick. Anything could have happened to contaminate the Excel data earlier than the signing. Had rows or columns been deleted or edited, Excel provides no way to tell.

The police seized several blood glucometers from the ward in question and presented the data on them as evidence. The police failed to seize at least two other glucometers that had been docked on the ward over the period of the alleged fabrications but which were (presumably) in other wards or being serviced when the police seized the ones they did. (Note that the database only shows where glucometers are docked, not where blood glucose tests are made — to know that, the test data needs to be related to patient/ward data.)

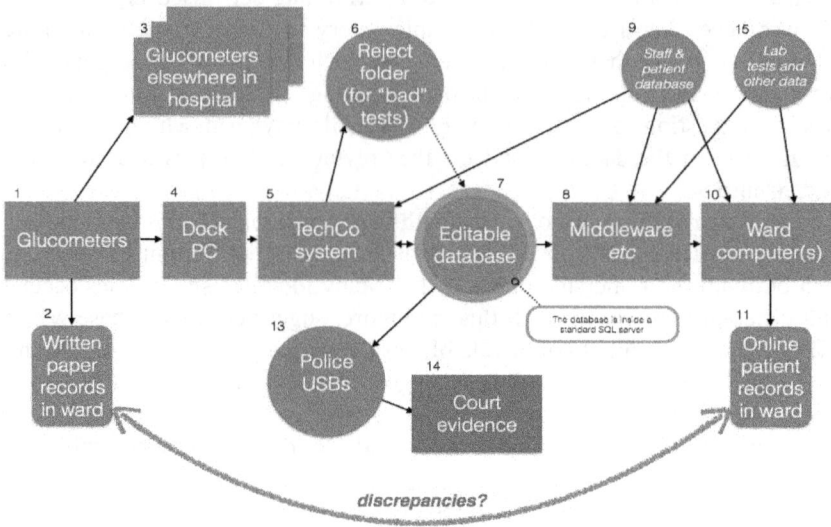

Fig. 2. Diagram presented in court (anonymized) originally sized as A4; the smaller reproduction here serves to indicate the complexity of the network, but note that Box 8 contains unknown further software. The basic problem was that there were discrepancies between the paper records (bottom left) and the final computer records (bottom right). Note that not all relevant systems are TechCo's (the "middleware" box may, and probably does, contain further complications). Numbers in the figure were used to cross reference this diagram to other expert evidence.

The police sent the seized glucometers to TechCo to confirm they worked correctly and to get their data. Unsurprisingly, TechCo said they worked correctly.

In fact, as figure 1 shows, glucometers move around the hospital. If a glucometer has a fault (e.g., a dead battery) would be returned for servicing and replaced by another. If a nurse needs a glucometer but cannot find one, they may borrow one from another ward. Glucometers may also get lost, perhaps at the back of a cupboard or sent off for repair. In the Excel data, glucometers were used almost hourly during the day on the ward in this story, but some glucometers in the hospital were not used at all for over 100 days and many not used at all for over a week — where were they? Where were *all* the glucometers that had *ever* been used on the ward when the police came? I do not think the hospital kept an inventory that tracked where glucometers were: if it had, it would have formed an essential part of their evidence.

It is possible that the alleged non-measurements are still sitting on a glucometer somewhere, but which is still waiting to be docked. Indeed, one XML file I got during the trial showed a 4 year gap between a measurement being taken and the data transferred to the database. The alleged incidents happened

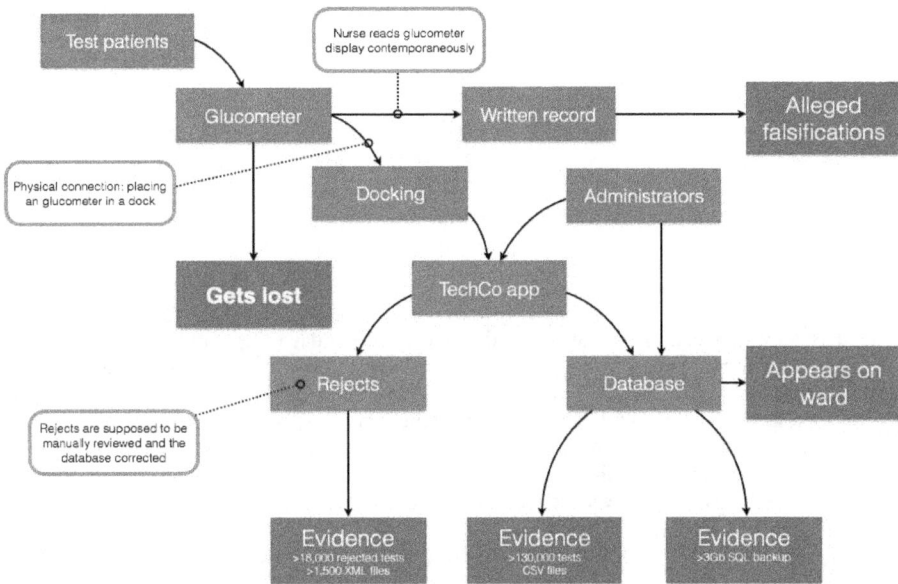

Fig. 3. Diagram presented in court (anonymized) to show key sources of evidence, though originally sized as A4 — for the present paper the details are less important than noting the complexity, with no end-to-end checking. The police made digitally signed copies of Evidence 1, 2 and 3, but signing occurred *after* the police had manually copied the data, so it could only be used to show the data had not changed after it had been collected. The digital signatures did not assure that the data was what it was claimed to be. This diagram does not show TechCo's modifications of data, which only became apparent after the diagram submitted to the court.

less than 4 years before the trial, so perhaps the missing data is still on its way — the alleged incidents did not happen that long before the trial!

There was much internal evidence that the databases were of very low quality, and of course there was the problem that there was no forensic route from the original SQL database to the Excel worksheet; worse, by TechCo's design, there was no reliable connection between the glucometers and the main database itself. There were many failure points, for example (see also figures 2 and 3):

- A glucometer may lose data itself;
- A glucometer may not be docked;
- A glucometer may be physically lost or returned for repair;
- Docking may fail, whether because of manual interference on the ward or by technical issues such as internet connectivity problems, unrecognized new servers and so on;
- TechCo's glucometers only store about 2,000 readings, yet the database shows they were used for nearly 5,000 tests: the data on the glucometers is

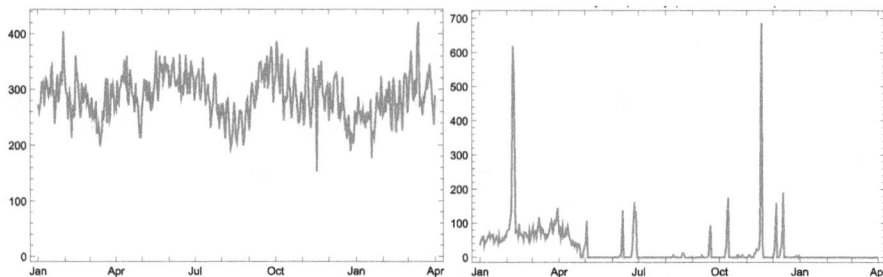

Fig. 4. Side-by-side comparison of accepted glucose tests per day (left), with rejected tests per day (right) over the same period. Successful tests per day closely track patient numbers, but rejects show no obvious pattern — one would assume rejects per day to correlate with the number of successful tests per day. In particular, note long periods of exactly zero rejected tests contrasting with brief periods of very high reject rates, almost double the number of accepted tests. A simple explanation is that some records of rejected tests were manually deleted, some may have been merged, and possibly some dates were arbitrarily corrupted.

not evidence whether nurses used the glucometers earlier than the glucometer records cover;

- … this list is not exhaustive.

Once docked, the data has a tortuous route through middleware and Tech-Co's own software. It can take days to get through. In particular, manual intervention is required for some data, but there was no evidence provided that any manual interventions had occurred. For example, if a glucose reading is rejected by the software, it is put into a holding folder and must be sorted out manually. Many patient IDs on the database were short-hands like 000, probably occurring because for the nurse it is far more important to get a measurement than worry about the technicalities of what happens after the glucometer is docked. Nobody had sorted out this poor data.

Such problems have not been reported on national databases to my knowledge (such as the US FDA's Maude), but there are peer-reviewed research papers that report identical problems at other hospitals with *the same TechCo devices* [so I shall not cite them]. At one hospital discussed in one paper, starting to check the database reduced poor practice such as using staff IDs for patient IDs. So what the prosecution should have concluded is not "this problem does not occur anywhere else" (i.e., the nurses must be being unprofessional) but rather "this problem routinely occurs elsewhere, but nobody worries about it enough to report it" (i.e., the nurses are behaving normally).

4.1 The collapse of the case

During the trial, the prosecution produced new evidence: an encrypted CD of gigabytes of XML files. This was the first evidence I had seen that had timestamps covering the period of the alleged incidents. (Though, as before, there was no digital signature or proof of provenance.)

The court adjourned while we analyzed this data. With effort the defence and prosecution then agreed a joint statement on the significance of this data, though some parts of the report, covering different features of the data, were written separately. The prosecution's case continued: the critical missing data was still not present.

I noted that the data had a very peculiar distribution, strongly suggesting data mismanagement, bugs or equivalent. See figure 4 for a simple example from my analysis.

TechCo was therefore called to be cross-examined over the joint report. During this cross-examination it became clear that this witness had visited the hospital before the police had seized a copy of the data. It emerged that TechCo had "tidied up" the database, and they had kept no records of what they had done.

The judge then made a ruling: the prosecution evidence was unreliable and had to be excluded. Asked what they wanted to do, the prosecution said they had no evidence and sat down. The judge ordered the jury foreman to clear the defendants as there was no case to answer. "Set the prisoners free!" he said.

My many other arguments for the unreliability of the evidence therefore never needed to be raised or cross-examined before the jury. For example, the trial did not get to exploring the consequences of the police seizing the wrong glucometers, of the hospital not having (or appearing not to have) any inventory for tracking glucometers. The court never explored the very poor data quality or reasons behind it.

The prosecution had implicitly gambled that any problem with their evidence could undermine the credibility of all of it. They ultimately needed to prove that *absent* data was caused *specifically* by deliberate nurse behavior, and they needed to prove that absence proved nurse fabrication rather than other possibilities. I would have liked to have shown, for instance, that the poor design of TechCo's equipment could provide no evidentially acceptable relation between absence and nurse behavior, but the confession that TechCo had manipulated data (and forgotten exactly what they had done) was enough to undermine all the prosecution evidence.

If the glucometers had been designed appropriately, they would have kept better records of successful (and failed) end-to-end blood test transmission. If they had, either the police would have had a very easy job (if the nurses had actually fabricated data) or the hospital would have easily known about their poor IT systems before the police turned up. Certainly if the TechCo database

This product is not for diagnostic use; all patient diagnostics should be based on results reported by the point of care instrument.

Fig. 5. Copy taken directly from TechCo's database system manual. It is worth noting that the nurses were doing what the manual says: they were using the point of care instrument (the glucometer) and writing down the results.

had included confirmed transfer and missing data information, the police would have known immediately the evidence was unreliable — though it will be recalled that they did not notice error flags and other indicators of poor quality in the data they actually had.

4.2 Technical discussion

The blood glucometers and related systems were not designed to be dependable. Or at least they were not designed to be dependable for any purpose other than taking measurements and immediately displaying results. The written evidence from TechCo strongly suggested that the software in them was not of high quality: for example, they wrote "[…] we should hopefully be able to confirm that this meter also had no corrupt records." But "hopefully" is not good enough!

It is interesting that there are research papers showing the TechCo glucometers are accurate blood glucometers. The prosecution understandably mentioned this, arguing that they were therefore good devices. As a matter of fact, accuracy was not relevant to the case. Regardless of whether the glucometers were accurate, the issue was *fabricated* readings not whether they were *accurate* readings. The relevant quality criterion for the case was whether the glucometers reliably transmitted test data to the hospital's patient record system. I could find no research exploring this aspect of their reliability. From the vague written evidence from TechCo, I suspected they were not very reliable at all in this regard, or at least they had been designed (by TechCo) so that TechCo could not answer this question definitively.

TechCo's database software has a warning "this product is not for diagnostic use" (reproduced in figure 5) — and if it is not good enough for diagnostic use, why was it used for evidence? Why did the hospital even have software that was not for diagnostic use managing glucometer databases? Rejected test data (e.g., with bad patient IDs) has to be edited with this product; if that isn't clinical use, what is it?

To be charitable, the database might have been intended for maintenance of the glucometers. For example, while I was double-checking I was correctly interpreting it, I plotted performance of the glucometers against battery voltage and temperature, data the glucometers record. There were interesting digitization effects, though I could not tell whether this was how the glucometers worked or whether it was an artifact of processing the data through TechCo's software, SQL or Excel.

Surely, if the data had not been used for clinical purposes, then the police should have been very cautious extrapolating to imply criminal clinical practice: they should have used independent evidence to establish the records were reliable. Given that the glucometers did not store all relevant data, the only independent evidence was the written patient records. The police assumed they were fabricated.

The police assumed the glucometers and hospital IT systems were completely reliable, even though they knew they must require human intervention. The management of the data was not questioned by the police, and there was no evidence submitted about day-to-day management of the data. Possibly nobody was managing it. On the contrary, some evidence says the TechCo database system crashed frequently —a problem the court never examined because the case collapsed before we needed to draw this issue to the court's attention.

To take any other view than "perfect IT" would imply considering the problems of managing the data and of questioning the reliability of TechCo systems. If the police suspected a problem, they would not have sent the glucometers to TechCo. I think the hospital, like the police, just assumed the data was perfectly reliable for a criminal investigation.

In a subsequent internal disciplinary hearing, the discredited police evidence was reused as if it was unproblematic. The disciplinary hearings presented a logical fallacy to support using it:

"in an internal disciplinary proceedings the burden of proof is a lower threshold than in criminal proceedings ...

The investigation looked on [the database] to verify if this blood glucose had been taken for this patient. This is not verified on [the database] ... [the hearing concludes that] patient did not have 9 blood sugar recordings checked [by this nurse] ... [etc]"

Indeed, words like "this is not verified on [the database]" is a recurring phrase in the disciplinary hearing transcripts. To be clear: there is no reliable database evidence to verify any fabricated tests or any other poor procedures. We know there is a lot of corrupted data, and there is a simple reason it is corrupted.

I submitted written evidence to this disciplinary hearing, including a full explanation. I quoted from the judge's ruling:

"Professor Thimbleby has shown that the chain has various breaks where the data can be lost. None of the data now relied on is original; it was all made after human intervention by [a TechCo employee] and he has no real recollection of what he was asked to do, what ID codes he was asked to consider, and did not note it at the time. All the material is at best edited. [The evidence] has lost significant amounts of data: but there is no way to tell whether the missing files were reintegrated into the [TechCo] database, in which case the Prosecution case might have force, or simply deleted, in which case it would not. I should exclude the evidence as being more prejudicial [...] and unreliable hearsay. [It] would serve only to suggest to the jury a conclusion they could not draw – namely, that absence in the searches meant those results had never been in [the database] or the reject folder."

Since my arguments were ignored, I wonder what other influences persuaded the hospital it was appropriate to use invalid computer evidence even a judge (on the basis of professional analysis and weeks' of detailed, critical discussions and cross examination) had rejected as unreliable and misleading? It is worth adding that the experts from the prosecution also fully agreed on the ruling.

In hindsight, somebody in the hospital ought to have ensured at the time and subsequently checked whether the police acted appropriately on original, un-contaminated evidence. And after the court case collapsed, somebody at the hospital should have checked why the case collapsed and made an informed decision whether disciplinary procedures should proceed on corrupt evidence.

At any stage, it would have been easy to compare a copy of the police evidence taken against a rolled-back hospital database. In fact, it seems bizarre the police did not take a signed and dated copy of the *actual time-stamped database and check it*, rather than just take what happened to be there after it had been corrupted and not check it. Unfortunately, a TechCo "expert" performed the transfer of data from the hospital to the police, probably reinforcing the naïve impression of infallibility.

5 Some recommendations

5.1 What should hospitals do?

5.1.1 A hospital collected data it did not monitor, so deleting or otherwise tampering with clinical data was not detected when it happened. Basic cybersecurity should have signaled TechCo's tampering (or any other unexpected changes) as and when it happened

5.1.2 A hospital procured equipment that unreliably recorded clinical proce-dures. The hospital should monitor operational data, and take steps to correct or manage problems when they occur. In this case, the hospital ignored the data it was collecting until the police seized it, and then it was too late.

5.1.3 A hospital should procure more dependable equipment and systems.

5.1.4 Hospitals should disable all IT features it is not using or not monitoring.

5.1.5   A hospital should not have clinical 'systems that are not designed for clinical use, nor should police use such systems except with proper caution. See figure 5.

5.1.6   Hospitals should engage regular external oversight to help avoid blindspots. IT is very complex, and it may be impossible to recognize one's own misunderstandings of it without external input. Note that the manufacturers are not the right people for any such oversight.

5.1.7   When police request data from a hospital, proper governance procedures should be adhered to. One of the problems with the case here is that the police obtained evidence without a court order, and the court had no idea what evidence the police had — hence the surprise of new evidence presented in court without any warning for the defence to interpret it. This wasted much court time.

5.1.8   Hospitals, police and courts should realize that the research literature does not tell the whole story of whether equipment is appropriate for clinical use. Clinical research papers focus on a very narrow aspect of dependability (e.g., whether measurements are clinically accurate), but real use is much more complicated (e.g., whether measurements get processed reliably in multi-vendor systems).

5.1.9   "Data fishing" is a serious problem. After the court case collapsed the hospital resumed disciplinary proceedings that had been suspended because of the criminal investigations. Using the police evidence the hospital argued that nurses had failed to take appropriate care of patients, a conclusion drawn from the discredited computer evidence. But one can fish data (especially bad data!) to support almost any case. One type of lack of care the nurse was accused of *also* happened over 1,000 times and affected all nurses on the same ward over the period I have data for, though note that I do not have the patient data to match against sliding scales of insulin. (The corruption of data may have *created* the "evidence" for this alleged incident and may also have inflated the 1,000 similar cases I can find.) Given the high frequency of such incidents, what this one nurse was accused is routine practice. I suspect the disciplinary hearing had no idea of such widespread practice, or that they were using discredited data. It is worth saying that if this lack of care was serious, then TechCo's system should have been configured to detect it *as and when it happened*; but TechCo's system was not used for any auditing (so far as I know). I infer the hospital did not worry what nurses did until it became a public issue.

5.1.10 Hospitals should routinely and regularly disclose to staff what data they are collecting and they should allow staff to see and, if necessary, to challenge it and the processes used to collect it. This means releasing data perhaps weekly if not daily. If data has no clinical role, then it should never be allowed to be used as if it had or might have done (which is what happened in this case).

5.1.11 Evidence discredited in the rigor of a criminal trial (and, in this case, very well summarized by the judge, see section 4.2) should be used to help seek the causes of the system failures, and should only be used with informed caution for disciplinary purposes.

5.1.12 Cybersecurity, improvement and culture change is at least a full time job. Avoiding predictable future problems and catastrophes will require dedicated staff and investment. Mature cybersecurity cannot be done alone, but requires collaboration across the healthcare sector and beyond — it should be a national priority, and local leaders need to be networking in this wider community just to stay up to date. This paper only discussed "simple" point of care equipment, but cybersecurity necessarily covers everything from wall-mounted emergency equipment, implants, medical apps, linear accelerators to PCs, all susceptible to hacking, ransomware, trojans and viruses — as well as all staff education (personal apps, phishing risks, etc).

5.1.13 Hospitals should also consider recommendations in all other sections in this paper (and elsewhere); hospitals should not work alone and be unaware of the activities and concerns of the wider community working to help improve IT and cybersecurity.

5.1.14 Best practice is, of course, that cybersecurity policies and implementations must be externally reviewed.

5.2 What should manufacturers do?

5.2.1 The very public discovery of VW's fraudulent IT to help their cars pass emission tests (Hotten, 2015) — which became public during the trial — should serve as a powerful reminder that IT is not just unreliable, but that it may be unreliable intentionally. VW's illegal emission levels are estimated to have contributed to tens of excess premature deaths. Manufacturers should adopt open source methods so that their code can be externally vetted.

5.2.2 TechCo testified that their equipment was CE marked, and therefore any problems *must* be the nurses' fault. With regulatory cover stories like that, there is little incentive for manufacturers to try harder. Manufacturers of clinical products should closely consider the quality of their programming, and have much better arguments for courts than that they have CE marks.

5.2.3 Hospital IT systems are very complex (through no particular fault of TechCo) and this complexity is not going to change any time soon. In view of the complexity, manufacturers must develop more defensive software — for instance with end-to-end checking, more logging and diagnostics, and with formal proofs of correctness (see section 5.5). Auditing needs to be provided, work and be used.

5.2.4 Installed software in hospitals that is not being actively managed (as happened here with TechCo's database) should be automatically reported, at least to the manufacturer who can then take remedial steps (such as reconfiguring it or notifying the hospital to audit it).

5.2.5 It was disappointing that no technical experts from TechCo wanted to appear in court, although they had provided much written evidence (though only for the prosecution it must be said). TechCo is based in a country outside of the UK's jurisdiction. Manufacturers should be eager to support investigations concerning their products.

5.2.6 Elsewhere I have written about the user interface of TechCo's systems [which I will not cite here to preserve anonymity]; the poor design of the user interface suggests that reliable operation was not a priority or perhaps not a competency for the manufacturer. The evidence presented in court further suggested the programming of the device and the database management software was substandard too —TechCo's written evidence says things like "it is not possible to say categorically that there were no corrupt records"; and, symptomatically, the TechCo database systems regularly crashed. Given the serious consequences of poor quality systems (patient harm, substandard care, pressure on staff — even prison) manufacturers should feel an obligation to put high quality professional effort into their products.

5.2.7 TechCo's systems have an "audit" feature. I beg to disagree; it certainly has a feature *called* audit, but it is manual and fallible, and generates documents that are not authenticated. Features should be named and implemented to support the conclusions most people (and courts) would reasonably draw from their names.

5.2.8 It may seem unfair to criticize manufacturers without offering any solutions. One easy thing to do, then, would be for all blood glucose tests (in fact, tests data from any equipment) to be assigned a serial number. Along with the glucometer ID, it would then be trivial to detect lost data (additionally, using digital signatures to circumvent cybersecurity problems). Lost data should be reported to the manufacturer's post-market surveillance team. It would then be easy for a hospital to use best efforts to respond and recover it.

5.3 What should regulators do?

The CE marking system is discredited (Cohen, 2012; Cohen & Billingsley 2011), and fixing it to be more effective for complex computer-based systems (devices, medical apps, etc) is essential, particularly as computer-based technology is ubiquitous and taking over healthcare. Healthcare IT systems (PC, tablet, embedded, point of care, etc) support patient care and it is therefore negligent if they are not developed using equivalent processes to the rigorous processes used in pharmaceutical development (Thimbleby *et al*, 2015).

There are rigorous process in pharma because we recognize that there may be side-effects and unknown variation in patients — analogous problems to IT and cybersecurity. Randomized controlled trials (RCTs) as used in pharma may not be essential for cybersecurity, but there are other methodologies such as formal methods where correctness is proved (see section 5.5). Such methods should be used, and should be shown to be used in any certified product. Formal methods are routine in aviation (where lives depend on them); they ought to be routine in healthcare too.

Manufacturers will complain that they do not know how to use formal methods for their complex products. Well, they should start making products that are simple enough for them to understand. Logically, if the manufacturers can only "hope" to understand their own devices (e.g., see quote above; and failing to use formal methods means relying on hope alone), then hospitals and nurses are very unlikely to understand them either.

The US FDA device regulation is only concerned with the patient; staff well-being and effectiveness clearly ought to be a consideration too. In Europe and the UK (whether or not their regulation is normalized) is much more closed, obscure and informal (e.g., the role of notified bodies); in our complex world of IT threats, regulation urgently needs opening up and becoming more responsive.

5.4 What should the police do?

5.4.1 Although not discussed in detail in this paper, the police management of data exposed numerous problems. For example, one piece of evidence says that the police found that their Excel crashed analyzing the data. It is surprising the police even tried using Excel to analyze the complex relations in such a large volume of data.

5.4.2 Evidence I was given included analysis spreadsheets, which made me wonder what other edits the police had made to what was claimed to be evidence: there was no clear management of the evidence. Despite the "forensic" methods the police claimed to use, they were not available with the evidence and did not help confirm provenance. I was never given any evidence with signatures.

5.4.3   The police assumed that data (or missing data) could be used to prove poor clinical practice. This encouraged the media, public, patients and relatives to assume the trial was about poor clinical practice and reinforced a view that patients were victims of poor care. But whether or not the hospital had poor clinical practice, the criminal case was much narrower. Even if the data had been reliable (which it was not), the connection between abstract data and clinical practice would have been tenuous at best. Indeed, even if the case had been proved, there had been no patient harm and it could not have said anything about quality of care.

5.4.4   So far as I know, the police never considered assessing the internal or external validity of their evidence. A cursory look at the database would have raised many questions about it. Instead, I think they probably never took a holistic view. Accepting IT evidence on faith is problematic.

5.4.5   This was a case where alleged data discrepancies were used in evidence, so I was surprised at typos affecting data presented in the prosecution evidence — this is ironic, as the prosecution case relied on the quality of data (I am not saying the typos were sufficient in themselves to discredit the evidence). Of course I may have made some typos in my own evidence that we did not spot. In fact, I used *Mathematica* to analyze the data and automatically generate reports, tables and diagrams, etc: insofar as I can program reliably (and I more than double-checked every result), this ensured my reports were factually accurate.

5.4.6   The police seized several glucometers. Other glucometers had been used on the ward, so not all of the relevant glucometers were seized. Had the trial proceeded, this would have become a criticism of the prosecution case — it is possible that the alleged fabrications are *still* stored on a misplaced glucometer somewhere.

5.4.7   What was the ward supposed to do when they lost their glucometers to the police? I wonder whether the police did a risk analysis (there is no evidence either way): in any case, removing glucometers off a ward puts patients at serious risk of harm, at much greater risk than any of the alleged incidents.

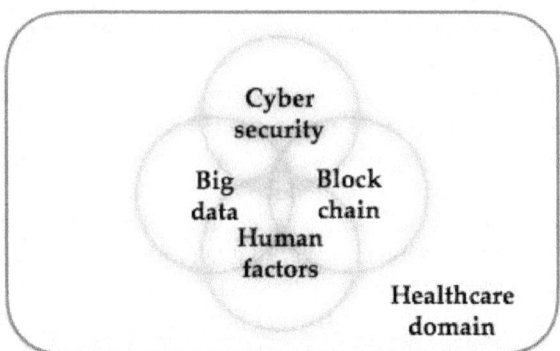

Fig. 6. Convergence of powerful research challenges, particularly when applied to the complexity, risks and priorities of healthcare. Healthcare has many opportunities, it needs IT to drive quality improvement — and new technical challenges (cybersecurity, big data and blockchain, etc) need relating to *actual practice and human limitations* through applying human factors embedded in the healthcare domain. See section 5.5.

5.4.8   There were surprising conflicts of interest. A technician from TechCo corrupted data, then selected the data from the hospital and handed it over to the police, and the police sent the glucometers to TechCo themselves to analyze and confirm whether they were functioning correctly. Independent experts should have been used throughout.

5.4.9   As TechCo's glucometers do not use an open architecture, independent experts should have been *required* to be present when the data was taken from the hospital and when any glucometer analysis was performed.

5.4.10 Knowing glucometers function correctly when analyzed tells you very little about how they might have performed during alleged incidents. (For example, they may have been serviced since the incidents.) The glucometers were not designed to answer such questions. The police should exercise more caution with test results, especially when undertaken by the manufacturer.

5.4.11 It should be routine for manufacturers to be required to disclose relevant quality control documents and risk analyses in support of any claims that their products work to specification (e.g., as required by ISO 14971, ISO 13485, etc).

5.5 *What should researchers do?*

5.5.1   A serious issue remains for researchers, the industry and regulators to address is that clinical trials alone are insufficient to justify the quality of computer systems or devices in normal use. The current peer reviewed literature is inadequate. We need *both*: clinical research (do things measure the clinical factors they claim?) *and* situated IT and HCI research of effectiveness in the real complexity of healthcare (will they be used correctly and is the data reliable?). Such research needs tying up "end to end": is the final data, however the clinicians summarize or interact with it, effective for clinical use and correctly based on true clinical data?

5.5.2   While cybersecurity research has a high profile, cybersecurity is only one aspect of the potential problems and vulnerabilities of medical devices and systems. More research is needed on end-to-end dependability, from HCI to networking and multi-vendor databases, interoperability, etc. Researchers should spearhead an analogous structure to the Information Sharing Analysis Organizations (ISAOs) established for cybersecurity.

5.5.3   Formal methods is a substantial research area that has resulted in many robust approaches to software development — widely used in aviation, for example. SPARK Ada is a good place for programmers to start (Barnes, 2003). Unfortunately, there is little connection between the formal methods community and healthcare, let alone healthcare IT. This gap urgently needs to be bridged. One of the many research problems is how to migrate large, complex, buggy software (as in blood glucometers and their networking, for example) into high quality software that works "well enough" — and increasingly more reliably — until it is rigorously correct.

5.5.4   Throughout this paper I have criticized the culture of assuming IT and data is perfect. In the UK, this culture is enshrined in law: the Criminal Justice Act 2003 created the presumption that IT works correctly (Mason, 2012; Mason, 2014). Computers are deemed to be "in order" and "properly set and calibrated." So we cannot simply blame the police or the hospitals when they just reflect the wider legal culture in which they operate, the absurd presumptions of the Act of Parliament being but a symptom. While Mason (*op cit*) gives a very professional discussion, including the problems of the inscrutability of proprietary systems (e.g., those that are not open source) and the imbalance between prosecution and defence scrutiny, the challenge to researchers is to create awareness and transformation of this absurd legal position.

5.5.5   Researchers always need resourcing, and the convergence of cybersecurity, healthcare risks and costs, big data and blockchain technology closely matches many national research funding priorities, to say nothing of digitalizing healthcare and increasingly relying on (unregulated? insecure?) apps. Human factors (e.g., human error) is often overlooked, for exactly the same reasons ("loss of situational awareness") that cybersecurity is overlooked by healthcare — people are too busy doing, urgent, hard complex jobs, and this distracts attention from longer-term, broader priorities that are not immediately visible. Figure 6 visualizes this opportunity.

6 Discussion

The big picture is that nobody seems to be fully aware of the complexity and risks of IT. This results in lax legislation, lax regulation and lax procurement, and in turn lax manufacturing since no useful standard of quality can be demanded by hospitals. Unawareness in turn results in lax management, and unnoticed inconsistencies between clinical care and its unreliable monitoring.

Prominent peer-reviewed papers such as Nichols (2011) "Blood glucose testing in the hospital: Error sources and risk management" reinforce the naïvety: "software ensures accurate documentation," "automation is the best prevention for errors," "smarter software is assisting with result documentation," etc. Even the recent UK *Making IT Work* report (Wachter, 2016) takes it for granted that computers work (with minor caveats on interoperability and usability) — hospitals just need to "digitalise" more, with another huge government investment (£4.2 billion) available to purchase IT. It takes it for granted there is appropriate IT that can be just purchased! The Wachter Report has a "digital maturity" scale, but it is about whether and to what extent a hospital has *adopted* levels of IT (and is paper free), not whether the IT is effective (and for what?) or even fit for purpose, which is just assumed. Every hospital is in good company, then, unwittingly "drifting into failure" (Dekker, 2011).

Ironically, if the hospital in the story here had been "paper free" (as the Wachter Report wants) there would have been no discrepancies to investigate; although there would have been no trial, the underlying IT problems would never have come to light. If we want a paper free health service, we also need to work out how to make it more reliable!

System failure only becomes apparent after there is a visible incident. In the case described here, something triggered the police investigation and that became "the incident," as discussed in this paper. In hospitals, reportable incidents usually involve patient harm or near misses of harm; in this case, thankfully, there was no patient harm but considerable staff harm. Nobody benefitted from the process.

With hindsight we can see many causes of the incident. All of them were avoidable; and avoiding only a few would have resulted in a much happier outcome. The wholly uncritical view of IT coupled with a remarkable unwillingness to consider alternative explanations for multiple IT problems related to tens of nurses are a textbook example of confirmation bias and cognitive dissonance (Tavris & Aronson, 2007); with so many nurses, underlying systemic factors, including management (Ball *et al*, 2013) and IT support (most of this paper) should have been obvious priorities to critically examine.

The causes, then, are many and complex, and other hospitals will have analogous but diverse complex IT problems. A key priority should be to have a mature cybersecurity strategy, which implies having an implementation of IT that permits having a workable strategy. Whatever the mess or causes of the mess — external hacking or just internal hacking, as here — effective procedures must be in place to detect, interpret and respond to unusual or unauthorized activity immediately. The regular crashing of critical systems should at least have been a warning sign. Hospitals need to tighten up their cybersecurity maturity. Good guidance on data management is available elsewhere (DSIWG, 2016).

The hospital had ignored evidence of behavior the police treated as criminal; that is, the recording of tests on the database had no day-to-day significance (so far as I can tell). In any case, the process of getting data from a glucometer into the patient records was so slow and unreliable it was of no clinical use. Having correct and timely blood glucose measurements is a clinical priority but "correct" use of a point of care device and its IT system that was, unknown to anybody, not working well was not an issue, and certainly was not an issue for the hospital.

When data was deleted, the hospital did not notice. The police apparently assumed the corrupted data was perfect, and adequate evidence to charge nurses. There were many reasons that the data was unreliable, but the simplest was that TechCo themselves had corrupted large parts of it. That fact alone was sufficient for the trial to collapse, and it left the prosecution with no admissible evidence.

Once the police investigation started, my impression was that the hospital felt unable to pursue any parallel investigation, and certainly they felt unable to help the expert witness in his enquiries (in hindsight I should have sought a court order). This missed early opportunities to uncover some of the systemic problems. A parallel cybersecurity investigation, done within the hospital, would have saved a huge waste of time and huge costs to the defendants.

It is too easy to blame people at the sharp end. Nurses are at the sharp end, and a witch hunt makes a compelling story. A witch-hunt involves human things we feel we understand, and we feel angry about poor patient care and a sense of betrayal by nurses being incompetent (if they were). For everybody concerned, if you get rid of a witch the problem is immediately and visibly solved! Unfortunately, on the other hand, complex IT is hard to write a gripping story about; it really isn't very interesting, and there is no quick fix. Dekker (2009) discusses these important issues in much more detail than we can here.

7 Conclusions

This paper explored a case where unconsciously accepted unmanaged IT complexity unnecessarily led to horrible outcomes. While the hospital had a core part to play in this story, the real villain in my opinion is TechCo. In principle they could produce better IT if they wanted to; they have had many years developing hospital equipment and must know very well how to do it: their equipment could have been much easier to use correctly in a real ward, and much easier to use to *dependably* monitor patient data and ward activity. TechCo's systems failed the hospital, its staff and its patients.

Nevertheless, the data corruption this case revolved around (the nurses' alleged falsifications, the use of staff IDs, TechCo's deletion of data, and the poor software that allowed it all to happen…) could have been detected and managed as soon as they occurred had the hospital had a mature approach to cybersecurity. Given that TechCo systems were used, the problems were unavoidable, but they could have been detected and managed. They need never have escalated to a criminal prosecution.

Given the widespread use of equally poor IT throughout healthcare, hoping for manufacturers to improve is less realistic than hoping for hospitals to prioritize mature cybersecurity. Fortunately, cybersecurity already has a high profile (thanks to fear of malicious hacking), and this paper adds one more reason to take it even more seriously. Luckily nobody was harmed as a result of any of the issues discussed in this paper, but a mature cybersecurity approach would also reduce unnoticed data problems of all sorts, and therefore would help reduce patient harm too.

The broader problem remains our culture of uncritical acceptance of IT, from legal, regulatory, procurement and other perspectives, especially for healthcare where billions is eagerly invested in more IT stuff. Our culture makes us all uncritically believe that IT and especially the *latest* IT is wonderful — don't we all want new things? (Of course, this is how companies stay in business.) But the reality is that behind the façade of superficial wonder, modern hospital IT is too complicated for its own good, for the good of patients, for the good of staff. Ironically, the newer IT is and the more exciting it seems, the less tested it is in the clinical environment.

This culture nurtures a lax approach to cybersecurity. It created the perfect environment (bad IT, bad IT management) for accepting a superficial explanation of alleged multiple nurse failures instead of exploring underlying causes.

Acknowledgements. EPSRC funded part of this work under grant [EP/L019272]. I am grateful to a huge number of anonymous people who helped with this paper and the background information and research. I am very grateful to the nurses who had the courage to defend their innocence.

References

Ball, J. E., Murrells, T., Rafferty, A. M., Morrow, E. & Griffiths, P. "Care left undone during nursing shifts: associations with workload and perceived quality of care," *British Medical Journal Quality & Safety*, **0**:1–10, doi:10.1136/bmjqs-2012-001767, 2013.

Barnes, J. *High Integrity Software: The SPARK Approach to Safety and Security*, Addison Wesley, 2003.

Cohen, D. "How a fake hip showed up failings in European device regulation," *BMJ*, **345**:e7090, doi: 10.1136/bmj.e7090, 2012.

Cohen, D. & Billingsley, M. "When it comes to medical devices, Europeans seem to get a worse deal than US patients," *British Medical Journal*, **342**:d2748 doi: 10.1136/bmj.d2748, 2011.

Davidson, J. "Cyberattacks on personal health records growing 'exponentially'," *Washington Post*, www.washingtonpost.com/news/powerpost/wp/2016/09/28/cyberattacks-on-personal-health-records-growing-exponentially, 28 September 2016.

Dekker, S. W. A. "Prosecuting professional mistake: Secondary victimization and a research agenda for criminology," *International Journal of Criminal Justice Sciences*, **4**(1):60–78, 2009.

Dekker, S. W. A. *Drift into failure: From hunting broken components to understanding complex systems*, CRC Press, 2011.

DSIWG, Data Safety Initiative Working Group, *Data safety guidance*, Safety Critical Systems Club, ISBN 978-1519533579, 2016. http://scsc.uk/paper_130/Data%20Safety%20(Version%201.3).pdf

Hotten, R. "Volkswagen: The scandal explained," BBC News, http://www.bbc.co.uk/news/business-34324772, 10 December 2015.

Koppel, R., Wetterneck, T., Telles, J. L., & Karsh, B-T. "Workarounds to barcode medication administration systems: Their Occurrences, Causes, and Threats to Patient Safety," *Journal of the American Medical Informatics Association*, **15**(4):408–423, 2008.

Mason, S. *Electronic Evidence*, Chapter 5, 3rd ed., Butterworths, 2012.

Mason, S. "Electronic evidence: A proposal to reform the presumption of reliability and hearsay," *Computer Law & Security Review*, **30**(1):80–84, doi: 10.1016/j.clsr.2013.12.005, 2014.

Nichols, J. H. "Blood glucose testing in the hospital: Error sources and risk management," *Journal of Diabetes Science and Technology*, **5**(1):173–177, 2011.

Tavris, C. & Aronson, E. *Mistakes were made (but not by me)*, Harcourt Inc: Florida, 2007.

Thimbleby, H., Lewis, A. & Williams, J. "Making healthcare safer by understanding, designing and buying better IT," *Clinical Medicine*; **15**(3):258–262, doi: 10.7861/clinmedicine.15-3-258, 2015.

Wachter, R. M. *Making IT Work:Harnessing the Power of Health Information Technology to Improve Care in England Report of the National Advisory Group on Health Information Technology in England*, Crown Copyright, 2016.

Data: Your Life in its Hands

Tom Adams, Paul Hampton

Mike Parsons

NATS Ltd

Abstract *This work is an examination of how safety-related data is currently being managed across the healthcare domain in the UK. Some areas where safety data can cause specific problems are highlighted and extant mitigations noted. Some case studies where data has been a contributory factor in the occurrence of harm are discussed. The work concludes with new proposed guidance material to supplement the existing safety guidance on Health IT Systems.*

1 Introduction

Data is at the heart of modern healthcare driving everything from consultations, investigations, lab tests, imagery, prescribing, and GP care. Some of this data is patient-specific for individual treatment and care, and some relates to population health (e.g. hospital infection rates). Some data is time-critical, for example lab results for acute infections, and some is related with identifying long-term health trends. There are now myriad forms of health data used in numerous ways and the volume will only increase with the government's initiative to make the NHS 'paperless' by 2020 (NIB 2014).

Although some data in healthcare is machine generated (for instance in imaging), some is still manually (human) generated. However much of this data is now held in some form of electronic record, and this is often shared across several databases, in multiple locations, and can be distributed for multiple purposes.

Traditionally there has been an assumption that important data is always checked at the point of treatment (e.g. identity of patient), and that little reliance is placed on the existing data quality, i.e. the health professionals will recognise and deal with any data problem, for instance by asking for a re-test for clearly wrong test results. However this position is now very hard to justify with increasing reliance on electronic data being used throughout the process without additional checks. Clinicians are increasingly being asked to do more in less time and are swamped with a flood of data that modern IT systems and integration between them is generating – it is no longer possible to check everything. The data and meta-data is now being used to direct the clinicians workflow, for example flagging what records need the clinician's immediate attention.

In many cases the data itself has some safety significance; and this is a particular problem where there is no sensible way of checking the data quickly (e.g. if patient allergy data is missing or incorrect then a potentially fatal drug could be administered in an emergency situation).

Apps on smartphones and tablets are a new problem as these are often not controlled to the same degree as traditional computer systems. This is not just a software issue, but also the data used by the apps can cause problems. Some of this data may be acquired by personal monitoring devices linked by radio, e.g. blood pressure or glucose meters, and fed directly into the apps [1] with little or no obvious checking of the data.

In some situations, electronic annotation systems are used to replace manual mark-ups (e.g. of patient notes, or diagnostic images), so there is no choice but to use the electronic record. Typically a clinician will mark up anything that requires follow up. With paper-based systems, a physically marked up image could not easily be separated from the annotation, but now, if an electronic annotation is lost, there may be no other way of detecting that a potentially critical diagnostic indication has lost its association with the image.

There are also new hospital systems that provide workflows that guide both patients and clinicians through complex diagnosis and treatment pathways. These may be different to old working methods and heavily rely on configuration data to create the correct workflow. One of the main reasons why such workflows and set processes are becoming common is that less skilled staff can be used who are only expected to "follow the process". The potential new hazard is that the less experienced staff don't realise the limits of their knowledge and may not spot incorrect data.

In summary healthcare has become heavily data-dependent: there is a *"Forest of data and we need to create a path through"* (Tom Adams).

---

[1] Note that some of these new monitoring devices or the apps themselves may be classed as medical devices, and so may be subject to specific regulations.

1.1 Properties of Data

Firstly it is important to identify what we think of as critical properties of medical data. Some data properties that can affect safety are:

Table 1. Critical Data Properties

Property	Description
Integrity	the data is correct, true and unaltered
Completeness	the data has nothing missing or lost
Consistency	the data adheres to a common world view, e.g. units
Continuity	the data is continuous and regular without gaps or breaks
Accuracy	the data has sufficient detail for its intended use
Resolution	the smallest difference between two adjacent values that can be represented in a data storage, display or transfer system
Traceability	the data can be linked back to its source or derivation
Timeliness	the data is as up to date as required
Verifiability	the data can be checked and its properties demonstrated to be correct
Format	the data is represented in a way which is readable by those that need to use it
Availability	the data is accessible and usable when an authorized entity demands access
Fidelity / Representation	how well the data maps to the real world entity it is trying to model
Priority	the data is presented / transmitted / made available in the order required
Sequencing	the data is preserved in the order required
Intended Destination/Usage	the data is only sent to those that should have them
Accessibility	the data is visible only to those that should see them
Suppression	the data is intended never to be used again
History	the data has an audit trail of changes
Lifetime	when does the safety-related data expire
Disposability / Deletability	the data can be permanently removed when required

It is clear that many of these properties apply in a medical context (for instance if lab test data is not complete when used in a consultation then an incorrect diagnosis could be made).

1.2 The SCSC Involvement

Data having safety properties is not unique to healthcare and other sectors have similar issues. To address the issue across all sectors, the Safety-Critical Systems Club has been involved in two areas:

I. It has sponsored an industry working group (the DSIWG [1]) to produce guidance in this area. The DSIWG has produced guidance documents, of which the latest is available at this symposium (DSIWG 2017). The aims of the group are *"To have clear guidance on how data (as distinct from software and hardware) should be managed in a safety-related context, which will reflect emerging best practice."* The group's objectives were set as follows:

1. Produce cross-sector guidance by the end of the year including a clear statement on handling of data as a separate component within safety related systems;
2. Produce a high level strategic plan by the end of the year for fuller adoption e.g. into existing standards or a new standard;
3. Influence standard updates currently in progress where we can;
4. Actively promote and disseminate objectives and outcomes of the initiative to the wider community, professional bodies, etc.

II. The SCSC has held two seminars on data safety, the first "How to Stop Data Causing Harm" (SCSC 2012) was held in December 2012, featured six talks from various sectors, and the second "How to Stop Data Causing Harm: What you need to know" was in December 2015 (SCSC 2015) with seven speakers covering Rail, Aviation, Water, and Health. (Slides available via the links in the references.)

1.3 Data Safety in the Literature

Several authors have been active in this field for some years, producing informative papers (Faulkner et al 2000), (Story & Faulkner 2001), (Story & Faulkner 2003), (Story & Faulkner Q4 2003), (Faulkner 2004), (Knight et al 2004), and (Inge 2011). One of the few standards that has specific guidance on data is DO 200A (RTCA 1998), which is concerned with aeronautical data such as runway lengths and navigational aid information. This identifies a number of aspects of data quality including Accuracy, Resolution, Confidence, Traceability, Timeliness, Completeness, and Format of the data. It also proposes a con-

---

[1] Data Safety Initiative Working Group

cept of Data Processing Assurance Level related to the Data Safety Assurance Level (DSAL) scheme proposed later. Edition 2 of IEC 61508 (IEC61508 2010) also makes reference to data but doesn't provide any real guidance. The new issue of the Def Stan 00-055 now mentions data in an annex specifically. Relevant work has also been conducted in the maritime navigational field.

1.4 Overview of the Paper

In the following sections the paper looks at the issue of data in healthcare and some real life case studies where data issues have arisen. The intention is to give concrete and readily identifiable examples of how data-related issues can manifest themselves and to show what practices the organisation adopted to mitigate the associated risks. The paper then shows how these issues can be classified into one of several different categories within the healthcare context. The paper concludes by presenting a proposal for a more structured approach to managing data safety risks and in particular, determining the mitigation methods and techniques that are recommended for managing those risks.

2 Data Safety Issues in Healthcare

2.1 Introduction

Data used in healthcare can be placed into three broad categories:

1. **Primary** – data used at the point of treatment to support the clinical and administrative processes. Primary data is generally time critical to some extent and includes data used in Primary Care settings (e.g. GP surgeries and hospitals) and Secondary Care settings (specialist consultants)
2. **Secondary** – data for clinical audit, population health, commissioning, research uses etc. Although this data usually has no immediate clinical impact, it can be used to determine treatment strategies and policies, decide on which drugs are effective, etc. with a consequential long-term impact.
3. **Reference** – reference data forms a third set and is used on a regular basis during treatment, but changes relatively infrequently. This data is traditionally held in books but now is often available in electronic form. Examples are the British National Formulary (BNF 2016). Some reference data can

be more time critical, e.g. drug alerts. There are many reference data sets available for different purposes.

2.2 Historical Context

Traditionally IT systems holding data in healthcare were rarely used to control anything that directly impacted patient health without professional intervention[1]. Usually with healthcare IT there is time to check results and a "person in the loop" is available (i.e. a trained healthcare professional) to supervise or advise; to make sense of the data and reject erroneous or misleading data. They will often check verbally with another professional if they think the data is wrong (e.g. by phoning the lab if they think a test result is strange.) Often there are still some paper records which can be used to cross-check electronic results. Clinicians and hospitals have a mature safety culture and some measures exist to manage IT safety. There are now standards in the NHS for the manufacture and deployment of Health IT Systems in a clinical context (SCCI0129 2016) (SCCI0160 2016) and these are considered useful and sensible. Hence, historically, the data in IT systems has not been directly making medical decisions.

However this situation is changing and data is increasingly being used to directly inform clinicians (e.g. via systems that synthesise an overall patient summary screen from multiple hospital systems) and many hospital functions are monitored and managed using multiple systems in a heavily data-dependant way.

Data can directly monitor, influence or control treatments, for instance electronic patient records / health records, patient dashboards, patient alerts, decision support systems, and of course, in medical devices. Clinicians therefore increasingly rely on system generated alerts or flags to indicate which patients, events or encounters require immediate attention.

2.3 Some Published Cases

The following section outlines some historical accidents in healthcare related to data.

---

[1] Note that medical devices (e.g. infusion pumps) are a special case and are covered by specific regulations.

Appointment Data

Three year-old Samuel Starr died in the arms of his parents as they read him his favourite stories at the local hospital (Guardian 2014). After successful heart surgery at nine months, Samuel should have had regular scans to see if his condition had worsened. But he didn't have any scans for 20 months, in part because of difficulties in organising the appropriate appointments on the hospital's new IT systems. During the migration to the new system, the appointment data for Samuel's follow-up scans had been lost. In the vast array of data used by hospitals it is easy to forget that even the most innocuous of data such as appointment data can have safety significance. The data in this case had 'fallen between the cracks' between the old and new systems. In general, a lost appointment is likely to be less critical than a lost medical record, such as an allergy record or discharge prescription, but nevertheless it is arguable that a better appreciation of the safety-related nature of the appointment data and the adoption of appropriate assurance techniques during the data migration might have led to an alternative outcome for Samuel.

Electronic Data Entry

The medical error that killed Genesis Burkett – a 24 week premature baby but with otherwise good prognosis (Huffington 2011) - began with the kind of mistake people often make when filling out electronic forms: a pharmacy technician unwittingly typed the wrong information into a field on a screen. Because of the mix-up, an automated machine at the hospital prepared an intravenous solution containing a massive overdose of sodium chloride — more than 60 times the amount ordered by a physician.

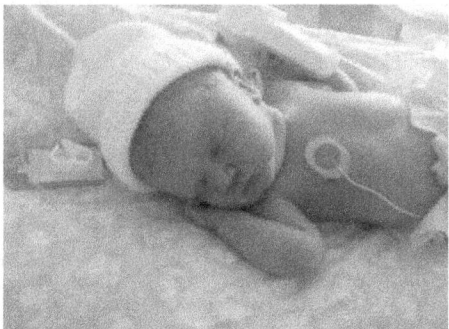

Fig. 1. Genesis Burkett

Although arguably a human error in this case it does illustrate that there is a *data supply chain* and in this case it was relatively short and direct with little intermediate checking, what was keyed by the clinician initiated a series of automated processes and procedures that led directly to the administration of the fatal dosage.

Loss of Availability of Data

In 2013, severe network problems at a major city hospital led to 709 patients having their hospital appointments postponed (DigitalHealth 2013). A significant number of these were chemotherapy appointments and hence safety-related and time-critical in nature. The problem was eventually tracked down to a system that kept a master record of clinician privileges to logon to clinical systems – this data had become corrupted, which effectively resulted in a loss of availability of key clinical systems.

Personal Perspective - 1

The following is a personal opinion and insight from a safety engineer who has worked at a major hospital on the organisation and its handling of data

The trust was fragmented and did not have joined up processes in respect to handling of IT safety risk. Safety incidents were occurring but these were not being communicated to the IT department. There was a lack of overall architectural vision and global understanding for the Trust. There was over-reliance on IT suppliers to provide advice and even pressure on suppliers to take ownership for the wider trust problems outside of the supplier's scope of supply.

Lack of overall oversight and vision meant that on some occasions a change was rolled out that was not compatible with a legacy system leading to safety incidents. In one case, the currency of data was properly tagged in the new IT system, but this flagging was omitted in the legacy system (because of presentational constraints) so there was risk that clinicians would be using old data on the legacy system. On another occasion, an upgrade to a prescribing system failed to identify that a data flow to the discharge system also needed to be updated as part of the change, with the effect that discharge letters did not contain the most current medication regime for the patient.

The Trust was subject to SCCI0160 but was not sufficiently skilled to follow the standards and relied on IT suppliers to provide guidance. They didn't have the correct underpinning processes, not least because of significant budgetary constraints. There was a preoccupation with risk of litigation when accidents did occur, this could be costly but they did not seem to want to invest in the risk management activities that could help prevent these from occurring in the first place.[1]

---

[1] This highlights that some hospital IT departments are not information management departments. IT department staff often try to understand the clinical requirements but are not as knowledgeable as the clinicians. The UK Department of Health publication "Making IT Work" (DoH 2016) proposes that there should be senior clinicians who are highly IT literate in every acute trust. Such people should be in a good position to check that the IT department has put in place a proper data migration plan because they are likely to understand risks better.

Communication of Results

An 18 year old man had been admitted into A&E with a suspected overdose with his mother having found an empty bottle of paracetamol that had been full the day before (Medical Error 2005). Blood tests to establish paracetamol and salicylate levels[1] were ordered and a student nurse was at the desk when the lab technician phoned in the results. She wrote the results in the message book. The salicylate level was negative but when it came to the paracetamol result, the technician said, "two" paused, and then, "one three". "two point one three" repeated the nurse, and put down the phone. She wrote "2.13" in the book, this value being significantly below treatment value. The technician didn't say whether this level was toxic and he didn't check whether the nurse had understood.

The actual reading was 213 showing high toxicity that required immediate treatment. The mistake was not discovered until two days later, by which time irreversible liver damage had occurred and the patient died a week later.

The real weakness was identified as the lack of safety checks in the system of communicating test results. In fact, no-one made a really big mistake. At least three people made a series of small ones, and the system failed to pick these up.

Prescription Errors

A frail 86-year-old lady, was suffering from rheumatoid arthritis (Medical Error 2005). Her rheumatologist started methotrexate as a disease-modifying drug and explained to her that she'd need to carry on taking methotrexate although, subject to periodic monitoring, the dose of the tablets could change.

The Senior House Officer (SHO) on duty that day wrote the discharge prescription, as "methotrexate 2.5mg once daily". The SHO didn't work for the rheumatologist and was merely covering for the ward where the lady had been an inpatient. The SHO had no rheumatology experience, and was unaware that this medication was usually given on a once-weekly basis. He had previously worked on an oncology unit where daily doses of methotrexate had been used. The SHO didn't notice the designation on the patient's drug chart that indicated that the medication was to be given weekly. The error bypassed the normal pharmacy checking arrangements in the hospital. Two weeks' medication was dispensed. The SHO didn't fill in any clinical or follow-up details on the discharge summary. This meant that no rheumatological follow-up appointments were established.

When the patient picked up her prescription, she was surprised to see that she needed to take her tablets more often than usual. She didn't say anything to the pharmacist as the hospital doctor had told her that her dosage might change.

---

[1] These levels can give an indication of Aspirin overdose.

After taking her methotrexate for about ten days she had suffered continuous nausea and vomiting which had not responded to starvation or over the counter medicines. Thinking she had a bug that was going round, she continued taking the methotrexate. The patient eventually saw her GP and made it clear that she felt unwell. Her GP noted that she was taking a daily dose of methotrexate which he investigated and the error was discovered. Some of the mitigations that the hospital considered were to:

- adapt the dispensing IT systems to design out opportunities for human error in prescribing, using default settings that incorporate warnings;
- design patient information leaflets and patient-held monitoring and dosage records.

A study (GMC 2012) by the General Medical Council in 2012 examined over 6,000 prescriptions issued at a range of GP surgeries in England. It looked at factors such as dosage, record keeping and giving patients appropriate check-ups to assess the impact of their medication. Researchers found prescription errors had been made for one in eight patients overall, and four in ten patients over 75 years of age. In summary it concluded that 1 in 20 prescriptions written featured an error. Of the errors, 42% were judged to be minor, 54% moderate and 4% severe.

Mixup of Records

A consultant urologist took needle biopsies from the prostate glands of two patients (Medical Error 2005). Each sample was labelled with the patient's details and sent to the pathology lab with a request form for analysis. The consultant pathologist noted that one sample contained cancer cells and the other did not and noted the findings on the back of each request form and later that day, the findings on each request form were dictated onto tape, attached to the corresponding request form and passed to a secretary, who put the details into the computer. The computer generated reports, which the consultant double-checked that the information in the reports matched the findings on the request form. One patient was told that his biopsy didn't find any cancer and he didn't need any treatment. The other was told that it showed adenocarcinoma of the prostate and radiotherapy was recommended.

During this process it transpired that the patient results had been mixed up so one patient's cancer had not been diagnosed and had not been receiving treatment, while the other had been receiving treatment for a cancer they did not have.

The National Patient Safety Agency (NPSA) produces quarterly reports on patient safety incidents. In 2008, an analysis of incidents from 1st June 2006 to 31st August 2008 (NPSA 2008) showed that 1,329 were due to patient identification errors. 14.5% of these resulted in some form of patient harm some of which was classed as severe. The results are summarised in the following table:

Table 1. Patient Identification Incidents June 2006-August 2008

Hospital Number	Number of Incidents	Example Report Text
Incorrect hospital number	560	*"Client admitted via A&E with incorrect name and hospital number resulting in incorrect medications, incorrect GP details and incorrect old medical notes. Client received incorrect medications treated for asthma when client is not asthmatic."*
Incorrect matching of hospital numbers	297	*"CT scan viewed for the wrong patient [same first and last names]. Patient had been admitted with suspected renal artery stenosis. Having reviewed the incorrect scan, doctor reported that scheduled angiogram /angioplasty was unnecessary but that patient had ovarian cancer."*
Patient has more than one hospital number	141	*"Discovered upon patient discharge from hospital, that patient details incorrect on discharge letter. Further investigation via electronic hospital records revealed that name had been spelled incorrectly, 2 different home addresses and 2 separate hospital numbers but same date of birth on different records, each containing her medical information. This lady has significant chronic health problems. Hospital doctors unable to find her records due to the mistake. Discharge letter not sent to her GP as secondary record did not have information required. Failure to identify this problem would have resulted in patient not receiving newly prescribed medications. The circumstances may have resulted in a significant risk to her health and life."*
Incorrect matching of NHS number	64	*"Blood samples processed incorrectly by laboratory staff. NHS number belonged to a different patient with the same surname but a different date of birth. GP surgery chasing result on patient. Specimen had been matched to twin brother. NHS number available but not used. Member of staff used DOB search."*
Incorrect NHS number	43	*"Request for patient incorrectly booked in with another patients details."*

A Personal Perspective – 2

The following is a personal opinion and insight from a safety engineer who has worked at a major hospital on the organisation and its handling of data

In summary there are significant data problems and problems with data governance. There are a number of supply chains at play both internally and with other Trusts and GP centres. The provenance and ownership is often not clear; in one case there is a feedback loop where the same data is exchanged back and forth with another Trust. Addressing information remains the most volatile. The lack of overall design oversight means that the dependency on data and the impact of change are not fully understood and this has led to safety issues even when IT suppliers have built system correctly to specification.

A key data issue for the Trust, and this is inherent in the nature of healthcare work, is duplicate data. Without knowledge of a patient, the A&E department may re-register a patient even though they may already be in the hospitals systems. These records need to be merged at some point and this is a safety-related activity as the information can contain information on blood groups and test results. Incorrect merges have historically led to harm in the past. Clinicians perform the merging process but there seems little added assurance of the data feeds that inform this merging process. There remains no overall top-down approach to management of safety data.

Some other examples of other issues with data in the hospital are: (i) Scan results were not presented in the correct sequence and so consultants could potentially have been misled as to which was the latest result;(ii) Uncontrolled change of the configuration data for a networking device led to network outage and loss of availability of mobile clinical systems and (iii) Asset change information as a result of an asset upgrade/replacement had not been disseminated, which led to the invalidation of the current configuration data.

Data Sequencing

Eight children were listed in an ENT ward for routine operations (Medical Error 2005). They were having grommets (to drain the ear and improve hearing), a tonsillectomy, an adenoidectomy or a combination of these operations. After the second operation, a ward nurse phoned the theatre to say that the third child on the list had eaten a sweet and was therefore not starved for surgery. The next child was sent down to theatre but without informing the surgeon or the scrub nurse of the change in the list. This child's tonsils were then inadvertently removed when the child was only due to have a grommet inserted.

The organisation has since been looking at standardisation for marking surgical sites and to introduce more checking on patient's identity.

Although this paper focuses on the UK healthcare sector, problems are often reported elsewhere. In 2011 the U.S. Food and Drug Administration acknowledged the following cases which are all data-related to some extent (Chicago Tribune 2011):

- A patient died after a computer network problem caused delays in transmitting a critically important diagnostic image.
- Vital signs from patient monitors disappeared from electronic medical records after being viewed by hospital staff.
- A patient died after getting therapy meant for someone else after a wrong name was entered electronically on a scan performed by radiologists

2.3 Formalising Data Safety Mitigations

In previous sections we saw a selection of real-life examples where loss of critical properties of data caused or contributed to harm. These are just a few of the many issues that were analysed, but these start to give some indication of the types of mitigations that could be employed to manage data safety risks. Taking the entire dataset of issues together, it is possible to categorise these issues more formally and state the types of mitigation typically employed in practice. With an understanding of the current practices the aim is to use this knowledge to inform more structured guidance and approaches to mitigating data safety risk.

2.3.1 Typical Data Issues

The following 13 areas provide a useful framework for classifying typical healthcare data issues[1]:

1. Fluidity
2. Reuse
3. Ageing
4. Transformation

---

[1] Note that there is no claim here that these are normative or exhaustive, but they do provide a structure for thinking about data issues and their mitigations.

5. Ownership
6. Archiving and Retrieval
7. Biasing
8. Defaulting
9. Sentinels
10. Aliasing
11. Dissociation
12. Masking
13. Falsification

1. **Fluidity**

The Problem: Hardware and software undergo specific product assurance and once assured may change relatively infrequently[1]. Where change is required to hardware or software, it can be carefully managed and the impact on the safety case appraised. This is not always the case for data, which is often much more fluid and sometimes not under formal change control; indeed the ease with which data can be changed is one motivation for the move towards data-driven healthcare systems. An example is a hospital prescribing engine which has rules for drug interactions; these are defined in local configuration files which are frequently updated. Timeliness is a key issue: time based sequences can be critical (e.g. timestamps on diagnostic images monitoring the progress of a tumour) and data received out of sequence, e.g. before patient registration. The fluidity of data also makes controlling changes difficult: consider how changes to core reference data might be managed across different inter-operating systems in conjunction with the associated system software upgrades while maintaining a live service. One increasing issue is the move from manufacturers of health systems making changes, to healthcare staff themselves making changes. Healthcare "self–service design" has become a big issue – users can do it themselves but this clearly introduces risks.

Known Mitigations: A key mitigation strategy is to take a 'solutions architecture' approach to data. This entails developing an architectural model for where data is used in the overall enterprise and establishing the relationships that exist between component parts e.g. identifying the interfacing contracts between systems and management strategies for message flow. Management of change can be handled through good governance and configuration management practices, ensuring change is properly impact as-

---

[1] Many manufacturers of health systems follow agile development processes and are typically releasing upgrades on a monthly basis. However, these upgrades should be subject to an assurance process.

sessed and ensuring all data stakeholders are identified and engaged in the risk and impact assessment and implementation of change.

2. **Reuse/Sharing**
    The Problem: Data used in one Healthcare system is often used in a different system or system context, for example, when exchanging data between Trusts. Just because data was valid for use in a particular system, it does not immediately follow that it can be reused again in a different system. Many considerations associated with data reuse are similar to those of software reuse, for example: similarity of requirements; similarity of role in system and similarity in required integrity / assurance level. One consideration that is different is that of timeliness: data that was valid for use in a particular system at a particular time is not necessarily valid for reuse in the same system at a different time. This is closely related to the issue of Semantic Interoperability (i.e. the data means the same across different systems). This is a big issue for sharing: sharing improves efficiency, and can give the right data at the right time, however there are greater risks due to misinterpretation (and also possibly multiple modifications). Shared data can cause huge issues (e.g. a GP may interpret the same coding differently to a lab technician). Another example might be a radiology report that is annotated differently and interpreted differently by different users. It is important that the context is known (e.g. blood pressure may be taken in different ways: standing up, lying down) and this context is recorded with the data. Single data items without context can be subject to misinterpretation and can subsequently be dangerous. There is also the risk that a user may think they are seeing the whole record, but in fact something is missing (or partial) and key data (e.g. allergies) are omitted. Note there are potentially big savings in avoiding repeated testing if data already collected could be trusted e.g. blood tests done by a GP are often repeated by a hospital. An example of poor reuse is the use of patient safety indicators derived from administrative data sources. These can be valuable tools for case identification and the monitoring of rates at a single organization but should not be used to compare rates across hospitals. In a report (PSNET Case 137 2006) the authors state that in data derived from administrative sources, more than half of the observed variation in risk-adjusted rates of postoperative infections seen across hospitals could be attributed to differences in coding practices and not actual outcomes.

    Known Mitigations: A solution is to use a globally available Summary Care Record or similar so the data is stored once in a single global trusted repository. OpenEHR (OpenEHR 2016) is way of standardising on common data values and this could potentially address semantic interoperability issues. Also use of Healthcare Information Exchanges / Digital Collaboration Platforms which transform data (i.e. map data to common values).

3. **Ageing**
The Problem: All safety-related data has a lifetime and this needs to be explicitly managed. For example, data relating to the current medication of a patient will age over time. Without explicit events to trigger updates to the data, the data can become out of date and possibly misleading. If a system records an anticoagulant as a patient's current medication that they no longer take, then this could mislead clinicians and affect the outcome in the treatment of an acute stroke patient. In many hospital systems records are updated in a way that preserves old entries, but it is normal practice for the most recent data entry to be assumed to be the correct one (whether actually current for the patient or not). Alerts are often set up, based on lab results, some of which are time dependent. Alerts have time expiry dates/times (e.g. for flu results vs. MRSA - flu dates will expire in a matter weeks, but MRSA is much longer). However expiry dates can be wrong. An important related issue is that notes tend not to ever be deleted, but filtered to show views. But of course, the filtering could be configured wrongly, or be executed wrongly producing a misleading result. Data associated with current best practices can also age and issues can ensue if these are not communicated to clinicians.

Known Mitigations: These include prompts to ask users to check that data is still current (e.g. for current medication and allergies). Other mitigation strategies involve refreshing, purging, deletion or alerting based on the data and its age.

4. **Transformation**
The Problem: Data is often filtered, mapped, fused or aggregated as it moves through systems, sometimes creating new data sets as a result. Data properties are not necessarily preserved by these processes. The main issues are loss of heritage, history and origin; data can also appear to become something else. The issues for safety are that the integrity may become lowered to the lowest common denominator and this may not be recognised. In healthcare, *Semantic Interoperability* is a huge issue. In healthcare there are many different data sets used to encode common clinical terms such as medical conditions, prescriptions, units, procedures etc. Data exchanged between systems using different or non-standard encodings will need interpretation and transformation to be meaningful, accurate and ultimately safe for use in the target system. Typical examples would be the encoding of lab results and imaging data. Also, changes of units could cause

issues. For example, in recent years the reporting of glycated Haemoglobin (an indicator used in the control of diabetes) has changed from a percentage figure to mmol/mol. Systems unaware of this change of units (or make errors in transforming between the units) could therefore fail to flag an individual with severe diabetes as the abnormal ranges will be different. Reference tables and data sets can be used to help interoperability but these change frequently and configuration control and distribution of functional updates across disparate interoperating systems can be a challenge.

Known Mitigations: Some medical information systems do transform data, but others try to avoid changing the data, acknowledging the safety issues that might result. Other mitigations are to use standard tables: ICD, OPCS, NHS Data Dictionary, access to lexicons/reference information, including descriptions along with codings, using common standards for data exchanges such as OpenEHR, and mapping to common codes via an interface engine. One good example here is SNOMED-CT (SNOMEDCT 2016). Primary care system manufacturers are now being contracted to recode their systems in SNOMED-CT which will enable data exchange without transforming the data. It has been used extensively in secondary care (but not in primary where suppliers use other coding standards).

5. **Ownership**

    The Problem: The transformation of data can result in a lack of clarity regarding who has ownership of, and responsibility for, the data (if anyone). It is important that responsibility for errors can be tracked, for example, to determine whether they were present in the initial data or whether they arose as part of the transformation process. In Healthcare, sometimes complex and lengthy data supply chains of data can be established or evolve over time. For example, data can be captured at many points of patient encounters and the information exchanged amongst a variety of Health IT Systems such as GP System, pharmacy, hospital and community systems as wells as on to secondary use systems. Another example relates to the requesting of lab and imaging tests for inpatients – typically ordered against a hospital consultant but checked and reviewed by a fluid team of more junior medical staff. This means that there can be a lack of ownership and provenance for data. Related to ownership, there is a debate around consent: patients can opt-out of sharing data which can then become a big issue for practice (management of consent is also a problem in that it adds a lot of complexity to IT system design). Signoffs/signouts and handovers are a part of normal life in a healthcare environment but present multiple opportunities for information to "drop between the cracks", as ownership and responsibility for data is unclear. In (PSNET Case 134 2006) multiple signouts caused information to be missed or used in error (see Figure 2, below).

Known Mitigations: The key mitigation for this type of issue is to take an enterprise architecture view of the data and its lifecycle. This will describe the data supply chains, the responsibilities for data acquisition, production and consumption and establish well defined contracts for data exchange. For consent related issues there are some mitigations in current use, for example showing that data is present but not actually showing it to the user (although just knowing a value is present may have implications, e.g. HIV test). Sometimes it is allowable to open up a medical record but to log who has had access to what.

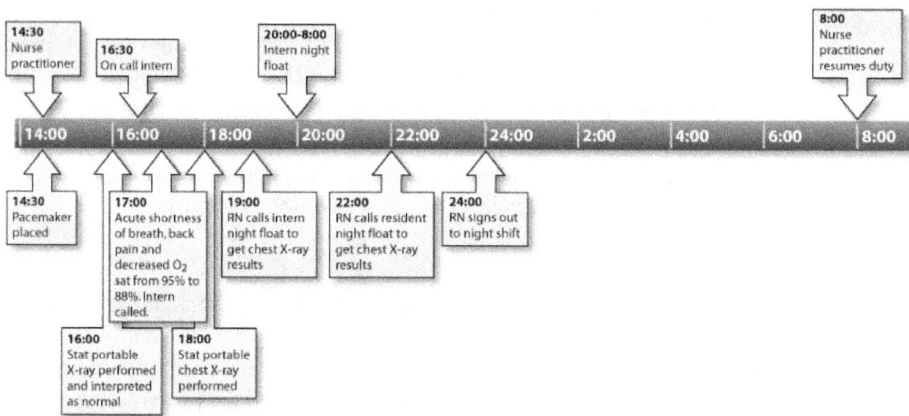

Fig. 2. Handoff Timeline[1]

6. **Archiving and Retrieval**

The Problem: Safety-related data needs to be available when required. There is thus a need to think about data accessibility over the complete system lifetime. It is also important to consider what properties of the data need to be preserved and how this affects the choice of storage medium. For example, it may be natural to focus on the availability of patient administration systems but availability of, say, a pathology system may be just as important as delays in processing a blood result could result in adverse patient outcomes.

---

[1] Reprinted with permission from AHRQ Patient Safety Network: Vidyarthi A. Triple handoff [Spotlight]. AHRQ WebM&M [serial online]. September 2006. Available at: https://psnet.ahrq.gov/webmm/case/134

Known Mitigations: Successful mitigation strategies involve establishing a clear enterprise wide understanding of the data lifecycle. This will involve understanding the criticality of access to data in all of a healthcare provider's systems and establishing plans for that data throughout its lifetime from creation through to deletion. In some instances it may be possible to add access to archived data through operational systems as part of the user interface. Such plans should include business continuity considerations – if data became unavailable, what contingency provisions would be in place to ensure the clinical business processes are not materially impacted? If data is lost, how could it be restored? What level of data loss is tolerable?

7. **Biasing**

The Problem: This is a systemic inaccuracy in data due to the characteristics of the process employed in the creation, collection, manipulation, presentation and interpretation of data. Hence this is usually unintentional distortion in the data set, which may be due to how the set has been selected or originated; it may also arise through the injudicious use of default data values. Biasing could be caused by faulty equipment, etc. but most medical equipment is subject to regular calibration which reduces the "window" of biased results, however calibration is potentially costly, involving specialist skills. There is now an issue with the new breed of mobile apps and associated sensors together with personal monitoring equipment, as these will not be subject to the same calibration regime. Ranges in reference data could also cause issues. Dropouts in the data, or loss of portions of data can be considered a biasing problem, as it is rarely independent of the data values themselves, e.g. in (NCBI 2002). Here the apparently common problem with clinical trials is the missing data caused by patients who do not complete the study. Possible reasons for patients dropping out include: death, adverse reactions, unpleasant study procedures, lack of improvement and early recovery. Missing data in a study due to dropouts may cause the analysis to be subject to a potential bias.

Known Mitigations: Biasing can be mitigated through statistical analysis of data (e.g. mean deviations) and trend analysis of results over time. Cross-checks across larger independent samples (e.g. nationally published statistics) can also help detect biasing at a local level. Biasing can also be closely linked to defaulting – if a drop down menu always displays the first entry then a user may forget to change that field and there will be biasing towards the default. This is resolved by forcing the user to make a selection in every case rather than relying on a default (as discussed in the next issue).

8.   **Defaulting**
The Problem: Many systems use default or initial values for data items; sometimes in data sets and sometimes embedded in software. Often these default values are designed to be neutral e.g. "0" or unrealistic e.g. "VOID" so as to avoid impacting the processing or ever being used. However this is not always the case and defaults may be used when in fact they should always be replaced by real data. There are therefore issues of default values being used by mistake. Clearly any default values in a clinical context, e.g. for a blood group, could cause harm if used without subsequent verification. Usually systems ask for explicit checking, for instance to confirm all known allergies (or to confirm that it was not possible to check allergies) on an e-discharge form. Use of electronic forms can cause defaulting to the previous values used on that form. As highlighted earlier, use of drop-down menus can lead to the wrong item being used if nothing is selected. In evidence given by the Royal College of Anaesthetists (Parliament 2015), particular dissatisfaction was expressed about the design of an incident reporting form from a system in widespread use to report patient incidents. The main problem is that it uses very limited drop down menus and has minimal space for free text. The drop down menus are mandatory and many do not offer appropriate clinical choices. Consequently reporting is very poor, with users forced to use inappropriate choices, leading to inaccurate descriptions of incidents. This instance could also lead to aliasing (see later).

Known Mitigations: Mitigations for defaulting are associated with reducing the cases where a default is used and increasingly forcing the user to consciously select a value. For example, rather than defaulting to the first item, the user is forced to select an item every time. A slightly weaker mitigation would be to default to the last selected value.

9.   **Sentinels**
The Problem: A sentinel or trigger value is a data value that is used to indicate a special action needs to be taken, typically indicating the end of a record or a data set, or that a particular action needs to be taken when it is encountered. The sentinel value should be one that is not allowable in the data set itself, however sometimes this is not properly considered and may use sequences of allowable values (e.g. five '9's). Sentinels can cause problems in several ways: i) where they are not recognised and so, for example, processing continues past the sentinel, and ii) where the data itself somehow contains the sentinel value and so the specific action takes place erroneously; iii) where the value does not apply to a patient with a specific condition (e.g. O_2 levels for a patient with existing COPD are knowingly lower than

normal) and should not trigger alerts. An example is where incorrect patient records are missing a sentinel value causing fields to become confused or in a worst case, two patient records to be merged as one. In current medical IT systems abnormal values are often used in messages to trigger alerts which appear on screen to make the clinician aware of something important.

Known Mitigations: The main mitigation strategies for sentinels is to discourage their use where possible, especially using encodings where a string sequence imparts special meaning which could easily be confused with real clinical data. Where record delimiters are required the preference is to use open standards for record construction where such delimiters are more universally known.

10. **Aliasing**
The Problem: This is an effect that causes different data to become indistinguishable when accessed; that is, there is only one item when there should be several. This could be due to the way the data is filtered, sampled, indexed, stored or retrieved. The data issues are typically related to loss of resolution leading to similar data appearing to be identical. For example, merging patient records as a result of multiple clinical visits could result in different medication events being erroneously treated as a single event. Also, updates of master records via multiple systems can result in aliasing problems with the last update overriding others. Aliasing can also occur in selections, in particular drop-down menus when many items in the list look similar or related (in fact in some cases there can also be duplicates), and often the first closely matching one is chosen. An example of this is reported in (PSNET Case 386 2016) where a clinician was entering the order for a CT scan; there was a long list of options that were labelled in many different ways with no clear distinctions (see Figure 3, below).

Known Mitigations: A number of techniques can mitigate the risk of aliasing. These include: ensuring the history of record merges is kept so the provenance of the current dataset can be checked, introducing robust duplicate identification support and the crosschecking of data with similar records held in other systems.

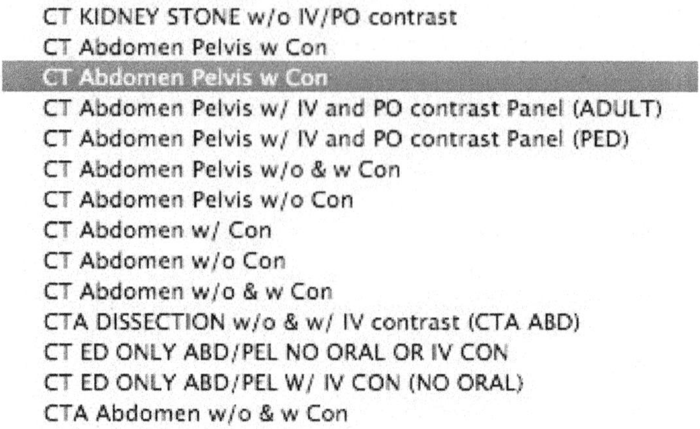

CT KIDNEY STONE w/o IV/PO contrast
CT Abdomen Pelvis w Con
CT Abdomen Pelvis w Con
CT Abdomen Pelvis w/ IV and PO contrast Panel (ADULT)
CT Abdomen Pelvis w/ IV and PO contrast Panel (PED)
CT Abdomen Pelvis w/o & w Con
CT Abdomen Pelvis w/o Con
CT Abdomen w/ Con
CT Abdomen w/o Con
CT Abdomen w/o & w Con
CTA DISSECTION w/o & w/ IV contrast (CTA ABD)
CT ED ONLY ABD/PEL NO ORAL OR IV CON
CT ED ONLY ABD/PEL W/ IV CON (NO ORAL)
CTA Abdomen w/o & w Con

Fig. 3. Example Picklist Resulting from Typing "CT abd" Into an Order Search[1]

11. Disassociation

The Problem: This effect is, in some senses, the opposite of aliasing; there are several records when there should only be one. This could occur, for example, if two records are created for the same individual using slightly different names (see Table 2, below). It could also arise if different systems use different indexing methods and the association between the indexes becomes corrupted. In Healthcare this is common as individuals will present themselves at different clinical settings such as GP, A&E, Community Hospitals, etc. and there will be multiple patient identifiers and often duplicate registrations even within the same system. This is a huge issue for healthcare. There can also be issues when different abbreviations are used for the same item, for instance "HB" or "HLB" used to refer to haemoglobin, with multiple entries in reference data. This can cause confusion and possibly lead to mistakes. There can also be a lack of a holistic view of a patient record; critical data can be held in different records in different systems and a single consolidated view cannot be presented. Seeing only a subset of the total dataset can lead to other issues such as 'assumption by omission' – clinicians may be tempted to assume absence of data such as an allergy in a dataset means the patient does not have any allergies. An example of using two disjoint records leading to a patient incident is (PSNET Case 382 2016): here a patient's new prescription was reflected in the outpatient medical record, which was on a different information tech-

---

[1] Reprinted with permission from AHRQ Patient Safety Network: Wears RL. Unintended consequences of CPOE [Spotlight]. AHRQ WebM&M [serial online]. October 2016. Available at: https://psnet.ahrq.gov/webmm/case/386

nology platform than the hospital system and did not communicate with it. At the time of hospital admission, the patient did not remember his new prescription. Unaware that the patient had started taking the new drug, the anaesthetist placed an epidural catheter through which a second drug was administered. The patient survived but required intubation and a brief intensive care unit stay.

Known Mitigations: Existing mitigations include defined patient merge systems; however merged records may not be handled by all systems. Also, it may not be possible to merge patients reliably without expert assessment. However mistaken merging can lead to the need for unmerging; this does happen occasionally and can be a very difficult problem to resolve as it may not be clear which records the data originally came from. Differing local and NHS numbers for the same patient are a big issue. One mitigation is to use a master Patient Administration System (PAS) (within a single organisation). Nationally the PDS (Personnel Demographic Service) based on a single record for a patient is preferable, and all health organisations in the UK should use this ultimately. Note that for sexual health systems aliasing is specifically allowed (also for mental health), i.e. patients can register under alternate or made-up names so as to ensure they remain anonymous (but systems may still be able to use the master NHS number in some situations).

Table 2. Example of how a hospital might write one patient's name 22 different ways[1]

J N Tobias	JTobias	Tobias, Jonah Noel
J Noel Tobias	Tobias J	TobiasJ
J Tobias	Tobias J N	TobiasJN
JNTobias	Tobias, J	Tobias J N
Jonah N Tobias	Tobias, J N	TobiasJonahN
Jonah Noel Tobias	Tobias, J Noel	TobiasJonahNoel
Jonah Tobias	Tobias, Jonah	
JonahNTobias	Tobias, Jonah N	

---

[1] Reprinted with permission from AHRQ Patient Safety Network: Koppel R. EMR entry error: not so benign. AHRQ WebM&M [serial online]. April 2009. Available at: https://psnet.ahrq.gov/webmm/case/199

12. **Masking**

The Problem: This issue can arise if a notable proportion of a data set is of a poor quality. This poor data can generate a lot of error reports which can then hide or mask errors from the system when handling the good quality data. This is a "wood for the trees" problem. Likewise, erroneous data can be filtered silently thus giving the impression of correct operation and data completeness. This category also includes cases where data is deliberately used in an erroneous manner in order to derive some benefit, for example, to use a field to carry data that it was not designed for. In such cases, actual errors in the data are therefore masked from proper error detection. In general, using a tool for which it was not intended can mask underlying erroneous behaviour IAEA (2001). Alert fatigue is also a case of error masking. It is accepted that alert fatigue does harm, and in this case (Allscripts 2016), potentially impacts clinical decisions. The article reports that there were alerts that clinicians frequently overrode as mere interruptions, or that they did not read for long enough to have any effect on their decisions.

Known Mitigations: The mitigation for masking is to address the root cause and improve, where possible, the quality of the data being produced. This may involve applying greater development rigour to the assurance of the system or replacing the system with another with better pedigree. Deliberate misuse, albeit with good intentions, can be addressed through training, awareness of system issues, understanding of data criticality and increased restriction and lock downs.

13. **Falsification**

The Problem: This issue arises where data held in paper or electronic format is manipulated (created, modified, deleted) in order to deliberately mislead potential consumers of that data. This might typically be done to give the impression that an event occured when in fact the event never actually took place (and vice versa, to show an event did not occur when in fact it did). This is surprisingly prevalent in healthcare but rarely causes a major issue. An example is where a clinician fabricates diagnostic readings such as blood glucose readings for a vulnerable patient to avoid (for whatever reason) having to take the reading – a route explored in the prosecution case in (Thimbleby 2017). A high-profile falsification case was reported in the Guardian, involving nurses making-up blood test results (Guardian 2015):

Fig. 4: Lauro Bertulano, Rebecca Jones and Natalie Jones have been struck off the nursing register by making up blood test results at Princess of Wales hospital, Bridgend

Known Mitigations: Typical mitigations for falsification revolve around techniques for "non-repudiation", that is, assuring that an individual cannot deny they did or did not take some action associated with the data. For example, digitally signing transmitted data, strong access controls and activity logging, maintaining sufficient audit records to 'replay' a transaction in its entirety.

3 Mitigating Data Safety Risks – New Guidance and Standards

At the time of writing (October 2016) the following guidance has been submitted as a formal supplement to the Implementation Guidance for the current NHS IT standards. The future aspiration is to incorporate the guidance into the healthcare standards themselves but it is acknowledged that this is a complex and time consuming process. Note also that the related international standard IEC 80001-1 (IEC80001 2010) is being extended to include health IT systems and manufacturers (currently only software and user organisations are covered). This means that this standard may eventually supersede the NHS standards and so this would become the target for guidance on data safety.

The content of this guidance is largely drawn from the work of the Data Safety Initiative Working Group (http://www.scsc.uk/groups.html) encapsulated in their Data Safety Guidance document DSIWG (2017). Full credit is therefore given to their work and contributions in informing this sector specific guidance.

3.1   Audience

This guidance is applicable to all involved in the establishment and operation of the Clinical Risk Management process for Health IT Systems.

3.2   Scope

This guidance is intended as a supplement to the Implementation Guidance for the following standards:

- SCCI0129 - Clinical Risk Management: its Application in the Manufacture of Health IT Systems (SCCI0129 2016) and
- SCCI0160 - Clinical Risk Management: its Application in the Deployment and Use of Health IT System (SCCI0160 2016).

3.3   Assumptions and Constraints

It is assumed that the reader is familiar with the SCCI0129 and SCCI0160 standards and their associated Implementation Guidance.

3.4   Guidance

3.4.1   Data Safety Issues

Data can give rise to risks which may be different to other system elements such as hardware and software. An awareness of data-related issues promotes a more comprehensive assessment of safety risk. Typical issues that arise in the Healthcare context have been covered in earlier sections and include:

- Fluidity
- Reuse
- Ageing
- Transformation
- Ownership
- Archiving and Retrieval
- Biasing
- Defaulting
- Sentinels
- Aliasing
- Dissociation
- Masking
- Falsification

3.4.2   Data Types

There are many different data types that can have safety implications. The Data Types section of (DSIWG 2017) provides a more comprehensive list. In this guidance, five of these types of data are discussed as presented in the following table. These are the data types where there is significant reliance in the Healthcare context and most likely to be encountered in Health IT Systems.

Table 1. Safety-related Data Types

Type	Description	Explanation	Typical containers
Verification	Data used to test and analyse the system	This is data comprising the test values and test data sets used to verify the system. It may include real data, modified real data or synthetic data. It includes data used to drive stubs, and any data files used by simulators or emulators.	Test data sets, Stub data, Emulator and Simulator files

Type	Description	Explanation	Typical containers
Infrastructure	Data used to configure, tailor or instantiate the system itself	Data used to set up and configure the system for a particular installation, product configuration, or network environment	Network configuration files, Initialisation files, Hardware pin settings, Network addresses, Passwords, etc. Messaging and interface engines
Performance	Data collected or produced about the system during trials, pre-operational phases and live operations	Data produced by and about the system during introduction to service and live service itself. Includes fault data and diagnostic data. This may be the results of various phases of introduction and may include trend analysis to look for long-term problems.	Field data, Support calls, Bug reports, Non-Compliance Reports, Defect Reporting and Corrective Action System data.
Dynamic	Data manipulated by the system during operations	This is the data processed or produced by the system which has end-user meaning. It may be displayed and used within the system or may be for transfer or distribution to other systems or downstream users. It is data that has some real domain meaning.	May be manipulated within the system in data structures or transferred into or out of the system through interfaces.
Justification	Data used to justify the safety position of the system	Data used to justify the safety position of the system. Data used to justify, explain and make the case for starting or continuing live operations and why they are safe enough. Often passed to external bodies (regulators, Health and Safety Executive, Independent Safety Auditors) for their review	Clinical Safety Case Report, Certification case, Regulatory documents, COTS Justification file, Design Justification file.

3.4.3   Data Properties

When considering how data can give rise to harm, it is more useful to define this as "where failure to preserve a critical property of data results in a hazardous system state that in combination with an external event can give rise to harm". For example, missing data from patient allergy records gives rise to a hazardous system state. Harm can occur if this information is then relied upon (the external event in the definition) – for example, administering penicillin when a record of an allergy to this drug has been lost.

A non-exhaustive list of typical data properties were given earlier in the document as listed below:

- Integrity
- Completeness
- Consistency
- Continuity
- Accuracy
- Resolution
- Traceability
- Timeliness
- Verifiability
- Format
- Availability
- Fidelity / Representation
- Priority
- Sequencing
- Intended Destination/Usage
- Accessibility
- Suppression
- History
- Lifetime
- Disposability / Deletability

In general, the objective of the risk management process is to assess which of these properties are critical to each data set in the scope of the analysis and establish mitigations to reduce the risk of occurrence. Not all properties will be critical in all cases. For example, precision of an INR (International Normalized Ratio) reading may be more critical than an individual's estimated weekly alcohol intake. The important point is to identify those properties that are significant for the data set in a given context of use and to ensure those properties are preserved with a sufficient level of confidence.

3.4.4   Assessing Applicability of Managing Data Risk

As data safety risk assessment is a relatively new discipline, Manufacturers and Health Organisations will naturally want to understand to what extent the guidance applies to them and how much effort will be required in addressing the risks. Patient safety is of course a key reason for investing effort in managing safety risk but the level of investment is also driven by the organisation's appetite for risk and the wider corporate level implications that may arise from patient safety incidents and accident.

The Organisational Data Risk (ODR) Assessment Form has therefore been developed as a useful tool to capture a high-level perspective of the risk posed to an organisation by data safety issues within a specific enterprise. The ODR is based on the ISO 31000 standard for risk management. The ODR is effectively a multiple choice questionnaire covering different aspect of data safety that in totality provide an overall holistic assessment of the corporate risk for the given enterprise.

The ODR firstly asks the assessor to establish the context of the risk assessment to ensure that the enterprise being considered and the scope of the assessment is well defined. The ODR questions will then establish:

- The risk tolerance of external stakeholders;
- The level of risk that is allocated to the organisation;
- The applicable regulatory environment within which the enterprise will operate;
- The maturity of the organisation in terms of their attitude to not simply risk, but specifically data-driven risks;

- Responsibilities for data ownership through the use cases of the system;
- The data-driven specifics about failure consequences and the issues raised by data complexity, boundary complexity and system complexity for the enterprise.

The ODR is intended to be a simple and straightforward method of determining and quantifying how exposed an organisation is to data safety risk for a given enterprise. If the level of effort required in managing risk is commensurate with the level of risk exposure, then the ODR can give an early indication of the scale of effort required to manage the organisational risk. The ODR itself is presented in an Appendix of the Data Safety Guidance (DSIWG 2017).

3.4.5   Managing Data Safety Risks

The following sections provide specific guidance for those Manufacturers and Health Organisations responsible for implementing SCCI standards. Each of the key requirements from the standards is presented along with specific guidance for meeting those requirements from a data safety perspective. Throughout, the guidance is supported by practical examples based on a Manufacturer building a new Health IT System and the Health Organisation that will be deploying it. These examples are by no means exhaustive, but should give insight into how to apply to guidance in practice.

3.4.6   Establishing the Context

The guidance in this section relates to the following requirements in the relevant SCCI standards from the perspective of managing data safety:

Table 2. SCCI Requirements relating to Establishing the Context

Standard	Section	Requirement
SCCI0129	4.2.1	The Manufacturer MUST define the clinical scope of the Health IT System which is to be deployed.
SCCI0129	4.2.2	The Manufacturer MUST define the intended use of the Health IT System which is to be deployed.
SCCI0160	4.2.1	The Health Organisation MUST define the clinical scope of the Health IT System which is to be deployed.
SCCI0160	4.2.2	The Health Organisation MUST define the intended use of the Health IT System which is to be deployed.
SCCI0160	4.2.3	The Health Organisation MUST define the operational environment and users of the Health IT System which is to be deployed.

From a data perspective, fundamental to fulfilling these requirements is establishing the context within which the use of data in system development, enhancement, introduction, integration or operation is occurring. This should establish the risk appetite: essentially, how much effort is devoted to making data risks as low as practicable. In turn, this will inform the nature and scope of assessments that are conducted during system development and, furthermore, its introduction into operational service.

Like other components of a safety-related system, the safety dependency of data is dictated by the context in which it is used and the causal links that become established where loss of one or more of the required properties can contribute to hazardous system states. For example, a given data set (say Configuration Data) could be used in a number of separate contexts such as:

- prototyping a system to demonstrate solution feasibility of a safety-related system;
- development testing of a safety-related system; or
- live operational use of a safety-related system.

In these cases, the same data set could be used but the context of its use changes the safety significance and therefore the level of assurance that it may require. Hence, not only is the type of data under consideration important but also when in the organisation's particular process lifecycle the data will be used and relied upon. It follows that the assessed integrity level of a data set is also predicated on where and when in the lifecycle the data set will be applied. It is recommended that these considerations are addressed in the Clinical Risk Management Plan by modelling the organisation's lifecycles and explicitly documenting where a specific data set will be used and therefore subject to further assurance techniques.

Example

A Manufacturer is building a new integrated health and social care system to support holistic care for community health services. The system supports clinical workflows for aspects such as referrals, tracking clinical encounters, appointment scheduling, outcome measures through to letter and report generation. The system follows a typical development lifecycle as a series of phases: Business modelling, requirements, analysis & design, implementation test and deployment.

The Manufacturer decides to build an early prototype to show to clients to help elicit requirements definition. To support this, the Manufacturer plans to create a test data set that comprises a typical range of scenarios that the system will encounter. This form of data is identified as Verification Data. The system also needs to be configured to support deploying Health Organisations' policies. This data is Infrastructure Data and for the prototyping phase, the Manufacturer plans to use largely default values.

In later phases when the system functionality is specified and the system is being built, the Manufacturer plans to create a test data set that will be key to demonstrating the correct functioning of the system and hence acceptance by the deploying Health Organisation. This still involves the use of Verification Data and Infrastructure Data but there will be far greater dependency on these data sets than the prototyping case.

The Manufacturer therefore documents the planned use of each of the data types during the entire delivery lifecycle in the Clinical Safety Management Plan.

The procuring Health Organisation will have a different perspective of the IT system that they will deploy into their organisation. They will already have many integrated systems in live operations and as part of establishing the context for the system's deployment they will need to consider many different types of data sets:

- Infrastructure Data: how the system will be configured in the specific environment;
- Verification Data: the test data sets to be used to support certain deployments such as integration testing and training;
- Dynamic Data: the data entered or fed into the system and the data presented to the user, generated in the form of reports or data passed to other systems.

Post acceptance, the procuring Health Organisation decides to run a series of user training sessions for clinicians. Once users are trained, the system will be integrated into live operations. The Health Organisation identifies the Infrastructure, Verification and Dynamic Data types to be used during these phases. The Health Organisation also realises that the system will form part of a data supply chain as a number of external organisations and departments within their own organisation engage in the procurement and use of safety-related data. For example, it will receive referral data from a number of other GP systems, it will receive outcome measures from hospitals and clinical data acquired from remote workers visiting patients in the community and from the patients themselves using the system's online portal. The system also produces data for other external systems such as electronic prescriptions for pharmacies.

The Health Organisation understands that it is important that when establishing the context that the roles within a data supply chain are clearly understood and defined in the Clinical Risk Management Plan:

- **The Commissioning User**: an organisation or unit of an organisation that has the need for the data;
- **The Data Provisioner**: the organisation or unit of an organisation that will fulfil that need for data;
- **The Data Acquirer:** the organisation employed by or collaborating with the Data Provisioner to carry out physical collection of data.

The Health Organisation sees that by using the new system it will become a Commissioning User as it will require and be a Consumer of data from GP's systems, hospital systems, systems used by remote worker in the community and the system's portal capturing data entered by the patients themselves, each of these acting as Data Provisioners. Those healthcare professional (and the patient themselves) gathering patient data through physical inspections and measurement are the Data Acquirers.

The Data Provisioner acts both as a Consumer (from the Data Acquirer) and Producer (to the Commissioning User) of data. Similarly, an organisation that augments data sets is both a Consumer and Producer of data in the supply chain. Data augmentation may be, for example, to add patient family history to a data set.

The Health Organisation defines the data supply chain relevant to the system including the roles and interfaces involved in its Clinical Risk Management Plan. This will therefore show where there are dependencies on Dynamic Data used and produced by the system.

Questions the Health Organisation will need to address when establishing the context are:

- Have all the dependent interfaces been identified?

- Have the roles of Commissioning User/Data Provider/Data Acquirer been established and acknowledged?
- What 'service levels' or contracts exist for the delivery of the data?
- What level of assurance do Data Providers/Data Acquirers provide for their data?

3.4.7   Risk Identification

The guidance in this section relates to meeting the following requirements in the relevant SCCI standards from the perspective of managing data safety:

Table 3. SCCI Requirements relating to Risk Identification

Standard	Section	Requirement
SCCI0129	4.3.1	The Manufacturer MUST identify and document known and foreseeable hazards to patients with respect to the intended use of the Health IT System in both normal and fault conditions.
SCCI0160	4.3.1	The Health Organisation MUST identify and document known and foreseeable hazards to patients in both normal and fault conditions through the introduction and use of the Health IT System

The Risk Identification phase for data comprises an analysis of all relevant data items and data sets and considering whether harm can arise if some or all of the properties for those data items/data sets were lost. As discussed earlier, this assessment must be conducted based on knowledge of the context and project phase in which the data is being used. For example, a data set used in a proof of concept system cannot give rise to harm but the same data set used in live clinical operations could potentially cause patient harm.

To support the hazard identification assessment it is sometimes useful to consider the use of guidewords. Guidewords help hazard identification by providing a structured qualitative set of expressions to be considered for each data property. For example, the following guidewords can be used when considering the integrity property of a data set: 'Correctness', 'truth', 'original', 'trustworthy', 'coherency', 'stability', 'perfect', 'unquestionable', 'faithful', 'certain', 'ordered'. A series of guidewords have been developed to support this form of analysis for data related hazard. (see HAZOP Guidewords in the 'Identify Risks' section of the Data Safety Guidance (DSIWG (2017)).

Example

The Manufacturer of the Health IT System decides that during the prototyping phase there is little safety dependency of the test and configuration data sets as no clinical decisions will be made based on their content; the data is simply being used to support the elaboration of requirements.

However, in later phases when the system functionality is specified and the system is being built, the Manufacturer will want to create a test data set that will be instrumental in demonstrating the correct functioning of the system. This still involves the use of Verification Data and Infrastructure Data but there is far greater dependency on these data sets than the previous case. For example, if the verification or configuration data is not sufficiently diverse or insufficiently models real world scenarios, it is possible that erroneous and unsafe functional behaviour is present in the system during live operation despite this system having passed factory and site acceptance testing.

The Health Organisation will likewise need to conduct risk identification relevant to their deployment context. Hazards arising from data sources that are to be delivered into the new system from existing systems need to be assessed for data risks.

From the Health Organisation's perspective, one key focus for hazard identification is in the use of Dynamic Data, ie. The data that will be delivered into the new system from existing system data sources, and the data presented to the user. For the interactions identified in the supply chain, the Health Organisation needs to consider the risks associated with loss of properties of the data it will receive. Questions the Health Organisation will need to consider and address more formally in the Clinical Risk Management Plan are as follows:

- Which data sets or items being received from other systems have properties (such as timeliness, completeness, consistency, fidelity etc.) that are significant to patient safety?
- What data presented to the user has properties (such as availability, format, resolution, etc.) that are significant to patient safety?
- What existing barriers or mitigations (physical, technical, procedural) exist to reduce the risk of loss of data properties?

3.4.8   Risk Analysis

The guidance in this section relates to meeting the following requirements in the relevant SCCI standards from the perspective of managing data safety:

Table 4. SCCI Requirements relating to Risk Analysis

Standard	Section	Requirement
SCCI0129	4.4.1	For each identified hazard the Manufacturer MUST estimate, using the criteria specified in the Clinical Risk Management Plan:
		- the severity of the hazard
		- the likelihood of the hazard
		- the resulting clinical risk
SCCI0160	4.4.1	For each identified hazard the Health Organisation MUST estimate, using the criteria specified in the Clinical Risk Management Plan:
		- the severity of the hazard
		- the likelihood of the hazard
		- the resulting clinical risk

In order to analyse risks and, more particularly, to align data safety with other risk management processes, there is a need to overcome problems stemming from the use of the term "likelihood" in a situations where there may be no quantifiable failure rates. The Data Safety Assurance Level (DSAL) has been developed to address this. The DSAL metric is not a statistical measure of likelihood, or a literal numeric measure of integrity. Instead, the DSAL is qualitative and determines the level of rigour required in demonstrating that critical properties of data are sufficiently well preserved. As such, DSALs share a common theoretical basis with concepts like System Integrity Levels and (Item / Function) Development Assurance Levels.

The table below presents a classification system allocating DSALs to safety-related data items using a function of likelihood and consequences. The system described below is not prescriptive and can be tailored depending on the context; the reasons for any such tailoring should, of course, be documented and agreed amongst relevant stakeholders. Furthermore, where this table suggests a low DSAL because (for example) a work-around is simple to implement it is important to ensure that the work-around (or similar) is actually implemented.

To make the analysis applicable to safety-related data, the "consequences" are subdivided further into five categories which impact on the level of risk:

- Proximity: how directly a data failure will lead to an accident;
- Dependency: how dependent the application is on the dataset;
- Detection: the likelihood of being able to detect a data failure prior to an accident;
- Prevention: the ability of the systems architect/developers to guard against errors;
- Correction: the ability of the system to work around or correct errors.

Table 5. DSAL Classification Matrix

		Likelihood of Data Causing Accident		
	Concern	High	Medium	Low
	Proximity	A known use of the data[6] is highly likely to lead to an accident.	A possible use of the data could lead to an accident.	All currently foreseen uses of the data could lead to harm only via lengthy and indirect routes.
	Dependency	Data is completely relied upon.	Data is indirectly relied upon.	Little reliance on data.
	Detection	Low or no chance of anything else detecting an error.	Some other people/systems are involved in checking the data.	Many other people/systems are involved in checking the data.
	Prevention	Difficult or impossible to guard/barrier against errors.	Possible to guard/barrier against errors.	Easy to guard/barrier against error
	Correction	Difficult or impossible to correct or workaround errors.	Possible to correct or workaround errors	Easy to correct or workaround errors.
Severity or impact of data related accident				
Negligible	Negligible harm. Negligible environmental impact.	DSAL0	DSAL0	DSAL0
Minor	Minor injury or temporary discomfort for 1 or 2 people. Minor environmental impact.	DSAL1	DSAL0	DSAL0
Moderate	An accident resulting in minor injuries affecting several people or one serious injury. Some environmental impact.	DSAL2	DSAL1	DSAL1
Major	A serious accident resulting in serious injuries affecting a number of people, or a single death. Major environmental impact.	DSAL3	DSAL3	DSAL2
Catastrophic	An accident resulting in several deaths. The accident could affect the general public or have wide and catastrophic environmental impact.	DSAL4	DSAL4	DSAL3

There are a number of instances where the system architecture could justify the movement of a dataset from one DSAL for another. For example there may be cases where multiple independent data items fulfil the same (or similar) usage and assessors may choose to reduce the DSAL requirements on each item to reflect the inherent redundancy (and associated risk reduction) that brings. Additionally, the same dataset may be reused by multiple functions with different levels of risk, in this case it would be recommended to assign the highest required level of integrity to the dataset so that it meets the requirements of the most demanding use case. As discussed, the implementation of a classification system is the responsibility of the Manufacturer or Health

Organisation, but any such manipulation of DSALs should be carefully considered and appropriately documented and agreed in the Clinical Risk Management Plan.

Example

The Manufacturer had previously identified phases where the use of specific types of data could give rise to hazards. In the first, the prototyping phase, the Manufacturer sees no use of the data that can give rise to credible clinical risk and assessing the DSAL for that data set as DSAL0. In the second phase of the development lifecycle, where Verification Data and Infrastructure Data is being used to demonstrate the correct functioning of the system, the Manufacturer considers that loss of any of the data properties of Integrity, Completeness, Consistency, Continuity, Format, Accuracy, Resolution, Timeliness, Availability, Fidelity/Representation, Sequencing, Intended Destination/Usage of this data could give rise to hazards. For example, if the verification data set selected is not representative of the eventual diversity experienced in practice, then it is possible that the system may contain latent software errors that could give rise to harm. However, the Manufacturer acknowledges that the system will be subject to further testing and trials in the clinical setting and so there will be other opportunities to detect errors in the system:

- the likelihood that the data use gives rise to an accident is Medium as other systems and processes are in place that would detect errors;
- the severity is Moderate; failings in the system could give rise to non-optimal treatment plans for a patient that might delay detection of a more serious condition or prolong the recovery for a known condition.

The Manufacturer therefore assesses these data types as DSAL1 in this particular context of use.

From the Health Organisation's perspective, the main focus for risk assessment is in the use of Dynamic Data. For the interactions identified in the supply chain the Health Organisation needs to consider the risks associated with loss of properties of the data it will receive and present to the user. Questions the Health Organisation will need to consider and address more formally in the Clinical Risk Management Plan are as follows:

- How likely is it that there would of a loss of the given data property?
- How would such a loss of a property be detected?
- How would such as loss be isolated to prevent further risks of harm?
- What recovery action would be required to resolve the issue to maintain patient safety?

In considering the receipt of outcome measures data received from a hospital, the Health Organisations considers that it is likely that some credible errors would not be readily detected by their new system; if the hospital system confused a result or there were errors in the precision of data then there would be few chances to catch these once received by the system.

- the Health Organisation assesses the likelihood of this loss of property as High;
- The impact of such errors, although not realistically likely to lead to death, could cause delays to treatment that result in serious injury and hence Moderate impact.

The data received from this data source is therefore classed as DSAL2 in this particular context of use.

3.4.9   Risk Evaluation

The guidance in this section relates to meeting the following requirements in the relevant SCCI standards from the perspective of managing data safety:

Table 6. SCCI Requirements relating to Risk Evaluation

Standard	Section	Requirement
SCCI0129	5.1.1	For each identified hazard, the Manufacturer MUST evaluate whether the initial clinical risk is acceptable. This evaluation MUST use the risk acceptability criteria defined in the Clinical Risk Management Plan.
SCCI0160	5.1.1	For each identified hazard, the Health Organisation MUST evaluate whether the initial clinical risk is acceptable. This evaluation MUST use the risk acceptability criteria defined in the Clinical Risk Management Plan.

Once all data items and/or data sets have been classified under the DSAL scheme, an evaluation is conducted as to whether the resulting classification is acceptable for the data at each relevant point of use in the project/service/operational lifecycle. Such an evaluation may consider what existing assurance and mitigations exist to manage the risk of loss of critical properties to the data item or data sets.

Consideration should also be given to any arrangements in place with other parties who may produce or consume the data. For example, if a data set bears a high DSAL then a producer of that data should be aware that this places an assurance obligation on them to ensure data properties are maintained to a level of confidence commensurate with the risk. Such obligations would ideally be embodied in the contractual arrangements with 3rd parties. If no such arrangement or assurance can be guaranteed then further mitigation will be required to reduce the DSAL for that dataset. For example, additional validation of received data or crosschecks with other independently held sources of the same data.

Example

The Manufacturer has determined there is some, albeit low, risk (DSAL1) associated with its use of data at a specific point of its lifecycle. The Manufacturer evaluates this risk and considers that the risks should be reduced further by taking some reasonably practicable steps.

Likewise, the Health Organisation has identified DSAL2 data and needs to ensure risks are reduced as low as reasonably practicable.

3.4.10 Risk Treatment

The guidance in this section relates to meeting the following requirements in the relevant SCCI standards from the perspective of managing data safety:

Table 7. SCCI Requirements relating to Risk Treatment

Standard	Section	Requirement
SCCI0129	6.1.1	The Manufacturer MUST identify appropriate clinical risk control measures to remove any unacceptable clinical risk.
SCCI0160	6.1.1	The Health Organisation MUST identify appropriate clinical risk control measures to remove an unacceptable clinical risk.

Where the DSAL of a data item/data set is evaluated as being too high then mitigations must be introduced to reduce the DSAL to an acceptable level. There are many methods for the treatment of data safety risks that will differ based on the type of data and the DSAL level. For example, some techniques may only be applicable to Dynamic Data as opposed to Configuration Data and the higher the DSAL, the more rigorous the technique should be in assuring that the data property is maintained.

Specific guidance on how data-related risks may be treated is presented in tables in the Data Safety Guidance (DSIWG 2017) based on the data type and the DSAL of the data. These methods and techniques represent good practice in mitigating the risks associated with safety-related data. The tables indicate where a particular method/approach is applicable to a given lifecycle data type, and for each Data Safety Assurance Level, whether the method/technique is:

- No recommendation for or against being used (-);
- Recommended (R);
- Highly Recommended (HR).

The lifecycle data types considered in the referenced version of the guidance are as follows:

- Verification (V);
- Infrastructure (I);
- Dynamic (D);
- Performance (P);
- Justification (J).

The methods/approaches are not intended to be prescriptive, but they should be sufficiently well-defined to allow interpretations to be applied to the given context in which the guidance will apply. Methods/techniques employed are expected to be more rigorously applied as the DSAL level increases. For example, the depth, level of coverage and effort/resources employed for analysis techniques must be proportionate to the DSAL - sampling may be appropriate for lower DSALs where full coverage will likely be expected for higher DSALs. Assurance methods and approaches must be considered for each stage of the data lifecycle as appropriate for the given DSAL. Strategies for dealing with large data sets must be fully justified with respect to the DSAL.

The Clinical Risk Management Plan must document:

- planned compliance with the tables;
- the interpretation for the given method/technique (e.g. depth of checking);
- justification in the case where a technique is not to be adopted.

The overall safety justification, the Clinical Safety Case Report, for the given project/service/operational context must then provide evidence of compliance against the plan.

It is important to note that the tables are not intended to be exhaustive. As well as only considering certain data types each table only considers one particular aspect of data safety. No claim is made that the collection of tables provides complete coverage either across the system lifecycle or across all possible approaches to one part of the lifecycle. Also, these tables merely identify techniques using a few key words; in almost all cases, further information on the technique should be readily available from a range of sources. Despite these limitations, it is hoped, however, that the guidance they contain will prompt considerations that lead to the use, and justification for the use, of an appropriate set of methods and approaches for any given system.

Example

Having decided that further risk reduction is necessary, the Manufacturer needs to select assurance methods and technique that are appropriate for DSAL1 data and in doing so demonstrate that reasonably practicable steps have been taken to reduce the risk. The Manufacturer therefore refers to the tables in the Data Safety Guidance (DSIWG 2017) for guidance. For DSAL1 Verification Data, the tables show that the following are recommended (R) or highly recommended (HR):

Table 8. DSAL1 recommendations for Verification Data

Ref	Technique	R/HR	Focus / Description
			System Design
SD1	Syntax Check	R	Semantic checking of data values and sequences based on defined rule

Ref	Technique	R/HR	Focus / Description
SD2	Sanity / Reasonabil-ity Checks	R	Dedicated processing implemented to check that data is within rea-sonable tolerances and/or logically/semantically consistent with what the data represents. For example, range checks, date checks, record counts, record sizes, special values (e.g. NaN) etc.
	Data Design		
DD1	Governance Model	R	A governance model is established that defines aspects such as data ownership, processing roles & responsibilities (who can do what to the data), processing authorisations and permissions (what can be done to the data) etc.
DD2	Data Flow Diagram	HR	To describe the data flow in a diagrammatic form.
DD3	Data Model	HR	To articulate how data is organised.
DD4	Client Sign-Off	R	
DD5	Configuration Man-agement	HR	The recording of the production of every version of every 'signifi-cant' deliverable and of every relationship between versions of the different deliverable.
DD6	Data Dictionary	HR	A data dictionary is a collection of descriptions of the data objects or items in a data model for the benefit of data users.
	Data Implementation		
DI1	Review / Inspection	HR	Manual review/inspection of data possibly involving data visualisa-tion tools.
DI2	Ground-Truth Check	R	Inspection against physical measurements (eg. lengths, positions, heights) taken in the real world.
DI3	Auditing	R	A period of comprehensive internal and external testing of the data quality process, where Data is verified according to its intended use and definition.
DI4	Authorisation	R	A security model is established to control who is authorised to cre-ate, view, edit, delete the data.
DI5	Authentication	R	Data is authenticated to validate its provenance.
DI6	Defined Confidence / Trust Levels	R	Criteria are established to provide an objective measurement of the confidence or trust in a given dataset.
	Test Data		
TD1	Using Informal / ad-hoc means	R	This is where data is generated by simple spreadsheets, or by simple scripts or programmes. It may also be legacy data or basic assump-tions. There is no formal checking or review of the method of gener-ation.
TD2	Using Manual means	R	Simple test data can be produced by manual means, although this may be prone to human error. However manual checking of a sam-ple of test data generated using tools is a useful verification method.
TD3	Using Initial Runs of New System	R	This method is often used where the system is breaking new ground and there is no prototype or legacy system to produce test data. This must be carefully used as initial operations can be very different to eventual usage, and so the test data suite must also evolve.

Ref	Technique	R/HR	Focus / Description
TD4	Derived from Real Data	R	Where real data is available this is usually a good basis for generating test data (e.g. by modification to increase the test space coverage). However there are potential issues of sampling and coverage, i.e. is the real data a representative sample?
TD5	Produced by Client	R	Ideally the client is involved in producing or at least checking the test data. The client will often know the data intimately and can highlight any issues quickly.
TD6	Client Sign-Off	R	Where possible, the client should formally agree and sign-off the test data as appropriate. This gives the system developer some confidence in the data and also some protection of the data is in fact incorrect or not representative.
TD7	Error Seeding	R	This is where errors are deliberately inserted into the dataset to demonstrate the effectiveness of data validation.
TD8	Data Reuse	R	Reusing data for one project that was created and thoroughly assured for another project. This can be effective but the read-across should be established
TD9	Feedback testing	R	To check output data by comparing it.
	Media - Paper		
MP1	Photographic Copies	R	
MP2	Scan to Electronic Format	R	
MP3	Indexing / Cataloguing	R	
	Media - Electronic		
ME1	Regular Refresh / Rewrite	R	Of magnetic media or flash memory. Life of a hard disk might be <3 years. Life of a properly-stored DVD might be 5 years. Life of a USB might be 10 years. Life of a magnetic tape might be longer if a clean tape drive was used and the tape was stored properly.
ME2	Suitable Physical Environment	R	Store media in a clean, low-humidity environment at a steady temperature, cool but not cold
ME3	Copies at Different Locations	R	Physically separate to cover natural disasters, accidental or malicious damage.
ME4	Backups / Duplication	R	Backups are essential. Frequency of backup is dependent on the rate of change of the data. The number of generations of backup to be kept should be commensurate with the impact of data loss.
ME5	Sample Restores	R	Sample restores should be performed at intervals to ensure that the backups are readable and retrievable.

From these tables the Manufacturer decides on a series of activities to implement the recommendations that are applicable to its particular endeavour. These activities are expressed as a series of requirements that can be placed on the Manufacturer's delivery organisation and tracked through to completion.

Table 9. Requirements established to implement DSAL1 recommendations for Verification Data

Ref	Requirement	Guidance Reference
R1	The verification data shall be carefully controlled in the Manufacturer's configuration management system. There shall be a configuration management plan that shall define who has responsibility for the data and who is authorised to create and amend it.	DI4, DI5, DD1, DD5
R2	The verification data shall be held on an industry standard fileshare that is regularly backed up with copies moved periodically to offsite storage. The Backup/Recovery plans shall include periodic sampling of restores.	ME1, ME2, ME3, ME4, ME5.
R3	The data shall be modelled as a series of patient 'journeys' that cover the entire lifecycle of data from first encounter through to archival and deletion of data. The complete set of journeys shall be chosen to exercise all the functionality of the system. The modelling shall include a data dictionary, data flow diagrams and a data model.	DD2. DD3, DD6
R4	To model data from external systems, the Manufacturer shall use manual data entry and spreadsheet based records to hold the data.	TD1, TD2
R5	The Manufacturer already has a set of clinical standing data that was used for another system and derived from real data. This data includes data such as encounter codes, clinical terms, consultant names, surgery and hospital addresses etc. and this shall be reused for this system. The Manufacturer's Clinical Safety Officer has reviewed the data and agreed on its suitability for reuse.	TD4, TD8
R6	Some of the verification data sets shall include errors deliberately inserted to check the effectiveness of data validation.	TD7
R7	The controlled verification data set shall be subject to review and analysis against defined confidence/trust criteria. Scripts shall be written to check for syntax and semantic consistency of the data and provide a basic sanity check. The scripts themselves shall be validated and verified before use.	SD1, DI1, SD2, DI2, DI6
R8	The project shall be subject to an internal delivery quality assurance audit.	DI3
R9	Data loaded from external system into the system and displayed to the user shall be crosschecked against the original source data, using manual spot-checks.	

Ref	Requirement	Guidance Reference
R10	The level of rigour employed in verifying all the above requirements shall be commensurate with the DSAL criticality and so an ISO9001 compliant quality management system shall be adopted.	All

The following guidance recommendations have not been adopted by the Manufacturer for the reasons given. Note that some may however become relevant in the future so actions are set, where appropriate, to review the applicability of the recommendation when the given condition is met.

Table 10. Justification for the non-adoption of recommendations

Ref	Requirement	Guidance Reference
E1	TD3	The data will be used before any initial run of the system.
E2	DD4, TD5, TD6	There is no contracted client at the moment as the system is a new developed so it will not be possible to get the client to create or signoff data.
E5	MP1, MP2, MP3	There are no paper based resources for this system

4. Conclusions

This paper has considered the widespread use of safety-related data within healthcare and focussed on data within hospital IT systems. Some potential safety issues have been identified, and although some mitigations are acknowledged to be in place and working, the changing use of IT, and consequential clinical dependence on it, presents new problems and challenges. To improve the situation the proposed new guidance for handling data in Health IT Systems is presented and discussed, together with a worked example. Also recommended is the use of user-centred design practices involving clinical users at all stage of the product life cycle to assist in the context, use and risk management of IT in the delivery of clinical practice.

A Personal Perspective – 3

The following is a personal opinion and insight from a safety engineer who has worked at a major hospital:

It is essential that health trusts appreciate the risks arising from incorrect data. Once they do they can challenge suppliers and their staff specifying systems to demonstrate/prove the mitigations put in place to address the risks due to data.

Acknowledgments   Thanks to Ian Bingham, David Lund, Sean White and the SCSC Data Safety Initiative Working Group (DSIWG).

References

Allscripts (2016), Fewer alerts reduce the amount of "think time" for clinicians, http://blog.allscripts.com/2016/03/10/fewer-alerts-reduce-the-amount-of-think-time-for-clinicians/, accessed October 2016

BNF (2016), British National Formulary, https://www.bnf.org/

Chicago Tribune (2011), Baby's death spotlights safety risks linked to computerized systems http://articles.chicagotribune.com/2011-06-27/news/ct-met-technology-errors-20110627_1_electronic-medical-records-physicians-systems, accessed October 2016

DigitalHealth (2013), http://www.digitalhealth.net/news/28932/glasgow-praised-for-it-crash-response, access Oct 2016

DoH independent report (2016) Making IT Work: harnessing the power of health information technology to improve health care in England, https://www.gov.uk/government/publications/using-information-technology-to-improve-the-nhs/making-it-work-harnessing-the-power-of-health-information-technology-to-improve-care-in-england , Department of Health accessed November 2016

DSIWG (2017) Data Safety Guidance, The Data Safety Initiative Working Group, http://www.scsc.uk/p133 (available Feb 2017)

Faulkner et al (2000) The Safety Management of Data-driven Safety-related Systems, 19th International Conference on the Reliability, Safety and Security of Critical Computer Application (SAFECOMP 2000), http://www.eng.warwick.ac.uk/~neil/papers/safecomp%202000.pdf, accessed Sep 2016

Faulkner (2004), Data Integrity – an often-ignored aspect of safety systems, (EngD thesis), http://wrap.warwick.ac.uk/1212/, accessed Sep 2016.

Guardian (2014) https://www.theguardian.com/society/2014/mar/05/samuel-starr-three-heart-death-nhs-operation-scan, accessed Oct 2016

Guardian (2015) Nurses jailed for falsifying stroke patients' records, https://www.theguardian.com/uk-news/2015/dec/14/nurses-jailed-for-falsifying-stroke-patients-records

GMC (2012) Investigating the prevalence and causes of prescribing errors in general practice: The PRACtICe Study (PRevalence And Causes of prescrIbing errors in general practiCe)

Huffington (2011) http://www.huffingtonpost.com/2011/04/06/hospitals-sodium-overdose_n_845689.html , accessed Oct 2016

IAEA (2001) Investigation Of An Accidental Exposure Of Radiotherapy Patients In Panama

IEC61508 (2010), http://www.iec.ch/functionalsafety/standards/page2.htm, accessed Oct 2016

IEC80001 (2010) IEC80001:2010 Application of risk management for IT-networks incorporating medical devices -- Part 1: Roles, responsibilities and activities

Inge (2011), Safe Data: Recognising the Issues, Safety-Critical Systems Club – Newsletter, Volume 21, Number 1. http://www.scsc.uk/news.html?v=21&n=1&a=b&pap=877, accessed Sep 2016.

Knight et al (2004), Specification and Analysis of Data for Safety-Critical Systems, 22nd International System Safety Conference, Providence RI, http://www.cs.virginia.edu/~eas9d/papers/issc.04.pdf, accessed Sep 2016.

Medical Error (2005), National Patient Safety Agency

NCBI (2002) Problems in dealing with missing data and informative censoring in clinical trials, https://www.ncbi.nlm.nih.gov/pmc/articles/PMC134476/#B4

NIB (2014) Personalised Health and Care 2020, November 2014.

NPSA (2008) Quarterly Data Summary 10 patient id errors from failure to use or check ID numbers correctly (NRLS-0827-F1-QDS-10)

OpenEHR (2016), http://www.openehr.org/, accessed October 2016

Parliament (2015), Written Evidence submitted by the Royal College of Anaesthetists (CCF0021) http://data.parliament.uk/writtenevidence/committeeevidence.svc/evidencedocument/public-administration-committee/nhs-complaints-and-clinical-failure/written/17463.pdf pg 1, accessed October 2016

PSNET Case 134 (2006), https://psnet.ahrq.gov/webmm/case/134/triple-handoff, accessed October 2016

PSNET Case 137 (2006), https://psnet.ahrq.gov/webmm/case/137/getting-a-good-report-card-unintended-consequences-of-the-public-reporting-of-hospital-quality, accessed October 2016

PSNET Case 382 (2016), https://psnet.ahrq.gov/webmm/case/382/falling-between-the-cracks-in-the-software, accessed October 2016

PSNET Case 386 (2006), https://psnet.ahrq.gov/webmm/case/386/unintended-consequences-of-cpoe, accessed October 2016

RTCA (1998), DO-200A, Standards for Processing Aeronautical Data http://www.rtca.org/onlinecart/product.cfm?id=238, accessed Oct 2013.

SCCI0129 (2016) Clinical Risk Management: its Application in the Manufacture of Health IT Systems, V4, 15/02/2013 (updated 05/05/2016)

SCCI0160 (2016) Clinical Risk Management: its Application in the Deployment and Use of Health IT Systems, V3, 15/02/2013 (updated 05/05/2016)

SCSC (2012) "How to Stop Data Causing Harm", http://www.scsc.uk/e209

SCSC (2015) "How to Stop Data Causing Harm: What You Need to Know", http://www.scsc.uk/e343

SNOMEDCT (2016), U.S. National Laboratory of Medicine, https://www.nlm.nih.gov/healthit/snomedct/index.html , accessed October 2016

Story & Faulkner (2001) The Role of Data in Safety-Related Systems, Proceedings of the 19th International System Safety Conference, Huntsville, AL, http://www.eng.warwick.ac.uk/staff/ns/papers/data-based%20systems%20paper.pdf, accessed Sep 2016

Story & Faulkner (2003) The Characteristics of Data in Data-intensive Safety-related Systems, Lecture Notes in Computer Science, Volume 2788, 396-409

Story & Faulkner (Q4 2003) Data – The Forgotten System Component?, Journal of System Safety http://www.eng.warwick.ac.uk/~neil/papers/forgotten%20component%20paper.pdf, accessed Sep 2016

Thimbleby (2017) Cybersecurity problems in a typical hospital (and probably in all of them), SCSC SSS'17, Feb 2017

Safety critical systems - A brief history of the development of guidelines and standards

Ron Bell

Engineering Safety Consultants (ESC) Ltd

London, UK

Abstract   *This paper provides an overview of the development of standards and guidelines for safety critical systems over the past 35 years. In the context of this paper "safety critical systems" refers to those systems that are intended to achieve, together with the other risk reduction measures, the necessary risk reduction to meet the required tolerable risk. The period covered by the paper is from the time that concerns were raised about the adoption of programmable electronic systems for implementing safety functions to today. It is essentially a personal account based on experience and reflections of the developments have taken place with to respect of guidelines and standards and is not intended in any way to be an authoritative account covering all industrial sectors.*

1 Introduction

During the 1970s the design principles for safety critical systems, in the context of manufacturing industry and the process sector, were based on qualitative criteria and methods to increase the safety integrity of a basic design were based on design methods which included:

    i.   Simplex or single-channel systems having no redundancy. This architecture would be regarded as a basic design having minimum safety performance.

ii.   Dual channel systems with redundancy for sensors and final elements (e.g. Contactors/valves). This architecture would be regarded as having medium safety performance.

iii.   Dual channel systems with redundancy for sensors and final elements with different forms of diagnostics for improving the diagnostic coverage of the critical elements. This architecture would be regarded as having a high safety performance.

iv.   Fault criteria ("no single or double fault would cause the system to fail to a dangerous state");

v.   Use of the concept of "Failure to safety" for specific failure modes (e.g. on failure of the power supply this would lead to a safe failure of the safety functions).

In the context of safety critical systems in the manufacturing sector the approach to safeguarding machinery using safety critical systems was based on BS 5304 "Code of practice: Safeguarding of machinery" and first published in 1975.   Design features in this code of practice were essentially qualitative in nature as was the risk-based approach to selection of the various methods. The essentials in this guide existed, with various revisions, until well into the 1990s where such guidance was then provided by European Standards such as BS EN 954-1:1997 "Safety of machinery. Safety related parts of control systems. General principles for design."   This latter standard was based on qualitative approaches to the design with a qualitative risk assessment method (i.e. risk graph). Both BS 5304 and EN 954-1 related to safety critical systems based on non-programmable elements.

Based on the above design methods various elegant designs with great ingenuity were developed and were primarily focused on dangerous failures arising from random hardware failures, and well established good practice, to avoid or control systematic dangerous failures. Often they were so ingenious it was quite difficult in those days to determine objectively what performance level had been achieved since quantitative analysis was not the norm.

One of the earliest approaches using quantified analysis to address dangerous random hardware was adopted by the Heavy Organic Chemicals Division of ICI in their development of a High Integrity Protective System (HIPS) for what we would now refer to as a major hazards plant (Stewart & Hensley (1971)). The approach adopted was to develop a quantified risk target based on process sector experience and to adopt a quantified approach to determining the Probability of Failure On demand (PFD) of the safety functions to be performed in respect of random hardware failures. This was truly ground-breaking territory in the industrial, non-nuclear, sector.

In the nuclear sector, several research documents were published which provided important building blocks for the future of functional safety. These were developed by the United Kingdom Atomic Energy Authority (UKAEA) through the Systems Reliability Service. Such publications included:

- The variability of failure rate data (Edwards);
- Defences against common-mode failures in redundancy systems; A guide for management designers and operators (Bourne et al.);
- Reliability and Protection Against Failure in Computer Systems. (Daniels)

During the early 1980s, both within the manufacturing sector and in the chemical process sector, increasing use was being made of computer-based systems for controlling equipment and plant. During this period, particularly in the manufacturing sector there was increasing use of programmable controlled robots. This raised major challenges as to how to safeguard robots and in the early days the safety functions to be performed were invariably based on hardwired solutions. That is, to cater for a dangerous failure of the programmable element, carrying out an interlocking function, a non-programmable interlocking arrangement was designed to act in parallel with the programmable element to carry out the safety function in the event of a dangerous failure of programmable element. The hardwired arrangement would invariably be based on electromechanical devices (e.g. relays, contactors).

In the period 1980-1990 there were many initiatives and developments relating to the safe use of Programmable Electronic Systems (PESs) although initially the focus was on safety critical software rather than safety critical systems. Some of these initiatives and developments are summarised in Table A.1.

Note: Table A.1 is not intended to be an authoritative catalogue of initiatives and developments during the 1980 to the mid-1990s but provide a snapshot of some of the key issues that are considered of importance to the author and the activities associated with IEC 61508.

2 Developments during 1980 - 1990s

The period 1980 to the mid-1990s established many of the fundamental building blocks and principles for what is now referred to as the discipline of functional safety. It was a period of intense activity and cooperation amongst many different groups within the UK and other countries in Europe and internationally.

At the beginning of this period there was excessive caution in the adoption of programmable electronic elements for safety-related systems and a high level of uncertainty as to how such technology could be used for safety applications and gain acceptance from safety regulators. At the end of that period there were much higher levels of confidence that reasonable solutions were available, albeit that for many specific areas of functional safety there were gaps that needed filling, and the filling of these gaps depended on professional judgement.

Some of the key building blocks established during the period 1980 to the mid-1990s are now embedded in accepted good practice and include:

- Effective management of functional safety;
- A safety lifecycle approach that covered all relevant phases from initial concept until final decommissioning;
- The concept of dangerous random hardware failures and dangerous systematic failures and the need to develop measures and techniques to combat both types of failure;
- The concept of a Safety Integrity Level (SIL) for the specified Safety Function that was required to deal with dangerous systematic failures;
- The concept of the Safety Function contains the specification for:
    - ✓ the functionality of what must be achieved to prevent the hazardous event; and,
    - ✓ the safety integrity (i.e. SIL) necessary to achieve the required target risk.

Note 1: during this period the distinction between a Safety Function and the Safety-Related System was not sufficiently distinguished. With hindsight, the adoption of a Safety Function was a significant building block in achieving functional safety.

- Quantified Target Failure Measures for a Safety Function having a specified SIL;
- Competence of persons involved in any safety life-cycle activity that had to be formally addressed and justified;
- The need for assurance measures for all safety life-cycle activities including Functional Safety Assessment, Functional Safety Audit, Verification and Validation. Such assurance measures are relevant to both hardware and software;

Note 1: it is interesting to note that in the context of IEC 61508 the concept of a Safety Case was not built into the developing standard. To some degree the Safety Case concept was perceived as a "UK" concept (whether this be a fact or not that was the perception within the Working Group developing IEC 61508). However, several UK industries developed and adapted the concept and the principles of the Safety Case to demonstrate their understanding and management of risks within their business. (Inge)

Setting the Target Risk for a specified hazardous event was an essential parameter for the design activity. The concept of a Tolerable Risk within the UK facilitated a quantified approach to the Target Risk within the legal framework. Also, the advent of quantified target failure measures (emerging in the drafts of IEC 61508) for the safety functions facilitated the verification of the design of a safety critical system based on quantified Target Risk criteria. The use of quantified risk criteria was, and still is, a sensitive issue for many countries (e.g. A maximum risk of death of 1 in 10^{-3} per annum to workers in any industry was suggested. (Tolerability)

- During this period both qualitative and quantitative risk targets were adopted even for consequences having life changing injuries.
- The need to base the required safety performance of a safety critical system, in respect of a specific safety function, on all the risk reduction measures and risk parameters related to the specified hazardous event. Although individual elements of this concept existed prior to the application of programmable electronic systems, implementing such systems demanded a more rigorous systematic and holistic approach.

3 IEC 61508

A key standard that emerged from the various activities during the 1980 to the-mid-1990s was IEC 61508. This eight-part standard was published during the period 1998-2000 (i.e. Part 1-7) and Part 0 (which was a very basic introduction to functional safety was published in 2005. IEC 61508 has the status of being a standalone standard (i.e. it can be used on its own) but it is also a Basic Safety Standard which means that other standards within IEC are required to comply with the requirements specified in that standard

3.1 Structure of IEC 61508

The overall title of IEC 61508 is 'Functional safety of electrical, electronic and programmable electronic (E/E/PE) safety-related systems'. The Parts are as listed in Table 1.

Table 1. The Parts of IEC 61508

Part	Title
0	Functional safety and IEC 61508[1]
1	General requirements
2	Requirements for electrical/electronic/programmable electronic safety-related systems
3	Software requirements
4	Definitions and abbreviations
5	Examples of methods for the determination of safety integrity levels
6	Guidelines on the application of parts 2 and 3
7	Overview of techniques and measures

Parts 1, 2, 3 contain all the normative requirements[2] (e.g. X *shall* be undertaken) and some informative requirements (e.g. Y *should* be undertaken). Parts 0, 5, 6 and 7 do not contain any normative requirements.

Parts 1, 2, 3 and 4 of IEC 61508 are IEC basic safety publications. One of the responsibilities of IEC Technical Committees is, wherever practicable, to make use of IEC 61508, in its role as a basic publication, in the preparation of their own sector or product standards that have E/E/PE safety-related systems within their scope.

IEC 61508 is both a standalone standard and can also be used as the basis for sector and product standards. In its latter role, it has been used to develop standards for the process, nuclear and railway industries and for machinery and power drive systems and has influenced the automotive sector (see 3.3). It will continue to influence, the development of E/E/PE safety-related systems and products across all sectors. This concept is illustrated in Figure 1.

The application of IEC 61508 as a standalone standard includes the use of the standard:

- as a set of general requirements for E/E/PE safety-related systems where no application sector or product standards exist or where they are not appropriate
- by suppliers of E/E/PE components and subsystems for use in all sectors (e.g. hardware and software of sensors, smart actuators, programmable controllers)

---

[1] Part 0 has the status of a Technical Report and is purely informative.

[2] In IEC standards, a normative requirement is prefaced by 'shall' and if that requirement is relevant in the application then it is necessary to comply with the requirement. A requirement prefaced by 'should' is informative and can be considered as a recommendation but is not normative in respect of compliance to relevant requirements in the standard.

- by system integrators to meet user specifications for E/E/PE safety-related systems
- by users to specify requirements in terms of the safety functions to be performed together with the performance requirements of those safety functions
- to facilitate the maintenance of the 'as designed' safety integrity of E/E/PE safety-related systems
- to provide the technical framework for conformity assessment and certification services as a basis for carrying out assessments of safety lifecycle activities.

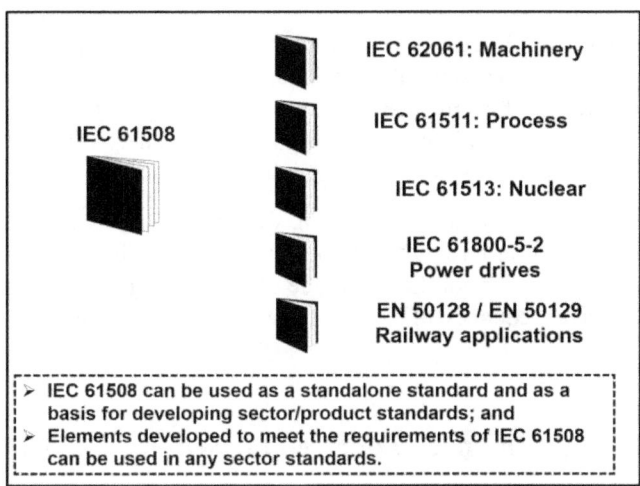

Fig. 1. Standalone and sector/product standards

Product or application sector international standards based on IEC 61508:

- are aimed at system designers, system integrators and users
- take account of sector-specific practice
- use terminology applicable in the sector to increase understanding for its intended users
- may specify constraints appropriate for the sector
- usually rely on the requirements of IEC 61508 for the design of subsystems.

3.2 Some of the key features of IEC 61508

- It enables the development of product and sector international standards, dealing with E/E/PE safety-related systems. This should lead to a high level of consistency (for example, of underlying principles, terminology etc.) both within and across application sectors; this will have both safety and economic benefits.

- It provides a method for the development of the safety requirements specification necessary to achieve the required functional safety for E/E/PE safety-related systems.
- It uses safety integrity levels (SILs) for specifying the target level of safety integrity for the safety functions to be implemented by the E/E/PE safety-related systems.
- It adopts a risk-based approach for the determination of the safety integrity level requirements.
- It sets numerical target failure measures for E/E/PE safety-related systems that are linked to the safety integrity levels.
- It sets a lower limit on the target failure measures, in a dangerous mode of failure, that can be claimed for a single E/E/PE safety-related system. For E/E/PE safety-related systems operating in:

  ✓ a low demand mode of operation, the lower limit is set at an average probability of failure of 10^{-5} to perform its design function on demand
  ✓ a high demand or continuous mode of operation, the lower limit is set at an average frequency of dangerous failure of 10^{-9} per hour.

3.3 Standards influenced / derived / based on IEC 61508

As indicated above, IEC 61508 can be used as the basis for sector and product standards and many standards have been influence or derived or based on IEC 61508. Several those standards are indicated below with the date they were originally published:

- **2003:** IEC 61511: Functional safety - Safety instrumented systems for the process industry sector.
  Note: The scope covers the requirements for the specification, design, installation, operation and maintenance of a safety instrumented system (SIF).

- **2003:** CENELEC, EN50128: Railway applications - Communication, signalling and processing systems - Software for railway control and protection systems 2000.

- **2003:** CENELEC, EN50129: Railway applications - Communication, signalling and processing systems - Safety related electronic systems for signalling.

Note: These CENELEC standards are railway-specific adaptation of IEC 61508. However, it is important to understand that they share common aspects but also differ in some key areas. (Braband, Hirao & Luedeke).

- **2005**: IEC 62061: Safety of machinery - Functional safety of safety-related electrical, electronic and programmable electronic control systems.
  Note: the scope covers the requirements and make recommendations for the design, integration and validation of safety-related electrical, electronic and programmable electronic control systems (SRECS) for machines.

- **2007**: IEC 61800-5-2: Adjustable speed electrical power drive systems - Part 5-2: Safety requirements – Functional.
  Note: The scope covers the requirements and makes recommendations for the design and development, integration and validation of safety related power drive systems (PDS(SR)) in terms of their functional safety considerations.

- **2010**: IEC 61784-3: Industrial communication networks - Profiles – Part 3: Functional safety fieldbuses - General rules and profile definition
  Note: The scope covers the safety communication layer (services and protocol) based on CPF 1 of IEC 61784-1 and IEC 61158 Types 1 and 9. It identifies the principles for functional safety communications defined in IEC 61784-3 that are relevant for this safety communication layer.

- **2011**: ISO 26262: Road Vehicles – Functional safety.
  Note: Scope is restricted to passenger cars. Work has started on Edition 2 of ISO 26262 (expected publication 2018). Edition 2 scope will now include motorcycles and truck/bus vehicles as well as new part specifically concerned with application of the standard to semiconductor devices.

- **2012**: IEC 61131-6: Programmable controllers - Part 6: Functional safety

- **2016**: IEC 61000-1-2: Electromagnetic compatibility (EMC) - Part 1-2: General - Methodology for the achievement of functional safety of electrical and electronic systems including equipment with regard to electromagnetic phenomena.

5 Concluding comments

The history of the development of guidelines and standards for what we now call functional safety have, simply put, three distinct phases:

- Phase 1 (1980 - 1990): Development and determination of the essential building blocks to achieve functional safety.

- Phase 2 (1990-2000): Refinements of the ideas developed and development of guidance based on emerging draft international draft standard IEC 61508.

- Phase 3 (2000-to date):
  - ✓ Implementation of the developed published standards in relation to safety critical products and the development of sector and product standards based on IEC 61508;
  - ✓ Adoption of the developed published standards by sectors that had been involved previously in their development. In such a situation, the migration to adoption of the standards was part of a planned process.
  - ✓ Adoption of the developed published standards by sectors that had not been involved previously in their development. In such a situation, the migration to adoption of the standards is a challenge to those sectors. Also, in such situations it interesting to know that the end user is often a key driver in requiring conformance to those standards.

In the context of IEC 61508, the revision process for IEC 61508 Edition 2 is just beginning. It is important that Edition 3 of IEC 61508 maintains the rigour of the performance levels (e.g. that is to meet the requirements of a specified SIL) but is also sufficiently flexible to meet the demands of those wishing to use the standard in new application areas (e.g. defence and aerospace) and where new Routes to compliance may need to be developed. Whatever Route is chosen there should be confidence that the performance level achieved is the same for all Routes.

Acknowledgments   In developing this paper, I am most grateful for the help given to me from the following people; Phil Bennett, Simon Brown, Brian Clarke, John Canning, Rainer Faller, Martin Goose, James Inge, Ken Simpson and David Ward.

References

Bell R (1986); Assessment Architecture and Performance of Industrial Programmable Electronic
Systems (PES) with Particular Reference to Robotic Safety. In Daniels, Safety and Reliability of
Programmable Electronic Systems, Elsevier Applied Science Publishers Ltd 1986 (ISBN 1-

85166-017-8).

Bourne et al. Defences against common-mode failures in redundancy systems; A guide for management designers and operators. Published by the United Kingdom Atomic Energy, Safety
and Reliability Directorate (1981)

Braband, Hirao & Luedeke; The Relationship between the CENELEC Railway Signalling Standards and the Other Safety Standards. Article in SIGNAL + SRAHT (95) 12/2003.

Competence: Safety, Competency and Commitment-Competency Guidelines for Safety-Related System Practitioners. Published by the IEE in 1999. Updated in 2007. Available from: http://www.theiet.org/resources/books/policy/comp-crit.cfm.

Daniels (1979); Reliability and Protection Against Failure in Computer Systems; published by the United Kingdom Atomic Energy Authority, National Centre of Systems Reliability.

Edwards (1979). The variability of failure rate data. Published by the United Kingdom Atomic
Energy Authority, Systems Reliability Service.

HSE PES; Programmable Electronic Systems in Safety-Related Applications:
1: An Introductory Guide;
2: General Technical Guidelines
Health & Safety Executive ISBN 0 11   883906 3 Crown Copyright

Inge http://safety.inge.org.uk/20070625-Inge2007_The_Safety_Case-U.htm and
Safety Critical Systems Club Newsletter; September 1993].

Managing competence #1: Managing competence for safety-related systems- Part 1: Key Guidance. (Managing competence #1). First published 2007. Available from:
http://www.hse.gov.uk/humanfactors/topics/mancomppt1.pdf

Managing competence #2: Managing competence for safety-related systems- Part 2: Supple-
mentary
Material. First published 2007. Available from:
http://www.hse.gov.uk/humanfactors/topics/mancomppt2.pdf

Out of Control: Why control systems go wrong and how to prevent failure. First published 1995.
Available from: http://www.hse.gov.uk/pUbns/priced/hsg238.pdf.

Tolerability: The Tolerability of Risk from Nuclear Power Stations¶ (1988 and Revised 1992). Available from: http://www.onr.org.uk/documents/tolerability.pdf.

IET- Safety Practices Report; A study of the computer-based systems safety practices of UK,
European and US industry (1989). Principal Author P. A. Bennett; ISBN 0 86241 700 0

SafeIT (1990)- A UK Government Consultation Document on the Safety of Computer-
controlled Systems;
SafeIT-1: The Safety of Programmable Electronic Systems
SafeIT-2: Standards framework.
Published by the Interdepartmental Committee on Engineering (ICSE)

R2P2: Reducing Risks, Protecting People: HSE's decision-making process. First published in 2001 ISBN 0 7176 2151 0. Available from: http://www.hse.gov.uk/risk/theory/r2p2.pdf.

Stewart & Hensley (1971); High Integrity Protective Systems on Hazardous Chemical Plants. Published by United Kingdom Atomic Energy Authority, Systems Reliability Service.

Appendix A: Table A.1: Some key initiatives related to the emerging concept of functional safety (1980 to the mid-1990s)

Item	Period	Projects & Initiatives	Key Outcomes
1	1980 - 1990	The Commission of the European Communities (CEC) funded a joint project between seven organisations. The project ran for 2 years from September 1983. These organisations already had programmes of research in the fields of Programmable Electronic Systems (PES's) and robotic safety. The CEC funding provided the opportunity to collaborate in the exchange of information which would, hopefully, lead to harmonization of approaches to assessment of PES's in this field across Europe. (Bell)	• Development of European approach; • Developing European network; • Identifying key challenges; • Identifying possible solutions.
2	1980 - 1987	Work began in the Health and Safety Executive (HSE), in the UK, on the development of guidance on the safe application of Programmable Electronic Systems. After several research projects and extensive collaboration and consultation with UK industry this led, in 1987, to the publication of guidance on "Programmable Electronic Systems in Safety-Related Applications" (HSE PES). This guidance was referred to as the "HSE PES guidelines" and are considered to be the first to be produced by a safety regulator covering industrial systems providing detailed guidance on the safe application of programmable electronic systems.	• Facilitated the adoption of programmable electronic technology for safety applications • The guidance was a major UK input into the Working Group developing the systems requirements of IEC 61508.
3	1980 - 1984	The UK proposed, through the relevant BSI Technical Committee, that an international standard be developed to deal with safety critical software. This led to an IEC Working Group being formed and active work on this project started in 1984. This work preceded the work on the systems aspects of functional safety. (see item 5).	• The publication of IEC 61508: Parts 1-7 (1998-2000) with respect to the specific sections concerning software.

Appendix A: Table A.1: Some key initiatives related to the emerging concept of functional safety (1980 to the mid-1990s) [Continued]

Item	Period	Projects & Initiatives	Key Outcomes
4	1985 - 1988	In 1985-1986 a UK proposal to the Advisory Committee On Safety (ACOS) of the IEC to develop an international standard on Programmable Electronic Systems (PES's) led to a Task Group to elaborate the proposal further. This led to an IEC Working Group being set up to develop an international standard on PES's.	• The Task Group elaborated the initial proposal and undeveloped basic contents. • The Task Group proposal was then voted on internationally and accepted. • This Task group document was the basis of the work undertaken by the IEC Working Group described in Item 5.
5	1988 - 2000	An IEC Working Group started work on and international standard title "Functional safety of Programmable Electronics Systems: Generic Aspects. The Software Working Group (see Item 3) and the Systems Working Group collaborated very closely with the intention of producing two standards, one for systems and one for software but after consultation on an early draft standard in 1989 it was agreed that one standard would be produced covering both systems and software requirements.	• The publication of IEC 61508: Parts 1-7 (1998-2000). • IEC 61508 was a generic standard to be used as a stand-alone but also to facilitate sector/product standards. • Safety Integrity Levels introduced in the 1989 draft.
6	1989	The IET (formerly the IET) published a document "A study of the computer-based systems safety practices of UK, European and US industry". This was funded by the UK DTI. (IET- Safety Practices Report). [IET- Safety Practices Report].	• In the context of standardisation, the overall view of the study was that a common international standard on safety related computer-based systems was supported within the UK, Europe and the USA.

Appendix A: Table A.1: Some key initiatives related to the emerging concept of functional safety (1980 two the mid-1990s) [Continued]

Item	Period	Projects & Initiatives	Key Outcomes
7	1989	DIN V 19250:1989: "Measurement and Control technology; Fundamental safety aspects to be considered for measurement and control equipment".	• This German standard had the concept of a Requirement Class (a precursor of the Safety Integrity Level (SIL)). The Requirement Class was a classification specifying the safety requirements needed to prevent and overcome specific types of dangerous failures. • The standard included a Risk Graph comprising eight Requirement Classes. (See item 8)
8	1990	DIN V VDE 0801:1990 "Principles for computers in safety-related systems".	• Specified required and recommended techniques for Hardware and Software (to a lesser extend) of computer systems for the DIN V 19250 risk levels. (See item 7)
9	1990 - 1990	A UK Government Consultation Document was published on the Safety of Computer-control Systems comprising two parts: SafeIT-1: The Safety of Programmable Electronic Systems. Covered several wide-ranging issues including that of standardisation. Indicated the advantages of all sectors following a common approach to standardisation but stressed the importance of the derivation of more specific sector standards from a generic standard. (SafeIT) SafeIT-2: Standards framework. Set out the proposed Framework Concept for the development of standards and elaborated Core Standards for the Framework. (SafeIT)	• The Standards framework influenced the development of IEC 61508. • The Safety-Critical Systems Club was conceived within the framework of the SafeIT initiative. The first Safety-Critical Systems Club Newsletter was published in November 1991.

Appendix A: Table A.1: Some key initiatives related to the emerging concept of functional safety (1980 to the mid-1990s) [Continued]

Item	Period	Projects & Initiatives	Key Outcomes
10	1991	Interim Def Stan 00-55/Issue 1 5 April 1991 – The procurement of Safety Critical Software in Defence Equipment Part 1: Requirements Part 2: Guidance	• Strong emphasis on the adoption of Formal Methods.
11	1991	Interim Def Stan 00-56/Issue 1 5 April 1991 – Hazard Analysis and Safety Classification of the Computer and Programmable Electronic System Elements of Defence Equipment. Issue 2 of 00-56 radically changed and became "Safety Management Requirements for Defence Systems" (13 Dec 1996). The emphasis evolved, certainly from Issue 3, to defence systems of all types and not just programmable.	• Claim limits introduce for dangerous systematic failures; • Claim limits based on Safety Integrity Levels (SILs); • Claim limit based on operational experience or where none existed based on qualitative failure criteria: SIL 1= Frequent; SIL 4 = Remote).
12	1988 & Revised 1992	Publication of HSE document "The Tolerability of Risk from Nuclear Power Stations". Suggested, for a risk of death for an individual worker, a risk boundary for what could be deemed just tolerable and one deemed unacceptable (i.e.10^{-3} per year). (SafeIT)	• Facilitated the establishment/demonstration of the Tolerable Risk. • This publication eventually led to the HSE publication "Reducing Risks, Protecting People". (R2P2)
13	1992	Education & Training Requirements for Safety Critical Systems IEE Public Affairs Board Report Number 12; January 1992	Recommendation: • Courses shall be developed and shall result in an accredited award (e.g. certificate).

Appendix A: Table A.1: Some key initiatives related to the emerging concept of functionalsafety (1980 to the mid-1990s) [Continued]

Item	Period	Projects & Initiatives	Key Outcomes
14	1994	MISRA (Motor Industry Research Association) 1994: Development of guidelines for vehicle-based software.	• MISRA Developed guidelines for vehicle based software. Based on principles from the emerging IEC drafts, including adoption of Safety Integrity Levels (SILs).
15	1995	Safety-Related Systems, Postgraduate Qualifications, Syllabus Proposals IEE Public Affairs Board Report Number 13; January 1995.	• Information about topics that post-graduate post-experience education and training safety-related and safety critical systems; • The topics were raised in modules based on the Safety Lifecycle.
16	1995 Updated 2003	"Out of Control-why control systems go wrong and how to prevent failure". Published by HSE (freely available to download). (Out of Control)	• The analysis of control system incidents underpinned the need for the concept of a Safety Lifecycle.

Appendix A: Table A.1: Some key initiatives relating to the emerging concept of functional safety (1980 to the mid-1990s) [Continued]

Item	Time period	Projects & Initiatives	Key Outcomes
17	1999 and updated in 2007. (Currently under review; new revision expected 2016)	Safety, Competency and Commitment- *Competency Guidelines for Safety-Related System Practitioners* Published by the IEE who managed the work to produce the guidelines which was carried out in collaboration with the British Computer Society (BCS). The UK health and safety executive commissioned the IET to manage the initial study that underpinned the development of the guidelines.	• Based on developing a Competence Profile, comprising three levels of competence, for each Task and Attribute. • Tasks are technical skills and knowledge; • Attributes are behavioural skills and knowledge.
18	2007	• Managing competence for safety-related systems- Part 1: Key Guidance. (Managing competence #1) • Managing competence for safety-related systems- Part 2: Supplementary material. (Managing competence #2) The guidance was issued by the Health and Safety Executive, the Institution of Engineering Technology and the British Computer Society.	• Guidance for those who are responsible for managing and assuring the competence of individuals and teams.

Balancing safety with rampant software feature-itis

Les Hatton

SEC, Kingston University and Oakwood Computing Associates Ltd

UK

Abstract *In the 30 years or so that we have been developing a safety methodology to accompany the growing presence of software in safety-related systems, understanding of the software development process itself hardly seems to have advanced. We still use in most part the same languages we did then but duly bloated to match the uncontrollable growth of software itself. We still teach entire paradigms without any basis in the scientific method whatsoever whilst the amount of software continues to grow alarmingly, particularly in the automotive industry. The result is an absence of any real forensic basis for understanding failure and subsequently avoiding it. This short essay looks at some of the reasons why, and demonstrates from recent results in information theory, that this stems from the fact that we are probably barking up the wrong tree.*

1. Introduction

30 years ago, 50,000 lines of code was considered pretty big. Ian Sommerville's magnum opus on Software Engineering had just appeared (Sommerville 1982, 2015); university computing departments formerly known as maths, electrical engineering, chemistry or indeed anything vaguely scientific were rushing to teach students the new ideas as computer science; the early versions of Unix had appeared embodying a new all-embracing language C; we taught compiler design, discrete mathematics and other hard subjects; the first twinklings of object-oriented design were stirring in the darkness beyond the Styx; we thought we might be able to prove programs were correct, perhaps even routine-

ly; we were beginning to realise that developing software for reliable systems was actually extraordinarily difficult; AND we also started measuring properties of software such as how many branches or how many *goto* statements it had, in order to make ambitious and entirely unsupported statements about the relationship of defects with these so-called metrics, (for which I must bear my own share of guilt).

Today, we encounter systems with 50 million lines of code, (Software Impact series, starting IEEE Software 2010). Ian Sommerville's 10th edition of Software Engineering has just appeared (Sommerville 1982, 2015); university computing departments are reconsidering returning to maths, electrical engineering, chemistry or indeed anything vaguely scientific to continue to pull in the punters[1]; Linux written in C along with countless open source packages runs the internet more or less (and very well too thank goodness); we mostly teach people how to do websites and play with SQL databases; we have given up with object-orientation and have renamed everything in terms of sporting analogies, so we have scrums and rushes and we have formed agile alliances and movements, and soon we may have television shows called "Programmer's partners" (I made that bit up); we haven't the faintest idea how to prove 50 million lines of code correct, (whatever that means when the requirements change more quickly than the verification time); we have realised that because nobody wants to write programs any more we are desperately short of programmers and now advertise via the BBC and other such paragons of science, "learn to code in a weekend"; AND we still extract endless largely meaningless metrics in the thin hope that they will correlate with something useful other than inside leg measurement.

What we still don't have is any systematic technology to understand why and how software based systems fail, so where did the wheels come off the great Software Engineering Juggernaut, and why are we now wallowing in millions of lines of the stuff which needs updating (i.e. fixing) on an often daily basis, and most of which affects our lives?

2. The rise and fall of Software Engineering

The whole problem probably started, certainly in the UK, around the time we began to conflate Computer Science in universities with ICT in schools, (Information and Communications Technology also known as "Computer Skills") such as getting a spreadsheet to add two numbers; writing a sentence or two in a word-processor or designing a web page). ICT was intended to titillate the palates of our young, preparing them for careers in Computer Science. In truth, it

---

[1]   At GBP 9,000 – 12,000 per year in England and Wales for example.

probably bored them senseless. Perhaps worse, ICT spread into management offices. Even the CEO had a computer on his or her desk and could fiddle about with a spreadsheet, no doubt thinking how jolly easy it is to change things. In the army, they say there is nothing so dangerous as an officer with a map. In business, it is a CEO in charge of a spreadsheet.

Meanwhile Computer Scientists were largely having far too good a time inventing new stuff to lay down the basis of a scientific and engineering methodology based on measurement-based systematic improvement for software development. If a structural engineer says "if you build a horizontal beam that long, it will break", they have decades of measurement-based material science to back it up. If a software developer says "If I add that feature, it will probably cause serious unintended consequences", they have nothing to back it up other than their experience, so it's very easy for a CEO aka spreadsheet jockey, to override them.

We therefore seem to have come full circle and arrived at a paradox. We are rushing to produce a new generation of "coders" as a proxy for proper engineering to crank the stuff out, whilst we are simultaneously overwhelmed by millions of lines of code in systems such as the formerly humble automobile, much of which in my opinion we do not fully understand, given the levels of reported failures. We have rampant feature-itis – not everywhere it's true, but in too many uncomfortable places and it seems to me that the distinction between what we used to know as safety-related software development, and the systematic pumping out of millions of lines of code for the latest in-car gimmick or whatever, has blurred to the point where there seems almost no distinction.

3. The growth of feature-itis

Software is mostly sold like soap powder as part of a fashion-based industry. As I have many times discovered to my cost in commercial software development, the moment you release a brand new version of your software with lots of ravishing bells and whistles, the first thing the users tend to say is, "Why can't it do this ...?". I have always felt that this is tantamount to buying a television and then wondering why it can't tumble-dry. It is symptomatic of the widely held delusion that software is infinitely malleable, whilst maintaining its integrity. One out of two isn't bad I suppose...

This has led to rapid growth in many software-producing industries as suppliers compete to please end-users with new features. The bottom line is that most systems are now growing in source code terms at around 20% per year with a surprisingly small variance, (van Genuchten and Hatton (2012)). This is roughly equivalent to a doubling in size every 42 months. No wonder we are wallowing in it and Fred Brooks' prescient quotation from Ovid made 40 years ago, "add little to little and you finish up with a big pile" (Brooks 1975) is perennially true.

I didn't expect to see this in safety-related systems however. I thought we would have a little more sense and that our burgeoning safety bureaucracy with its safety cases, safety management, too-often impenetrable safety vocabulary and plethora of standards such as IEC 61508[1] would have acted as a brake on vaulting over-ambition. It hasn't. Take the automobile for example. When I first started working in and around this industry in the early 1990s, there were around 100,000 lines of assembler in a car. 20% a year for say 23 years gives 12.8 million lines give or take a useless feature or three, and that's about where we are today, (Mossinger, 2010), except that it's mostly C now. It is true of course that it is not all safety-related but a significant amount is and the demarcation line between the bits which are safety-related and the bits which are not remains fuzzy to say the least, particularly as they tend to be distributed throughout multiple shared processors.

As a result of this unbridled creativity, we now have in 2016, the first reported death due to automatic driving when a Tesla system could not distinguish a large truck from a light sky, (Tesla (2016)); there have been 149 recalls involving tens of millions of cars (PopSci, 2016) and including unintended acceleration incidents, braking failures, engine cut-outs and numerous other features specifically related to software; we have drive-by hacking of cars via the stack of internet-related software we are now serving up to drivers for their comfort and "safety" (BBC 2013); and we have the software cheat, software designed solely to mislead, courtesy of Volkswagen and probably numerous others. In fact as I write this part on 9th September 2016, the BBC news had two items: 1) GM were recalling 4.3 million vehicles because of a software defect which might cause airbags not to deploy, a situation that may already have led to one death. It is expected to cost about half a billion dollars, (GM 2016). 2) In contrast, the success of a robot arm in improving surgical precision was reported, so clearly some systems not so prone to feature-itis are benefiting, (BBC 2016). I sincerely hope that the designers of the robot arm are not considering internet-stack upgrades so the surgeon can play the latest hot-off-the-shelf exercise in mindless violence in between patients, to keep their reactions up to speed.

Apart from the odd billion dollar pay-off to avoid further litigation, (Toyota (2012)), the legal profession has remained relatively quiet about software, largely because they are in the dark about it, just like the rest of us. The probable reason for this is that the almost complete lack of any empirical basis to software engineering means that expert witnesses can violently disagree with each other without perjuring themselves, (30 years ago or so, the courts even went so far as to comment on this in *Saphena Computing v. Allied Collection Agencies* (1985) – plus ca change ...).

---

1   http://www.iec.ch/functionalsafety/

4. Software development in court

A particularly high profile example of software problems in automobile systems is Toyota, (NASA (2011), Barr (2013)). These reports are very detailed and thorough. However, it becomes apparent that there are areas of coding and coding practice which are crying out for more substantial measurement support. For example, the use of global variables is an important issue in these reports as a dangerous practice. It is likely that many embedded system engineers would agree, but how dangerous are they and in what context are they dangerous? Everybody will have an example which may have affected them but this does not constitute a quantitative predictive system. Tim Hopkins and I checked this in the NAG library (Hopkins and Hatton (2008)), and there was no statistically significant relationship between the presence of global variables and the presence of defect after 25 years of use. Although hard real-time systems and scientific subroutine libraries are not the same beast, there is precious little sharing of good experimental data and analysis in any computationally reproducible form on which we could build. It could be argued that there are better, more modern alternatives, in that they make more sense to an experienced engineer, but where is the quantitative evidence that they are "better"?

There are other examples quoted in the reports associated with coding rules but again, even though we can compare their transgression rates with other systems, it tells us very little about any reliability implications. In fact, predicting reliability generally is pretty difficult, (Littlewood and Strigini (1993)). Reading the above reports as an engineer, I agreed entirely with the verdict but not with some of the arguments presented to support it.

(Note that Toyota have since settled with the US Justice Department for a reputed 1.2 billion dollars (Toyota (2012)). That would have paid for a lot of testing.)

5. Why is measurement-based improvement so difficult?

A good question. If we consider this in terms of control-process feedback, one important component of an answer seems to be that of disjoint timescales. In the time it takes to understand the failure modes of a process so that it can be changed beneficially, it is very important that the process has not changed significantly whilst analysis is in progress. This is rarely the case in software development, where there is such a great creative outpouring of paradigms, methodologies and other distractions that empirical support for these ideas simply never happens.

There is another and much more fundamental problem. What is it that we have to measure to bring about beneficial feedback? In control process feedback, the general idea is that we identify systematic modes of failure of the resulting product and then we change the process which produced it, either to mitigate or to remove such failure modes. This might be through either an improvement in the product itself and/or by use of some design methodology such as redundancy. Unfortunately, this relies crucially on successful diagnosis of failure, notably the ability to relate cause to effect in numerical terms so that we can quantify the risk of failure of certain practices.

In this we seem to have failed abysmally. Raking through a few hundred thousand lines of code and perhaps a lot more, trying to pin down certain kinds of failure turns out to be extremely difficult, as demonstrated by NASA (NASA (2011)), who tried but failed to find a specific software defect which led to the Toyota unintended acceleration failure, whilst later, Barr (Barr (2013)) did. Since it is much easier to see source code than systems behaviour, (which we view through the very dark glass of testing and the even darker glass of user experience), we have become fixated with source code metrics – how many lines, how many decisions, how many loops, how many variables (global or otherwise) – in pursuit of the dream that we might be able to predict defects from some combination of these measurements. In the process, we have spawned derivative candidate measures such as function-points. Indeed most of the literature on software defects seems to be attempting to fit some kind of statistical model based on lines of code, number of decisions, fan-in fan-out or whatever. It was hoped that this might allow us to define what we mean by software quality and be able for example to distinguish good code from bad code, which would at least be a start.

Not only have we failed, but it seems from recent theoretical studies that we cannot succeed in this endeavour. Indeed, software systems are one of a very wide class of discrete systems, (systems built from components, which are themselves build from indivisible discrete pieces), which conserve Hartley-Shannon information (Hatton (2014), Hatton and Warr (2015)). This conservation principle gently acts underneath the frenzied creativity of programmers, whatever language they are using and whatever application area they are working within, to force the length distribution of components – functions, subroutines, packages and the like – into the same characteristic canonical length distribution, (shown as Figure 1 for 80 million lines of C). Its shape is very interesting and comprises of a rather complex sharply unimodal distribution of lengths with a characteristic and extraordinarily accurate power-law for longer components and is the implicit solution of:

$$t_i = A.exp(-\frac{1}{2t_i}).(1 - a_i/t_i)^\beta.exp(\beta\frac{a_i(1 - a_i/t_i + 1/2t_i)}{t_i(1 - a_i/t_i)}).a_i^{-\beta}$$

where A, β are undetermined Lagrangian constants, t_i is the length of the ith component and a_i is its frequency, (which turns out to be proportional to its unique alphabet).

In other words, our fixation with the size of components and their possible relationship with defects is almost certainly misplaced. Indeed, we do not seem to have any real control over the length distribution, instead the conservation principle takes over, as it operates at all scales.

This has a number of important implications. First of all, the average size of components is conserved around the mode of this distribution. Second, it would appear that the distribution of defects across software systems is purely statistical. When defects are randomly distributed amongst component sizes with this distribution, this has the non-intuitive property that a large percentage of software components in any system, something like 75% will never exhibit any defects, and leads to the frequently-observed phenomenon we know as *defect clustering*.

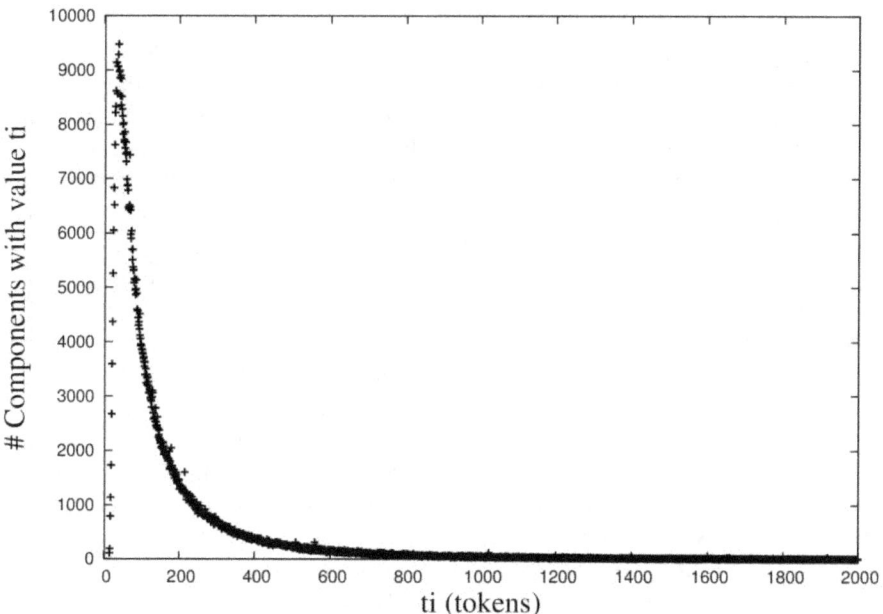

Fig. 1: The canonical length distribution of all software demonstrated in a large open source population of C.

On the face of it, this sounds like a vision of the Holy Grail and if only we could understand how these apparently defect-free majority were built, we might conceive of building a (near) perfect system. Unfortunately, they look just like the rest and it appears to be due simply to the random spreading of defects amongst components which have the canonical length distribution mentioned above. Whilst there is some evidence that we can exploit the resultant defect clustering by identifying a defect and then utilising the increased probability of finding another one, understanding exactly why a component has been defect free for a long time appears to be as futile as understanding why some people never win the lottery by studying people who previously have. It is a purely statistical effect caused by the underlying conservation principle - the reason that the majority of components never show any defect appears to be simply because the rest do have defects and they have to be somewhere.

6. Where now?

It is easy to criticize but we must be constructive. Society depends so much on software now that it would be foolhardy to blunder on regardless, although I see no immediate end to the feature explosion in sectors like the automobile industry unless there is a corresponding backlash from the end user. It costs a billion dollars or so on an increasingly regular basis, but the car-industry seems to be profitable enough to take the hit[1], and as even embedded software becomes more and more easy to change over the internet, updates come increasingly often, as there is little or no logistical barrier. This may be very well for the almost daily Android software updates, but the relationship between "little and often" and safety-relationship has yet to be explored.

The emerging results from information theory are a little more worrying, although they do explain why our efforts at producing predictive metrics of defect from properties of source code and the processes that produced it, have not yielded the reliability gains we might have hoped for. In summary, it seems to me therefore that *if we are to make progress in improving the safety and reliability of systems containing software, we are rather more likely to make it by pursuing the traditional engineering virtues of redundancy and designing for failure than we are by pursuing studies of the fabric of software, which, as far as information theory is concerned, is just a bunch of symbols.*

References

BBC (2013) http://www.bbc.co.uk/news/technology-23443215, accessed 11-Oct-2016.

---

[1]   "A **billion here, a billion** there, pretty soon, you're talking real money", as the celebrated Republican Senator Everett Dirksen was alleged to have said. Never a truer word ...

BBC (2016) http://www.bbc.co.uk/news/health-37246995 accessed 09-Sep-2016

Barr M, (2013) http://embeddedgurus.com/barr-code/2013/10/an-update-on-toyota-and-unintended-acceleration/ accessed 10-Sep-2016.

Brooks F.P. (1975) "The mythical man month", Addison-Wesley, ISBN 0-201-00650-2

GM (2016) (http://www.reuters.com/article/us-gm-recall-idUSKCN11F2AH, accessed 09-Sep-2016)

Hatton (2014) "Conservation of Information: Software's hidden clockwork?", IEEE Transactions on Software Engineering, 40 (5), p. 450-460.

Hatton L. and Hopkins T. (2008) "Exploring defect correlations in a major Fortran numerical library", http://www.leshatton.org/NAG01_01-08.html

Hatton L. and Warr G. (2015) "Protein Structure and Evolution: Are they Constrained Globally by a Principle Derived from Information Theory?", PloS ONE, doi:10.1371/ journal.pone.0125663

IEEE Software (2010) "The Software Impact columns", ed Michiel van Genuchten and Les Hatton, 2010-

Littlewood B. and Strigini L (1993) "Validation of ultrahigh dependability for software-based systems", CACM, 36 (11), p. 69-80.

Mossinger J. (2010) "Software in Automotive Systems", IEEE Software, 27 (2), p. 2-4.

NASA   (2011)
http://www.nhtsa.gov/staticfiles/nvs/pdf/NASA_FR_Appendix_A_Software.pdf, accessed 10-Sep-2016

PopSci (2016) http://popsci.com/software-rising-cause-car-recalls, accessed 29-Sep-2016

Sommerville I. (1982,2015) "Software Engineering", Pearson, ISBN 1292096131 (1st edition 1982, 10th edition 2015)

Tesla (2016) https://www.theguardian.com/technology/2016/jun/30/tesla-autopilot-death-self-driving-car-elon-musk, accessed 29-Sep-2016.

Toyota (2011), (http://www.wsj.com/articles/
SB10001424052702304256404579449070848399280, accessed 10-Sep-2016)

Van Genuchten M. and Hatton L. (2012) "Compound Annual Growth Rate for Software", IEEE Software 29 (4), p. 19-21.

AUTHOR INDEX